Advanced Photocatalysts for Energy Conversion and Environmental Applications

Advanced Photocatalysts for Energy Conversion and Environmental Applications

Editor

Ruowen Liang

MDPI • Basel • Beijing • Wuhan • Barcelona • Belgrade • Manchester • Tokyo • Cluj • Tianjin

Editor
Ruowen Liang
Fujian Province University
Key laboratory of Green Energy
and Environment Catalysis
Ningde Normal University
Ningde
China

Editorial Office
MDPI
St. Alban-Anlage 66
4052 Basel, Switzerland

This is a reprint of articles from the Special Issue published online in the open access journal *Molecules* (ISSN 1420-3049) (available at: www.mdpi.com/journal/molecules/special_issues/E8W3J98H52).

For citation purposes, cite each article independently as indicated on the article page online and as indicated below:

LastName, A.A.; LastName, B.B.; LastName, C.C. Article Title. *Journal Name* **Year**, *Volume Number*, Page Range.

ISBN 978-3-0365-8871-1 (Hbk)
ISBN 978-3-0365-8870-4 (PDF)

© 2023 by the authors. Articles in this book are Open Access and distributed under the Creative Commons Attribution (CC BY) license, which allows users to download, copy and build upon published articles, as long as the author and publisher are properly credited, which ensures maximum dissemination and a wider impact of our publications.

The book as a whole is distributed by MDPI under the terms and conditions of the Creative Commons license CC BY-NC-ND.

Contents

Fulin Wang, Zhenzhen Yu, Kaiyang Shi, Xiangwei Li, Kangqiang Lu and Weiya Huang et al.
One-Pot Synthesis of N-Doped NiO for Enhanced Photocatalytic CO_2 Reduction with Efficient Charge Transfer
Reprinted from: *Molecules* 2023, 28, 2435, doi:10.3390/molecules28062435 1

Leunam Fernandez-Izquierdo, Enzo Luigi Spera, Boris Durán, Ricardo Enrique Marotti, Enrique Ariel Dalchiele and Rodrigo del Rio et al.
CVD Growth of Hematite Thin Films for Photoelectrochemical Water Splitting: Effect of Precursor-Substrate Distance on Their Final Properties
Reprinted from: *Molecules* 2023, 28, 1954, doi:10.3390/molecules28041954 17

Deling Wang, Erda Zhan, Shihui Wang, Xiyao Liu, Guiyang Yan and Lu Chen et al.
Surface Coordination of $Pd/ZnIn_2S_4$ toward Enhanced Photocatalytic Activity for Pyridine Denitrification
Reprinted from: *Molecules* 2022, 28, 282, doi:10.3390/molecules28010282 31

Haiyan Jiang, Jiahua He, Changyi Deng, Xiaodong Hong and Bing Liang
Advances in Bi_2WO_6-Based Photocatalysts for Degradation of Organic Pollutants
Reprinted from: *Molecules* 2022, 27, 8698, doi:10.3390/molecules27248698 45

Jinhua Xiong, Xuxu Wang, Jinling Wu, Jiaming Han, Zhiyang Lan and Jianming Fan
In Situ Fabrication of N-Doped ZnS/ZnO Composition for Enhanced Visible-Light Photocatalytic H_2 Evolution Activity
Reprinted from: *Molecules* 2022, 27, 8544, doi:10.3390/molecules27238544 69

Jian Li, Xiaojia Wang and Yunyin Niu
M-Carboxylic Acid Induced Formation of New Coordination Polymers for Efficient Photocatalytic Degradation of Ciprofloxacin
Reprinted from: *Molecules* 2022, 27, 7731, doi:10.3390/molecules27227731 81

Linzhu Zhang, Lu Chen, Yuzhou Xia, Zhiyu Liang, Renkun Huang and Ruowen Liang et al.
Modification of Polymeric Carbon Nitride with $Au–CeO_2$ Hybrids to Improve Photocatalytic Activity for Hydrogen Evolution
Reprinted from: *Molecules* 2022, 27, 7489, doi:10.3390/molecules27217489 93

Parnapalle Ravi and Jinseo Noh
Photocatalytic Water Splitting: How Far Away Are We from Being Able to Industrially Produce Solar Hydrogen?
Reprinted from: *Molecules* 2022, 27, 7176, doi:10.3390/molecules27217176 108

Yayun Wang, Haotian Wang, Yuke Li, Mingwen Zhang and Yun Zheng
Designing a 0D/1D S-Scheme Heterojunction of Cadmium Selenide and Polymeric Carbon Nitride for Photocatalytic Water Splitting and Carbon Dioxide Reduction
Reprinted from: *Molecules* 2022, 27, 6286, doi:10.3390/molecules27196286 133

Zhi-Yu Liang, Feng Chen, Ren-Kun Huang, Wang-Jun Huang, Ying Wang and Ruo-Wen Liang et al.
CdS Nanocubes Adorned by Graphitic C_3N_4 Nanoparticles for Hydrogenating Nitroaromatics: A Route of Visible-Light-Induced Heterogeneous Hollow Structural Photocatalysis
Reprinted from: *Molecules* 2022, 27, 5438, doi:10.3390/molecules27175438 152

Article

One-Pot Synthesis of N-Doped NiO for Enhanced Photocatalytic CO$_2$ Reduction with Efficient Charge Transfer

Fulin Wang [1], Zhenzhen Yu [1], Kaiyang Shi [1], Xiangwei Li [1], Kangqiang Lu [1], Weiya Huang [1], Changlin Yu [2,*] and Kai Yang [1,*]

[1] School of Chemistry and Chemical Engineering, Jiangxi University of Science and Technology, Ganzhou 341000, China
[2] School of Chemical Engineering, Guangdong University of Petrochemical Technology, Maoming 525000, China
* Correspondence: yuchanglinjx@163.com (C.Y.); yangkai@jxust.edu.cn (K.Y.)

Abstract: The green and clean sunlight-driven catalytic conversion of CO$_2$ into high-value-added chemicals can simultaneously solve the greenhouse effect and energy problems. The controllable preparation of semiconductor catalyst materials and the study of refined structures are of great significance for the in-depth understanding of solar-energy-conversion technology. In this study, we prepared nitrogen-doped NiO semiconductors using a one-pot molten-salt method. The research shows that the molten-salt system made NiO change from p-type to n-type. In addition, nitrogen doping enhanced the adsorption of CO$_2$ on NiO and increased the separation of photogenerated carriers on the NiO. It synergistically optimized the CO$_2$-reduction system and achieved highly active and selective CO$_2$ photoreduction. The CO yield on the optimal nitrogen-doped photocatalyst was 235 µmol·g^{-1}·h^{-1} (selectivity 98%), which was 16.8 times that of the p-type NiO and 2.4 times that of the n-type NiO. This can be attributed to the fact that the nitrogen doping enhanced the oxygen vacancies of the NiOs and their ability to adsorb and activate CO$_2$ molecules. Photoelectrochemical characterization also confirmed that the nitrogen-doped NiO had excellent electron-transfer and separation properties. This study provides a reference for improving NiO-based semiconductors for photocatalytic CO$_2$ reduction.

Keywords: NiO; nitrogen doping; photocatalysis; reduction of CO$_2$

1. Introduction

The rapid development of current society has increased the consumption of non-renewable fossil fuels. Human beings have to face the problem of energy shortages, and the resulting large emissions of CO$_2$ are also an important cause of global warming [1–4]. Currently, using sustainable solar energy to photocatalytically reduce CO$_2$ in high-value-added products is a promising way to simultaneously solve the greenhouse effect and the energy crisis [5,6]. Therefore, it is very important to design and synthesize photocatalysts with low pollution, high efficiency, and low cost [7,8].

As an environmentally friendly transition-metal-oxide semiconductor, NiO has excellent conductivity, good chemical stability, and non-toxicity, and it has broad application prospects at the nanoscale [9,10]. At the same time, it is considered to be a semiconductor that can be used for CO$_2$ photoreduction due to its sufficiently negative conduction band position, fast hole mobility, and high charge-carrier concentration [11]. However, due to the high recombination degree of photogenerated carriers, the separation efficiency of electrons and holes in the reaction process is low, which greatly weakens the reactivity [12]. In addition, wide-band-gap NiO semiconductor catalysts can only use about 3–5% of solar ultraviolet light, resulting in the low efficiency of the photocatalytic reduction of CO$_2$, limiting the application of NiO in photocatalysis [13]. Therefore, NiO is often used as a

co-catalyst to improve photocatalytic performance and encourage the efficient separation of photoelectrons and holes [14]. For example, NiO can significantly improve the photocatalytic hydrogen-production performance of $SrTiO_3$, TiO_2, Nb_2O_5, Ga_2O_3, and other photocatalysts [15]. However, the activity was generally low in the reported photocatalytic reduction of CO_2 by NiO [16,17]. Therefore, NiO is usually modified by different methods to improve the photocatalytic performance [18].

Since NiO has suitable conduction band (CB) and valence band (VB) positions, it often forms heterostructures with many semiconductors. Zhang et al. [19] prepared an S-type BiOBr/NiO heterojunction. The experiment showed that the layered structure of BiOBr/NiO increased the light-absorption and charge-separation performance, and it improved the redox ability of BiOBr/NiO. In addition, the NiO-layered porous-sheet structure was conducive to the adsorption of CO_2, exposing abundant active sites for CO_2 photoreduction, thus achieving excellent CO_2 photoreduction performance. Moreover, Park et al. [20] prepared a single-layer hollow-sphere photocatalytic material (h-NiO-NiS) of NiO and NiS by partially replacing O with S on NiO hollow spheres. The construction of this heterojunction greatly enhanced the CO_2-adsorption capacity and increased the transfer of excited electrons from the NiS to the surface along the hollow spheres. The efficient transfer of electrons led to the prolongation of the photogenerated charges' recombination times, which further increased the conversion of CO_2 to CH_4.

Moreover, charge separation can be increased by adjusting the electronic structure of NiO, thereby improving its CO_2 photoreduction activity. Xiang et al. [21] constructed ultrathin NiO nanosheets with different oxygen-vacancy concentrations to achieve efficient CO_2 photoreduction performance. Density functional theory calculations and CO_2-temperature programmed desorption experiments confirmed that moderate oxygen vacancy concentrations achieved a strong combination of the material surface with CO_2, enhanced the adsorption and activation of CO_2, and encouraged effective charge transfer. By contrast, the excessive oxygen-vacancy content reduced the binding affinity of the CO_2; thus, the appropriate regulation of oxygen-vacancy content is an effective means to achieve a NiO electronic structure that is suitable for CO_2 photoreduction. In addition, the construction of a ternary bridging structure is also an important method to increase the separation of photoelectrons and holes. For example, Park et al. [22] introduced reduced graphene oxide (rGO) into the NiO-CeO_2 p-n heterostructure, which accelerated the separation and transfer of photogenerated electrons, and the surface of the material accumulated electrons more easily, thus improving the photocatalytic activity of the CO_2 multi-electron reduction.

In addition, heteroatom doping is an effective method with which to adjust the electronic structures of catalysts and has been extensively studied [23–26]. However, compared with anion doping, cation doping produces more harmful electron–hole recombination centers. Because oxygen and nitrogen show similar chemical, structural, and electronic characteristics, such as polarizability, electronegativity, coordination number, and ionic radius, when other elements (such as N 2p) with higher potential energy than O 2p atomic orbitals are introduced, new VBs instead of O 2p atomic orbitals can be formed, resulting in smaller E_{bg} without affecting the CB level, thereby improving the visible-light response [27]. Therefore, non-metallic-element-N doping is a preferable way to improve the photocatalytic CO_2-reduction effect of NiO. Furthermore, it is also important to choose the appropriate doping method. The molten-salt method of element doping is an efficient and low-cost method because its molten-salt liquid environment can make the element distribution more uniform, and the treatment process before and after the reaction is very simple [28].

In this research, NiO semiconductor catalysts with different nitrogen-doping contents were prepared using a molten-salt calcination method, and the CO_2-reduction activity was tested in a bipyridine ruthenium/triethanolamine heterogeneous catalytic system excited by different wavelengths of light [29]. The phase composition, band structure, optical properties, and surface morphology of the doped NiO semiconductor were researched through a series of characterizations. The enhancement mechanism of the photocatalytic

2. Results and Discussion

2.1. Phase Structure

As shown in Figure 1a, all of the samples corresponded to standard NiO (JCPDS PDF#47-1049), and no impurity phase was detected via XRD. The diffraction peaks at 2θ = 37.2°, 43.3°, 62.9°, 75.4°, and 79.4° corresponded to the (111), (200), (220), (311), and (222) crystal planes of the NiO, respectively [30]. In addition, the doping of the N significantly enhanced the crystallinity of the sample, which was more conducive to the migration and separation of photogenerated charges [31]. By enlarging the range of 2θ = 41–45° (Figure 1b), it was found that the doping of N made the (200) crystal plane of the N-NiO-x shift by a small angle. This is because the radius of the N was different from those of the Ni and the O. After the N doping into the lattice of the NiO, the Ni–O bond became compressed and stretched to a certain extent, resulting in a change in the crystal-plane spacing, which showed the shift in the crystal plane's diffraction angle macroscopically [32]. The average particle diameters of the samples calculated by the Scherrer equation are shown in Table S1. It was found that the calculated results of the NiO and N-NiO-2 were similar to the results of the SEM (Figure 2a,b). The particle diameter of the pure NiO was smaller and more uniform than that of the N-NiO-2. The doping of the N made the NiO agglomerate and the particle diameter increased.

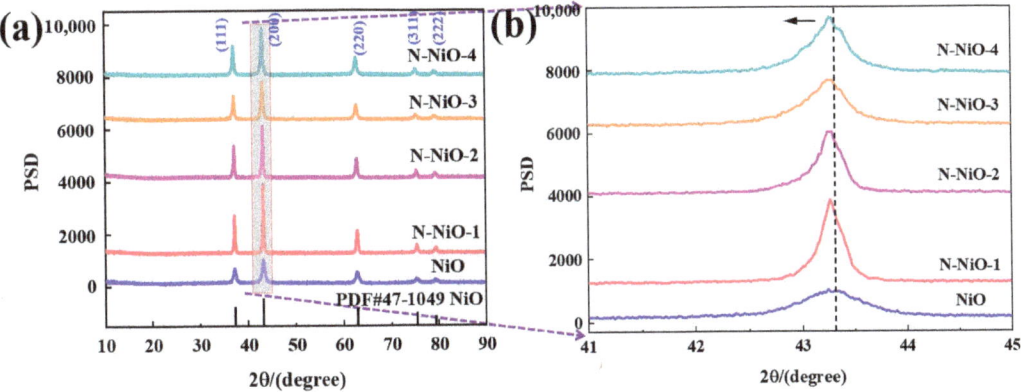

Figure 1. (a) The XRD patterns of the prepared samples; (b) local amplification diagram.

2.2. Microstructure

The microstructure information of the NiO and N-NiO-2 were collected using SEM and TEM. As shown in Figure 2a, the NiO appeared in the form of nanospheres, and the particles were evenly distributed. After the introduction of the N-element doping, the surface of the sample became irregular and agglomerated (Figure 2b). In addition, the TEM images showed that the N-NiO-2 was stacked in sheets and irregularly distributed (Figure 2c), which was similar to the SEM results. Furthermore, as shown by the high-resolution-TEM imagery in Figure 2e–f, it was found that there were lattice-fringe-spacing values of d = 0.22 and 0.24 nm in the N-NiO-2, which corresponded to the (200) and (111) crystal planes of the NiO, respectively. No lattice fringes of impurity phases were detected, indicating that the N doping did not form impurity phases on the surface of the NiO. It is worth noting that the formation of oxygen defects in the sample macroscopically showed the edge of the defect band [33]. In addition, the element-mapping spectra in Figure 2g–j show that the Ni, O, and N elements were uniformly distributed without impurity elements.

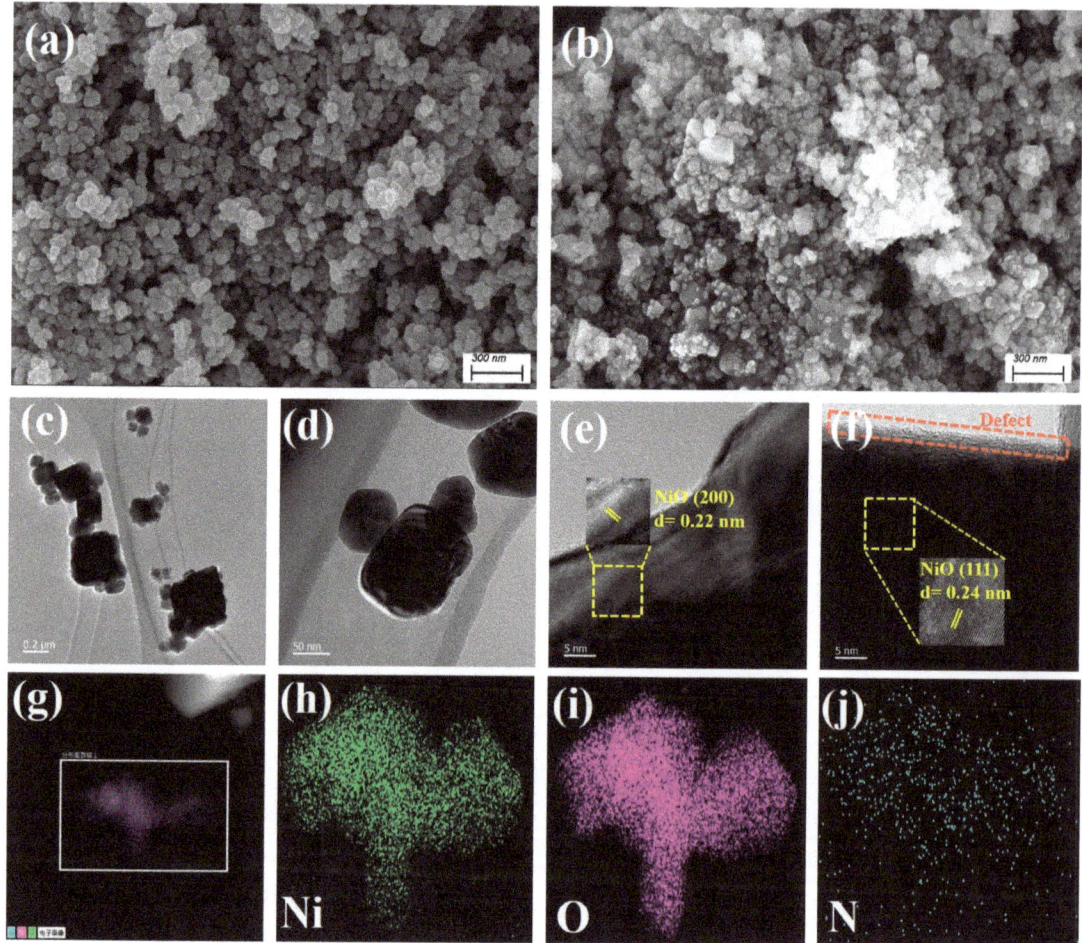

Figure 2. (a) SEM images of NiO and (b) N-NiO-2; TEM (c,d), HR-TEM (e,f), and element-mapping spectra of N-NiO-2 (g–j).

2.3. Optical Properties

The optical absorption spectrum was used to characterize the absorption characteristics of the sample to different wavelengths of light. In general, the larger the maximum absorption wavelength, the wider the spectral response of the semiconductor, but this causes the narrowing of the band gap of the semiconductor, which may further lead to a reduction in the redox performance in the photocatalytic process [34]. Therefore, it was necessary to balance the excitation wavelength and redox performance of the light-excited semiconductor. As shown in Figure 3a, the DRS showed that the maximum absorbance of all the samples was concentrated within a range of 200–350 nm. However, after the N doping, the original black NiO was transformed into yellowish brown N-NiO-x (Figure S1); thus, the absorption of the N-NiO-x in the visible range was weakened. In addition, the absorption peaks of the N-NiO-x at about 390 nm and 470 nm were attributed to the N 2p band introduced by the N doping [35]. The weak absorption band around 600 nm belonged to the defect band [36]. The other absorption peaks at 380–500 nm and the peak around 720 nm correspond to the NiO itself [37].

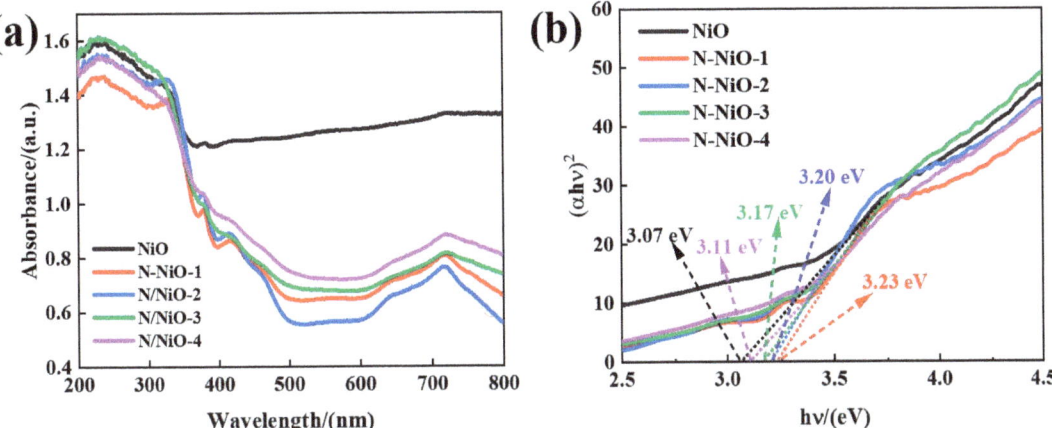

Figure 3. (a) Optical absorption properties of samples and (b) corresponding $(\alpha h\nu)^2$-$h\nu$ curves.

The band gap of the sample can be calculated according to the Kubelka–Munk equation [38]:

$$(\alpha h\nu)^{1/n} = A(h\nu - Eg), \qquad (1)$$

where α is the absorption coefficient, $h\nu$ is the light energy, A is a constant, Eg is the band gap, the direct band-gap semiconductor n is 1/2, and the indirect band-gap semiconductor n is 2. According to the literature, NiO is a direct band-gap semiconductor, and n is 1/2. Through drawing a $(\alpha h\nu)^2$-$h\nu$ diagram and linearly fitting the curve from the intercept to estimate the Eg of the sample, the results were obtained and they are shown in Figure 3b. It can be seen that the optical absorption of the NiO weakened after the introduction of the N doping into the NiO lattice; on the other hand, the doping of N made the band-gap value of the NiO change from 3.07 to 3.23 eV, and the wider band gap improved the reduction performance of the NiO.

2.4. Surface Chemical States

In the XPS full spectra of the N-NiO-2 shown in Figure S2a, Ni and O elements were present, and no obvious N element was found, which may have been due to the low doping amount. In the C 1s spectrum (Figure S2b), the peaks at 284.8 eV, 286.2 eV, and 288.8 eV corresponded to the C-C, C-O, and C=O of the external carbon source, respectively. Figure 4a corresponds to the energy spectrum of the Ni element. The characteristic peaks of the NiO at the binding energies of 853.6 and 872.0 eV corresponded to Ni $2p_{3/2}$ and Ni $2p_{1/2}$, respectively, corresponding to Ni^{2+}. In addition, the binding energies of 860.6 and 870.7 eV corresponded to the satellite peaks of Ni 2p [39]. However, compared with the NiO, the Ni 2p characteristic peak of N-NiO-2 shifted 0.59 eV in the direction of increased binding energy, indicating a decrease in the electron-cloud density of the Ni element [40]. This may have been due to the fact that the electronegativity of N is larger than that of Ni, and electrons are more easily attracted by the N element. In addition, it can be seen in Figure 4b that the O 1s were fitted to the three peaks of O_I, O_{II}, and O_{III} with binding energies of 529.2 eV, 531.2 eV, and 531.9 eV, respectively, corresponding to Ni-O lattice oxygen, the hydroxyl oxygen of the adsorbed water on the sample surface and oxygen defects, respectively [41–43]. Compared with the NiO, the oxygen defects of the N-NiO-2 increased from 4.9% to 10.4% (Table 1). The increased oxygen defects were more conducive to electron capture, thereby promoting the separation of photogenerated charges [44]. Moreover, the binding energy of 400.0 eV (Figure 4c) corresponded to the N 1s peak, indicating the successful doping of the N element [45].

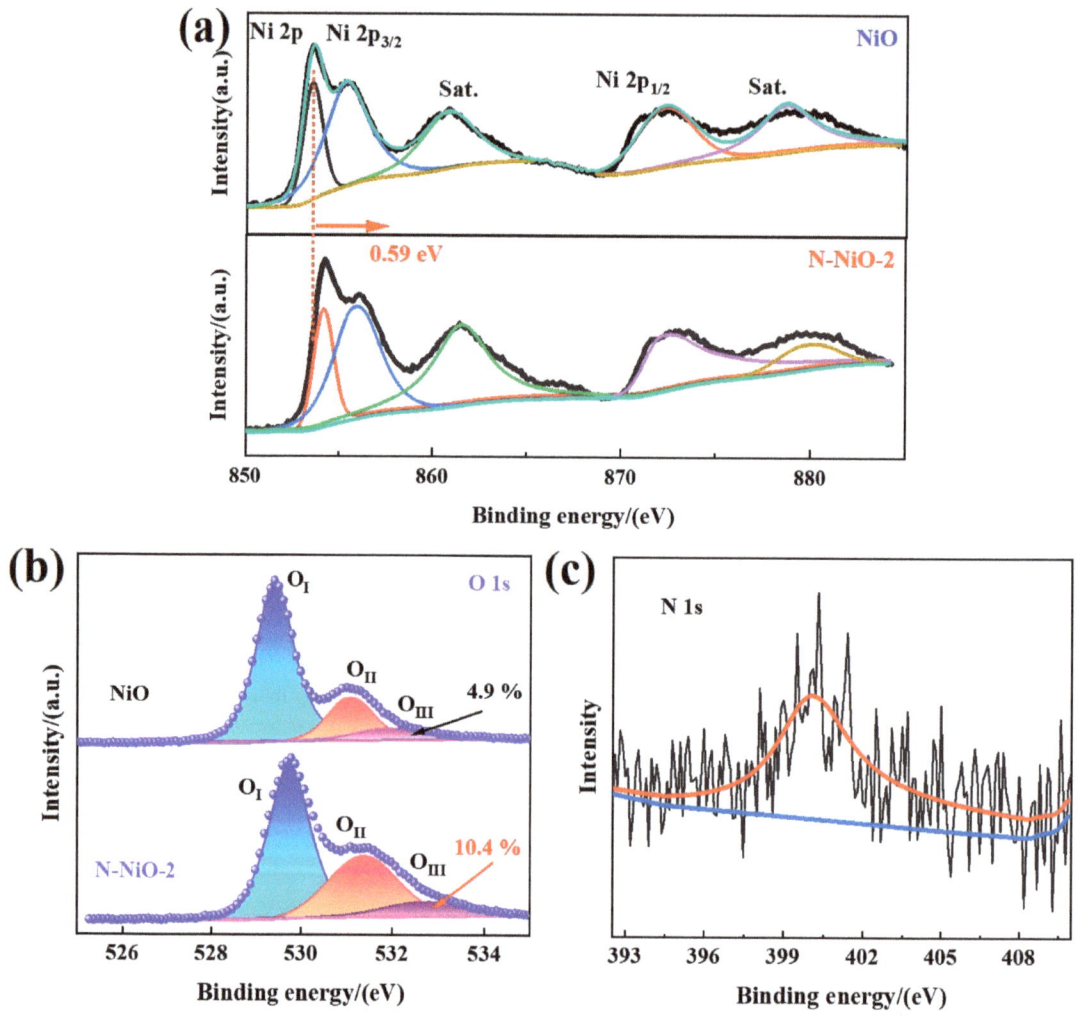

Figure 4. (a) X-ray photoelectron spectra of NiO and N-NiO-2: (a) Ni 2p, (b) O 1s, and (c) N 1s.

Table 1. The XPS fitted peak area and oxygen-defect ratios on NiO and N-NiO-2.

Samples	Oxygen Species				
		Ni–O	O–H	OV	OV Ratio
NiO		64,587	44,182	5635	4.9%
N-NiO-2		67,586	73,293	16,352	10.4%

2.5. CO_2-Photoreduction Performance

Using a LED lamp as the light source, the prepared samples were tested for CO_2-photoreduction activities. As shown in Figure 3, it was found by liquid chromatography and gas chromatography that the product had no substances other than CO and H_2. As shown in Figure 5b, the T-NiO exhibited extremely low CO_2-reduction activity under 365 nm of light, with a CO yield of 14 μmol·g^{-1}·h^{-1} and a selectivity of 39%, while the prepared NiO exhibited higher CO yield (95 μmol·g^{-1}·h^{-1}) and selectivity (82%) under molten-salt conditions. Furthermore, when N-doping was introduced into the NiO, the

CO yield increased to 235 µmol·g⁻¹·h⁻¹ and the selectivity increased to 98%. As shown in Figure 5a, with the increase in the N content, the yield and selectivity of the CO increased gradually and reached its maximum on the N-NiO-2. In addition, in order to research the photon-utilization rate of the prepared samples, the activity tests were carried out under 420-nanometer and 550-nanometer light sources, and the results are shown in Figure 5c. In order to investigate the necessary conditions of the reaction system in the catalytic process, a control experiment was also carried out. It can be seen from Figure 5d that only trace products were detected under the conditions of no Ru, no catalyst, the use of N_2 instead of CO_2, no light, and no TEOA, indicating that the CO did arise from the reduction of CO_2 in the system, not from the decomposition of the catalyst, and any changes in the reaction system greatly affected the catalytic activity.

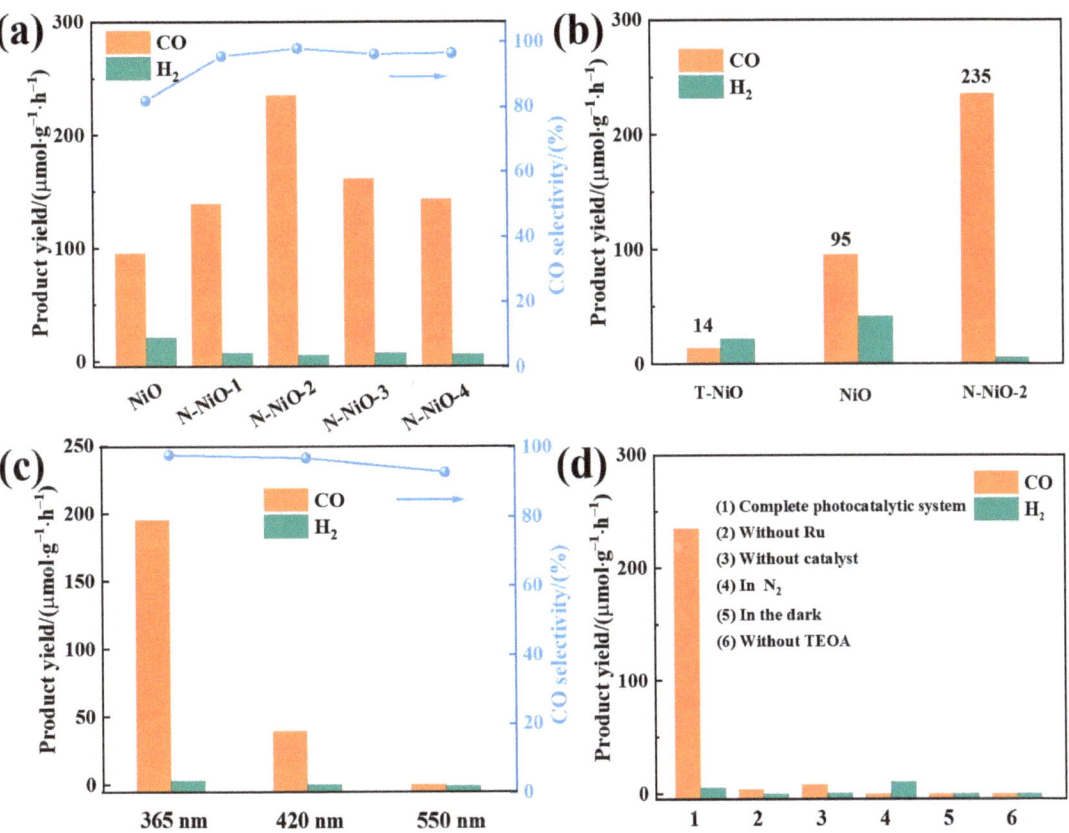

Figure 5. (**a**) Product yields and CO selectivity of NiO and N-NiO-x at 365 nm; (**b**) product yields of T-NiO, NiO, and N-NiO-2 at 365 nm; (**c**) product yield and CO selectivity of N-NiO-2 at different wavelengths; (**d**) product yields of N-NiO-2 under different reaction conditions.

In order to further explore the light-utilization efficiency of the system under the irradiation of different wavelengths of light, the 2-hour CO yield of the N-NiO-2, the calculated apparent quantum efficiency (AQE) value, and the optical power of the corresponding wavelength were determined, and they are listed in Table 2. It can be seen from the table that the AQE reached 2.4% at 365 nm, indicating that the activity corresponded to the energy of the light.

Table 2. Optical power, CO yield, and AQE of the N-NiO-2 at different wavelengths under 2 h of illumination.

Wavelength (nm)	Optical Power (mW)	CO Yield ($\mu mol \cdot g^{-1} \cdot h^{-1}$)	AQE (%)
365	456.4	235.5	2.4
420	358.7	48.2	0.5
550	59.2	1.3	0.06

2.6. Evaluating the Separation Performance of Photogenerated Carriers

In order to explore the photogenerated charge-separation abilities of the prepared samples and investigate the resistance during charge transport, photoelectrochemical tests were carried out. As shown in Figure 6a, the N-NiO-x exhibited an enhanced photocurrent response, indicating that the N doping increased the separation of the photogenerated carriers [46]. On the other hand, the electrochemical impedance spectra of the samples (Figure 6b) showed that the doping made the charge-transfer resistance of the NiO smaller, so that the electrons participated in the catalytic reaction more efficiently [47]. In addition, the laws of the photocurrent and the impedance were consistent with the activity law, which also indicated that the charge transfer was the decisive factor in the catalytic activity. Furthermore, the photoluminescence spectra of the prepared samples are shown in Figure 6c. At the excitation wavelength of 250 nm, all the samples showed emission peaks at about 400 nm, which came from the composite luminescence of the photogenerated carriers. The fluorescence-response values of the N-doped samples were weakened, indicating that the degree of recombination of the photogenerated electrons and holes reduced, resulting in more effectively separated electrons, improving the catalytic performance of the photocatalytic reduction system [48].

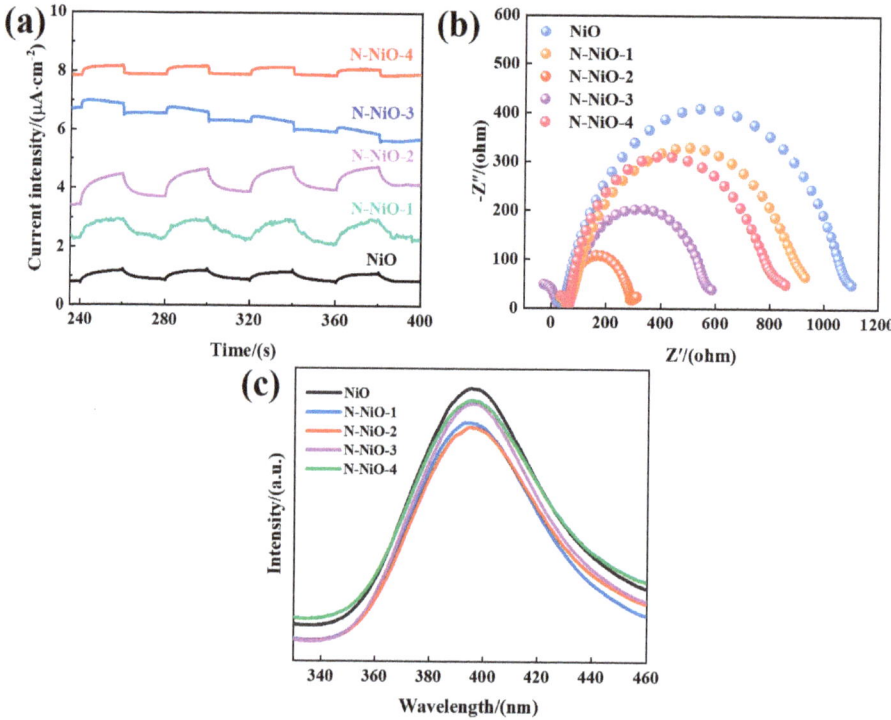

Figure 6. (a) Photocurrent-response curves; (b) electrochemical-impedance spectra; (c) steady-state PL spectra (excitation wavelength: 250 nm).

The EPR spectra of the samples at room temperature (Figure 7) revealed the presence of defects in the samples. The Lorentzian linear resonance peaks at g = 2.002 indicated the presence of unpaired electrons in the samples [49–51]. The results of the EPR and XPS showed that there were oxygen defects in the samples. The higher Lorentz resonance signal indicated that the N-NiO-2 had more oxygen defects than the NiO, which indicated that the molten-salt system effectively introduced oxygen defects into the NiO, and the presence of N could further increase the formation of oxygen defects. The presence of oxygen defects formed an electron-capture trap in the semiconductor, which encouraged the separation of electrons and holes [52]. The CO_2 combined with the accumulated electrons in the defects and was reduced to support the improvement in the reaction performance.

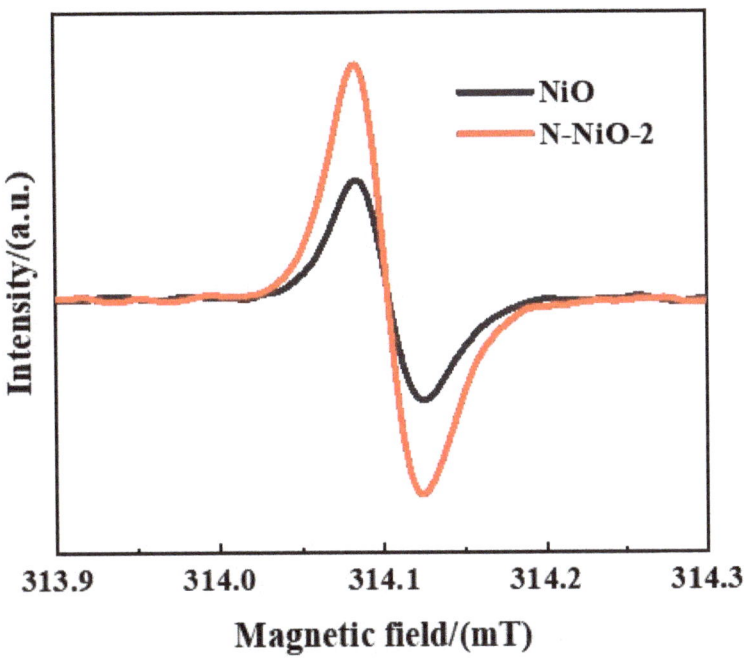

Figure 7. The EPR spectra of NiO and N-NiO-2.

2.7. Energy-Band Structure

In order to study the energy-band structures and redox properties of the prepared samples, the flat band potentials of the samples were tested via photoelectrochemical Mott–Schottky (M–S) analysis. As shown in Figure 8a, the T-NiO exhibited the characteristics of a p-type semiconductor [53], while the NiO in a molten-salt environment (Figure 8b) exhibited the characteristics of an n-type semiconductor [54], indicating that the molten salt encouraged the transformation of the NiO semiconductor type. The surface of the T-NiO itself was rich in holes, so it showed p-type characteristics. The reduction environment in the molten-salt atmosphere encouraged the formation of oxygen vacancies on the surface of the NiO, further enriching it with surface electrons to realize electron doping. Furthermore, the flat-band potentials of the NiO and N-NiO-2 were −0.75 V and −0.85 V (vs. Ag/AgCl pH = 7), respectively, which corresponded to −0.55 V and −0.65 V (vs. NHE pH = 7), respectively, as shown in Figure 8b,c. In general, the conduction band of the n-type semiconductors was about 0.1 V more negative than the flat-band potential [55], so the conduction band of the N-NiO-2 was reduced from −0.65 V to −0.75 V (vs. NHE pH = 7). This indicates that the N doping reduced the conduction-band position of the NiO and enhanced the reducibility of the reaction. In addition, the valence-band position of the

N-NiO-2 was 2.45 V, according to the band-gap diagram obtained through the optical absorption spectrum.

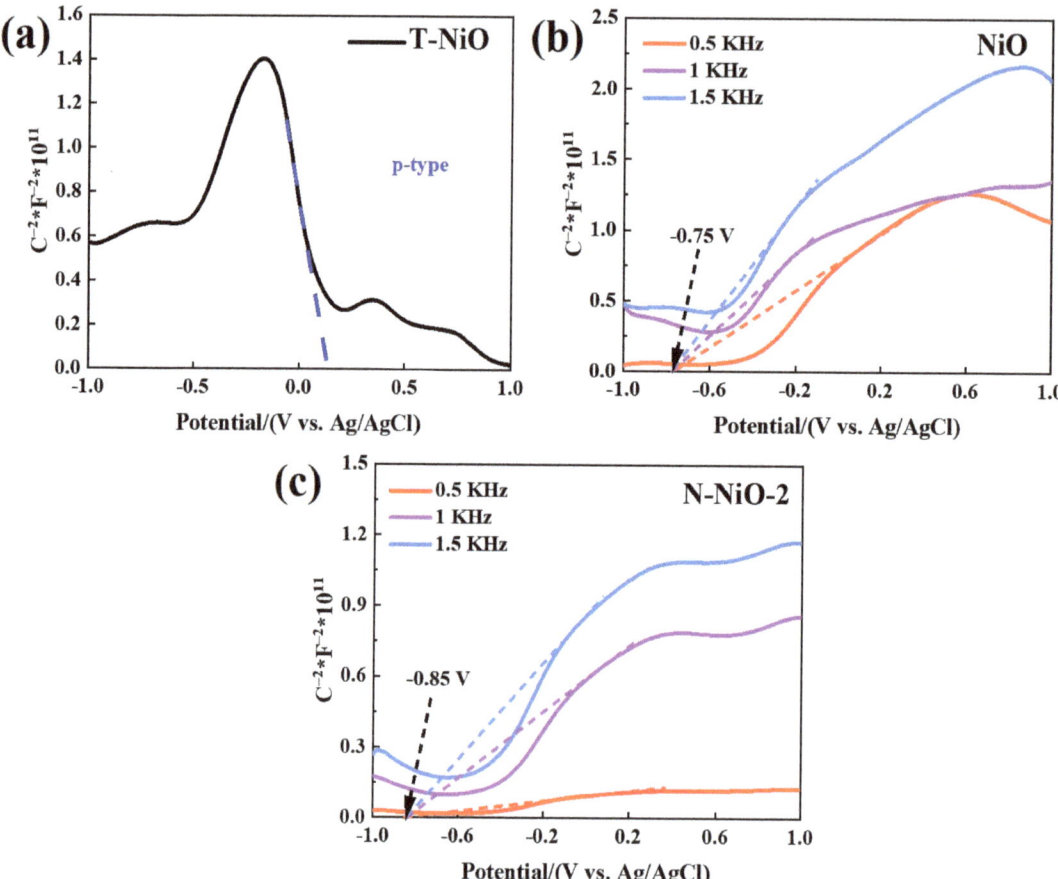

Figure 8. The M–S diagrams of T-NiO (**a**), NiO (**b**), and N-NiO-2 (**c**) semiconductors.

2.8. Possible Reaction Mechanism

Figure 9 shows the physical adsorption isotherms of the CO_2 on the prepared samples. Compared with the pure NiO, the adsorption capacity of the N-NiO-x materials for the CO_2 increased first and then decreased with the increase in the N content, and it reached a maximum with the N-NiO-2 sample, which was consistent with the order of reactivity. These results show that the N doping increased the adsorption of CO_2 on the surface of the NiO, and the combination of electrons with CO_2 on the surface of the NiO facilitated the photoreduction performance of the CO_2, indicating that the adsorption of CO_2 was the decisive factor in the activity.

The possible mechanism of the whole reaction is shown in Figure 10. Under illumination, the N-NiO-x semiconductor became excited, and it produced electron–hole pairs ($e^−–h^+$). Subsequently, the excited electrons in the conduction band of the N-NiO-x transferred to the defect energy level and accumulated there. The holes accumulated in the valence band were consumed by triethanolamine (TEOA), and the TEOA oxidized to diethanolamine and glycolaldehyde. At the same time, $Ru(bpy)_3^{2+}$ activated to the excited state of $Ru(bpy)_3^{2+*}$ under light irradiation and was then quenched by the TEOA to form $Ru(bpy)_3^+$. Subsequently, the electrons of the $Ru(bpy)_3^+$ were transferred to the conduction band of the NiO, and further accumulated at the defect level, and the Ru(bpy)$_3^+$ was oxidized to the initial state, Ru(bpy)$_3^{2+}$. Furthermore, CO$_2$ molecules combined with the excited-state electrons accumulated at the N-NiO-x defect level and protons in water and converted into the product, CO.

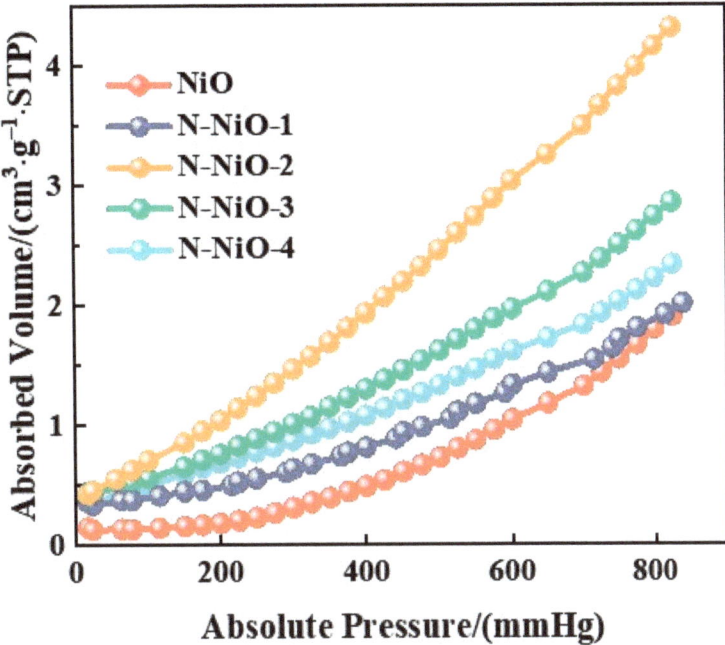

Figure 9. CO$_2$-adsorption isotherms of NiO and N-NiO-x.

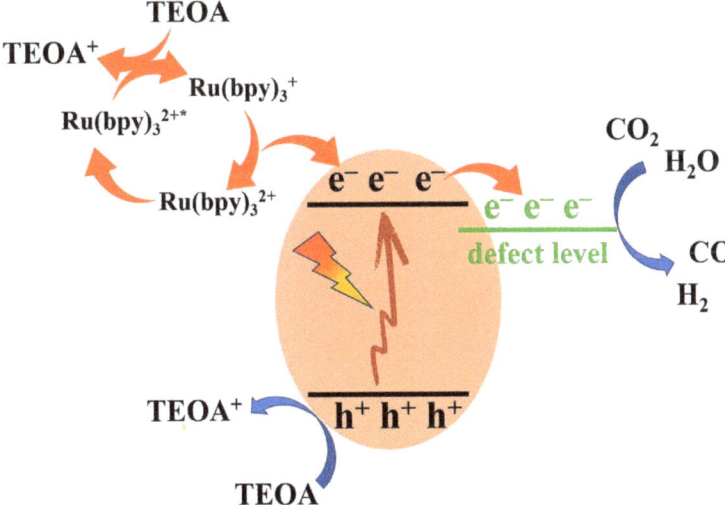

Figure 10. Possible photoreduction mechanism of CO$_2$ on N-NiO-x (* represents the excited state).

3. Experimental Section

3.1. Materials

The used chemicals were nickel nitrate hexahydrate ($Ni(NO_3)_2 \cdot 6H_2O$, Sinopharm Chemical Reagent Co., Ltd., Shanghai, China), sodium hydroxide (NaOH, Shanghai Aladdin Biochemical Technology Co., Ltd., Shanghai, China), anhydrous lithium chloride (LiCl, Shanghai McLean Biochemical Technology Co., Ltd., Shanghai, China), potassium chloride (KCl, Xilong Science Co., Ltd., Shantou, China), urea (CH_4N_2O, Shanghai McLean Biochemical Technology Co., Ltd., Shanghai, China), $[Ru(bpy)_3]Cl_2 \cdot 6H_2O$(Shanghai McLean Biochemical Technology Co., Ltd., Shanghai, China), triethanolamine (TEOA, Xilong Science Co., Ltd., Shantou, China) and acetonitrile (MeCN, Xilong Science Co., Ltd., Shantou, China). All chemicals were analytically pure and could be used directly without purification after purchase.

3.2. Synthesis of Precursor $Ni(OH)_2$

Precursor $Ni(OH)_2$ was prepared by a simple precipitation method: a total of 10 mmol $Ni(NO_3)_2 \cdot 6H_2O$ was dissolved in 40 mL of deionized (DI) water under magnetic stirring, after which 20 mmol of NaOH was added when the solid was completely dissolved. After 30 min of stirring, the precipitate was collected by filtration, washed once with 10 mL of deionized (DI) water and anhydrous ethanol, and dried at 60 °C for 8 h to obtain precursor $Ni(OH)_2$.

3.3. Synthesis of NiO

The 5 mmol of precursor $Ni(OH)_2$ was fully ground with 2.7 g of LiCl and 3.3 g of KCl and then calcined at 400 °C for 3 h. After the reaction, the bulk was fully dissolved in DI water and filtered, after which it was washed several times with DI water and ethanol alternately before drying at 60 °C for 8 h to obtain NiO. For comparison, traditional P-type NiO (T-NiO) was obtained by directly calcining $Ni(OH)_2$.

3.4. Synthesis of N-NiO-x

With urea as the nitrogen-doping source, excessive urea was added to the reaction to reduce the effect of volatilization. The preparation process of nitrogen-doped NiO was as follows: a total of 5 mmol of precursor $Ni(OH)_2$, m g urea (m = 0.2,0.3,0.4,0.5), 2.7 g of LiCl, and 3.3 g of KCl were fully ground and then calcined at 400 °C for 3 h. The bulk after reaction was fully dissolved in appropriate DI water and filtered, after which it was washed several times with DI water and ethanol alternately and dried at 60 °C for 8 h to obtain N-NiO-x (x is 1, 2, 3, 4). Nitrogen and oxygen contents over N-NiO-2 were determined by inert-gas-fusion technique using a nitrogen-and-oxygen elemental analyzer (LECO Corp., TC-436AR, St. Joseph, USA). The carbon content was obtained by carbon–sulfur analyzer. The ratio of C over N-NiO-2 was 0.02 wt% (probable error), which was negligible compared with 2.03 wt% N and 19.21 wt% (Table S2).

3.5. Photocatalytic CO_2 Reduction

In this research, the catalytic performances of the samples were evaluated for CO_2-reduction activity. The light source was an 80-watt LED lamp (illumination wavelengths were 365 nm, 420 nm, 550 nm, Zhenjiang Yinzhu Chemical Technology Co., Ltd., Zhenjiang, China). Typically, 30 mg of the catalyst, 5 mg of $[Ru(bpy)_3]Cl_2 \cdot 6H_2O$ (denoted as Ru), 3 mL of MeCN, 2 mL of H_2O, and 1 mL of TEOA were added to a 50-milliliter quartz reactor. Before the start of the reaction, the reactor was first vented with pure CO_2 for 30 min in the dark, in order to make the reaction system reach the adsorption saturation of CO_2; next, 1 mL of gas was extracted every 2 h under illumination and injected into a chromatographic system (H_2 and CO were detected by thermal-conductivity detector and flame-ionization detector, respectively).

The CO selectivity was calculated using the following formula:

$$\text{CO Selectivity } (S_{CO}) = \frac{Y_{CO}}{Y_{CO} + Y_{H2}} \tag{2}$$

where Y_{CO} and Y_{H2} represent the yields of CO and H_2, respectively.

Furthermore, the optical powers at different wavelengths were measured via an optical power meter, with a probe area of 1×1 cm^2 to contact light. The light-irradiation area was 2.5×2.5 cm^2. The apparent quantum efficiency (AQE) was calculated using the following formula:

$$AOE = \frac{2 \times the\ number\ of\ evolved\ CO\ molecules}{N} \tag{3}$$

$$N = \frac{E\lambda}{hc} \tag{4}$$

where
N: the number of incident photons;
E: the accumulated light energy in the given area (J);
λ: the wavelength of the light;
h: Planck's constant (6.626×10^{-34} J·s);
c: the velocity of light (3×10^8 m·s^{-1}).

3.6. Characterizations

The phase structure of the material was measured using X-ray diffraction (XRD, Cu Kα, λ = 0.15406 nm, Bruker D8 Advance). The microstructure and element distributions of the prepared samples were evaluated using scanning-electron microscopy (SEM, FESEM ZEISS sigma 500, Oberkochen, Batenwerburg, GER), transmission-electron microscopy (TEM, JEM-2100F), and energy-dispersive X-ray spectroscopy (EDX). The X-ray photoelectron spectra (XPS, Thermo Fisher, K-Alpha, Waltham, MA, USA) were examined to study the chemical states of the elements. The UV–Vis diffuse-reflectance spectra (DRS, Shimadzu UV-2600, Kyoto, Japan) were examined using BaSO$_4$ as the reference standard, in order to study the optical absorption properties of the samples. The vacancy-defect state in the photocatalyst was analyzed with electron paramagnetic resonance (EPR, Bruker ER200-SLC, Billerica, MA, USA) measurement at room temperature. The CO$_2$ adsorption at 273 K under ice–water-mixture conditions was studied on an automatic physical adsorption instrument (ASAP 2020, Norcross, Georgia, USA). Steady-state fluorescence (PL) spectra detected the reintegration of exposed electron–hole pairs at an excitation wavelength of 250 nm with a fluorescence spectrometer (FLS 980, Edinburgh, Scotland). Photoelectrochemical measurements were carried out in a three-electrode system on an electrochemical workstation (Shanghai Chenhua CHI-660E, Shanghai, China) using 0.1 mol/L Na$_2$SO$_4$ or 0.1 mol/L K$_3$Fe(CN)$_6$/K$_4$Fe(CN)$_6$ buffer solution as the electrolyte solution, Ag/AgCl as the reference electrode, Pt wire as the auxiliary electrode, and indium-tin-oxide conductive glass (ITO) as the working electrode (10 mg of the sample was dissolved in 3 drops of ethanol, including 10 µL of nafion solution, after which the solution was subjected to ultrasound for 40 min to completely disperse the sample, with an effective loading area of 0.25 cm^2).

4. Conclusions

In summary, NiO semiconductors doped with non-metallic nitrogen were successfully prepared using a molten-salt method. Compared with the p-type NiO, the reduction performance of the n-type NiO was improved. Furthermore, the photoreduction of CO$_2$ by the n-type NiO was more efficient with N doping. The improvement in the photocatalytic performance of the NiO semiconductor doped with non-metallic nitrogen was mainly due to three factors: (1) the molten-salt atmosphere increased the transformation of the p-type NiO to n-type and the conduction-band position met the potential requirements for CO$_2$

reduction, thus enhancing the reduction performance; (2) the nitrogen doping increased the adsorption and activation of CO_2 on the surface of the NiO semiconductor, and realized the rapid conversion of CO_2; 3) the defect-energy level induced by the oxygen defects increased the transfer and separation of electrons, and the CO_2 obtained electrons at the oxygen defects more easily and reduced. This research provides a new reference for solving the insufficient reduction performance of p-type NiO, as well as a new control method for inhibiting the photogenerated charge recombination of n-type NiO semiconductors.

Supplementary Materials: The following supporting information can be downloaded at: https://www.mdpi.com/article/10.3390/molecules28062435/s1, Table S1: Average particle diameters of the samples; Table S2: The contents of C and N obtained by carbon-sulfur analyzer and nitrogen-oxygen elemental analyzer; Figure S1: Photos of NiO and NiO-x; Figure S2: XPS survey spectra (a) and C 1s spectra (b) of N-NiO-2; Figure S3: TCD (a) and FID (b) of gas chromatogram; (c) Liquid chromatogram.

Author Contributions: Conceptualization, F.W. and K.Y.; methodology, Z.Y.; XRD, DRS, SEM and TEM analysis, X.L. and K.S.; XPS analysis, K.L.; resources, W.H.; data curation, K.L.; writing—original draft preparation, F.W.; writing—review and editing, K.Y. and C.Y.; supervision, K.Y. and C.Y.; funding acquisition, K.Y. and C.Y. All authors have read and agreed to the published version of the manuscript.

Funding: This research was funded by the National Natural Science Foundation of China (21962006, 22272034), Jiangxi Provincial Academic and Technical Leaders Training Program—Young Talents (20204BCJL23037), Program of Qingjiang Excellent Young Talents, JXUST (JXUSTQJBJ2020005), Ganzhou Young Talents Program of Jiangxi Province (204301000111), Postdoctoral Research Projects of Jiangxi Province in 2020 (204302600031), Jiangxi Province "Double Thousand Plan" (Yang Kai, Hou Yang), Guangdong Province Universities and Colleges Pearl River Scholar Funded Scheme (2019), Jiangxi Provincial Natural Science Foundation (20224BAB203018, 20212BAB213016), and the Foundation Engineering Research Center of Tungsten Resources High-Efficiency Development and Application Technology of the Ministry of Education (W-2021YB003).

Institutional Review Board Statement: Not applicable.

Informed Consent Statement: Not applicable.

Data Availability Statement: Not applicable.

Conflicts of Interest: The authors declare no conflict of interest.

Sample Availability: Samples of the compounds are not available from the authors.

References

1. Tang, J.; Guo, R.; Zhou, W.; Huang, C.; Pan, W. Ball-Flower like NiO/g-C$_3$N$_4$ Heterojunction for Efficient Visible Light Photocatalytic CO$_2$ Reduction. *Appl. Catal. B-Environ.* **2018**, *237*, 802–810. [CrossRef]
2. Chen, S.; Yu, J.; Zhang, J. Enhanced Photocatalytic CO$_2$ Reduction Activity of MOF-Derived ZnO/NiO Porous Hollow Spheres. *J. CO2 Util.* **2018**, *24*, 548–554. [CrossRef]
3. Han, C.; Zhang, R.; Ye, Y.; Wang, L.; Ma, Z.; Su, F.; Xie, H.; Zhou, Y.; Wong, P.K.; Ye, L. Chainmail Co-Catalyst of NiO Shell-Encapsulated Ni for Improving Photocatalytic CO$_2$ Reduction over g-C$_3$N$_4$. *J. Mater. Chem. A* **2019**, *7*, 9726–9735. [CrossRef]
4. Hiragond, C.B.; Lee, J.; Kim, H.; Jung, J.W.; Cho, C.H.; In, S.I. A Novel N-Doped Graphene Oxide Enfolded Reduced Titania for Highly Stable and Selective Gas-Phase Photocatalytic CO$_2$ Reduction into CH$_4$: An in-Depth Study on the Interfacial Charge Transfer Mechanism. *Chem. Eng. J.* **2021**, *416*, 127978. [CrossRef]
5. Bie, C.; Zhu, B.; Xu, F.; Zhang, L.; Yu, J. In Situ Grown Monolayer N-Doped Graphene on CdS Hollow Spheres with Seamless Contact for Photocatalytic CO$_2$ Reduction. *Adv. Mater.* **2019**, *31*, 1902868. [CrossRef]
6. Wang, L.; Zhu, B.; Cheng, B.; Zhang, J.; Zhang, L.; Yu, J. In-Situ Preparation of TiO$_2$/N-Doped Graphene Hollow Sphere Photocatalyst with Enhanced Photocatalytic CO$_2$ Reduction Performance. *Chin. J. Catal.* **2021**, *42*, 1648–1658. [CrossRef]
7. Neațu, Ș.; Maciá-Agulló, J.; Garcia, H. Solar Light Photocatalytic CO$_2$ Reduction: General Considerations and Selected Bench-Mark Photocatalysts. *Int. J. Mol. Sci.* **2014**, *15*, 5246–5262. [CrossRef] [PubMed]
8. Mao, J.; Li, K.; Peng, T. Recent Advances in the Photocatalytic CO$_2$ Reduction over Semiconductors. *Catal. Sci. Technol.* **2013**, *3*, 2481. [CrossRef]
9. Kong, X.; Lv, F.; Zhang, H.; Yu, F.; Wang, Y.; Yin, L.; Huang, J.; Feng, Q. NiO Load K$_2$Fe$_4$O$_7$ Enhanced Photocatalytic Hydrogen Production and Photo-Generated Carrier Behavior. *J. Alloys Comp.* **2022**, *903*, 163864. [CrossRef]

10. Jiao, Z.F.; Tian, Y.M.; Zhang, B.; Hao, C.H.; Qiao, Y.; Wang, Y.X.; Qin, Y.; Radius, U.; Braunschweig, H.; Marder, T.B.; et al. High Photocatalytic Activity of a NiO Nanodot-Decorated Pd/SiC Catalyst for the Suzuki-Miyaura Cross-Coupling of Aryl Bromides and Chlorides in Air under Visible Light. *J. Catal.* **2020**, *389*, 517–524. [CrossRef]
11. He, L.; Zhang, W.; Liu, S.; Zhao, Y. Three-Dimensional Palm Frondlike Co_3O_4@NiO/Graphitic Carbon Composite for Photocatalytic CO_2 Reduction. *J. Alloys Comp.* **2023**, *934*, 168053. [CrossRef]
12. Prajapati, P.K.; Singh, H.; Yadav, R.; Sinha, A.K.; Szunerits, S.; Boukherroub, R.; Jain, S.L. Core-Shell Ni/NiO Grafted Cobalt (II) Complex: An Efficient Inorganic Nanocomposite for Photocatalytic Reduction of CO_2 under Visible Light Irradiation. *Appl. Surf. Sci.* **2019**, *467–468*, 370–381. [CrossRef]
13. Kamata, R.; Kumagai, H.; Yamazaki, Y.; Sahara, G.; Ishitani, O. Photoelectrochemical CO_2 Reduction Using a Ru(II)–Re(I) Supramolecular Photocatalyst Connected to a Vinyl Polymer on a NiO Electrode. *ACS Appl. Mater. Interfaces* **2019**, *11*, 5632–5641. [CrossRef] [PubMed]
14. Tahir, M.; Tahir, B.; Amin, N.A.S.; Muhammad, A. Photocatalytic CO_2 Methanation over NiO/In_2O_3 Promoted TiO_2 Nanocatalysts Using H_2O and/or H_2 Reductants. *Energy Convers. Manag.* **2016**, *119*, 368–378. [CrossRef]
15. Hong, W.; Zhou, Y.; Lv, C.; Han, Z.; Chen, G. NiO Quantum Dot Modified TiO_2 toward Robust Hydrogen Production Performance. *ACS Sustain. Chem. Eng.* **2018**, *6*, 889–896. [CrossRef]
16. Lan, D.; Pang, F.; Ge, J. Enhanced Charge Separation in NiO and Pd Co-Modified TiO_2 Photocatalysts for Efficient and Selective Photoreduction of CO_2. *ACS Appl. Energy Mater.* **2021**, *4*, 6324–6332. [CrossRef]
17. Chen, W.; Liu, X.; Han, B.; Liang, S.; Deng, H.; Lin, Z. Boosted Photoreduction of Diluted CO_2 through Oxygen Vacancy Engineering in NiO Nanoplatelets. *Nano Res.* **2021**, *14*, 730–737. [CrossRef]
18. Haq, S.; Sarfraz, A.; Menaa, F.; Shahzad, N.; Din, S.U.; Almukhlifi, H.A.; Alshareef, S.A.; Al Essa, E.M.; Shahzad, M.I. Green Synthesis of NiO-SnO_2 Nanocomposite and Effect of Calcination Temperature on Its Physicochemical Properties: Impact on the Photocatalytic Degradation of Methyl Orange. *Molecules* **2022**, *27*, 8420. [CrossRef]
19. Wang, Z.; Cheng, B.; Zhang, L.; Yu, J.; Tan, H. BiOBr/NiO S-Scheme Heterojunction Photocatalyst for CO_2 Photoreduction. *Sol. RRL* **2022**, *6*, 2100587. [CrossRef]
20. Park, B.H.; Kim, M.; Park, N.K.; Ryu, H.J.; Baek, J.; Kang, M. Single Layered Hollow NiO–NiS Catalyst with Large Specific Surface Area and Highly Efficient Visible-Light-Driven Carbon Dioxide Conversion. *Chemosphere* **2021**, *280*, 130759. [CrossRef]
21. Xiang, J.; Zhang, T.; Cao, R.; Lin, M.; Yang, B.; Wen, Y.; Zhuang, Z.; Yu, Y. Optimizing the Oxygen Vacancies Concentration of Thin NiO Nanosheets for Efficient Selective CO_2 Photoreduction. *Sol. RRL* **2021**, *5*, 2100703. [CrossRef]
22. Park, H.R.; Pawar, A.U.; Pal, U.; Zhang, T.; Kang, Y.S. Enhanced Solar Photoreduction of CO_2 to Liquid Fuel over RGO Grafted NiO-CeO_2 Heterostructure Nanocomposite. *Nano Energy* **2021**, *79*, 105483. [CrossRef]
23. Wang, X.; Wang, J.; Li, Y.; Chu, K. Nitrogen-Doped NiO Nanosheet Array for Boosted Electrocatalytic N_2 Reduction. *ChemCatChem* **2019**, *11*, 4529–4536. [CrossRef]
24. Jaiswal, R.; Bharambe, J.; Patel, N.; Dashora, A.; Kothari, D.C.; Miotello, A. Copper and Nitrogen Co-Doped TiO_2 Photocatalyst with Enhanced Optical Absorption and Catalytic Activity. *Appl. Catal. B-Environ.* **2015**, *168–169*, 333–341. [CrossRef]
25. An, L.; Park, Y.; Sohn, Y.; Onishi, H. Effect of Etching on Electron–Hole Recombination in Sr-Doped $NaTaO_3$ Photocatalysts. *J. Phys. Chem. C* **2015**, *119*, 28440–28447. [CrossRef]
26. Qian, B.; Chen, Y.; Tade, M.O.; Shao, Z. $BaCo_{0.6}Fe_{0.3}Sn_{0.1}O_{3-\delta}$ Perovskite as a New Superior Oxygen Reduction Electrode for Intermediate-to-Low Temperature Solid Oxide Fuel Cells. *J. Mater. Chem. A* **2014**, *2*, 15078. [CrossRef]
27. Wang, W.; Tadé, M.O.; Shao, Z. Nitrogen-Doped Simple and Complex Oxides for Photocatalysis: A Review. *Prog. Mater. Sci.* **2018**, *92*, 33–63. [CrossRef]
28. Luo, L.; Wang, S.; Wang, H.; Tian, C.; Jiang, B. Molten-Salt Technology Application for the Synthesis of Photocatalytic Materials. *Energy Technol.* **2021**, *9*, 2000945. [CrossRef]
29. Liu, W.; Mu, Y.F.; Yao, S.; Guo, S.; Guo, X.W.; Zhang, Z.M.; Lu, T.B. Photosensitizing single-site metal− organic framework enabling visible-light-driven CO2 reduction for syngas production. *Appl. Catal. B-Environ.* **2019**, *245*, 496–501. [CrossRef]
30. Yang, J.; Wang, Z.; Jiang, J.; Chen, W.; Liao, F.; Ge, X.; Zhou, X.; Chen, M.; Li, R.; Xue, Z.; et al. In-Situ Polymerization Induced Atomically Dispersed Manganese Sites as Cocatalyst for CO_2 Photoreduction into Synthesis Gas. *Nano Energy* **2020**, *76*, 105059. [CrossRef]
31. Xiong, J.; Wang, X.; Wu, J.; Han, J.; Lan, Z.; Fan, J. In Situ Fabrication of N-Doped ZnS/ZnO Composition for Enhanced Visible-Light Photocatalytic H_2 Evolution Activity. *Molecules* **2022**, *27*, 8544. [CrossRef]
32. Zhang, J.; Wu, Y.; Xing, M.; Leghari, S.A.K.; Sajjad, S. Development of Modified N Doped TiO_2 Photocatalyst with Metals, Nonmetals and Metal Oxides. *Energy Environ. Sci.* **2010**, *3*, 715. [CrossRef]
33. Sarngan, P.P.; Lakshmanan, A.; Sarkar, D. Influence of Anatase-Rutile Ratio on Band Edge Position and Defect States of TiO_2 Homojunction Catalyst. *Chemosphere* **2022**, *286*, 131692. [CrossRef] [PubMed]
34. Sun, M.; Zhou, Y.; Yu, T. Synthesis of g-C_3N_4/NiO-Carbon Microsphere Composites for Co-Reduction of CO_2 by Photocatalytic Hydrogen Production from Water Decomposition. *J. Clean. Prod.* **2022**, *357*, 131801. [CrossRef]
35. Wang, J.; Yin, S.; Komatsu, M.; Zhang, Q.; Saito, F.; Sato, T. Photo-Oxidation Properties of Nitrogen Doped $SrTiO_3$ Made by Mechanical Activation. *Appl. Catal. B-Environ* **2004**, *52*, 11–21. [CrossRef]
36. Senobari, S.; Nezamzadeh-Ejhieh, A. A P-n Junction NiO-CdS Nanoparticles with Enhanced Photocatalytic Activity: A Response Surface Methodology Study. *J. Mol. Liq.* **2018**, *257*, 173–183. [CrossRef]

37. Deng, C.; Hu, H.; Yu, H.; Wang, M.; Ci, M.; Wang, L.; Zhu, S.; Wu, Y.; Le, H. 1D Hierarchical CdS NPs/NiO NFs Heterostructures with Enhanced Photocatalytic Activity under Visible Light Irradiation. *Adv. Powder Technol.* **2020**, *31*, 3158–3167. [CrossRef]
38. Hu, X.; Wang, G.; Wang, J.; Hu, Z.; Su, Y. Step-Scheme NiO/BiOI Heterojunction Photocatalyst for Rhodamine Photodegradation. *Appl. Surf. Sci.* **2020**, *511*, 145499. [CrossRef]
39. Chen, J.; Wang, M.; Han, J.; Guo, R. TiO$_2$ Nanosheet/NiO Nanorod Hierarchical Nanostructures: P–n Heterojunctions towards Efficient Photocatalysis. *J. Colloid Interf. Sci.* **2020**, *562*, 313–321. [CrossRef] [PubMed]
40. Liang, R.; Wang, S.; Lu, Y.; Yan, G.; He, Z.; Xia, Y.; Liang, Z.; Wu, L. Assembling Ultrafine SnO$_2$ Nanoparticles on MIL-101(Cr) Octahedrons for Efficient Fuel Photocatalytic Denitrification. *Molecules* **2021**, *26*, 7566. [CrossRef]
41. Hao, X.; Cui, Z.; Zhou, J.; Wang, Y.; Hu, Y.; Wang, Y.; Zou, Z. Architecture of High Efficient Zinc Vacancy Mediated Z-Scheme Photocatalyst from Metal-Organic Frameworks. *Nano Energy* **2018**, *52*, 105–116. [CrossRef]
42. Naik, K.M.; Hamada, T.; Higuchi, E.; Inoue, H. Defect-Rich Black Titanium Dioxide Nanosheet-Supported Palladium Nanoparticle Electrocatalyst for Oxygen Reduction and Glycerol Oxidation Reactions in Alkaline Medium. *ACS Appl. Energy Mater.* **2021**, *4*, 12391–12402. [CrossRef]
43. Nosaka, Y.; Nosaka, A.Y. Generation and Detection of Reactive Oxygen Species in Photocatalysis. *Chem. Rev.* **2017**, *117*, 11302–11336. [CrossRef]
44. Qi, K.; Liu, S.; Qiu, M. Photocatalytic Performance of TiO$_2$ Nanocrystals with/without Oxygen Defects. *Chin. J. Catal.* **2018**, *39*, 867–875. [CrossRef]
45. Qian, J.; Bai, X.; Xi, S.; Xiao, W.; Gao, D.; Wang, J. Bifunctional Electrocatalytic Activity of Nitrogen-Doped NiO Nanosheets for Rechargeable Zinc–Air Batteries. *ACS Appl. Mater. Interfaces* **2019**, *11*, 30865–30871. [CrossRef] [PubMed]
46. Guo, F.; Wang, L.; Sun, H.; Li, M.; Shi, W. High-Efficiency Photocatalytic Water Splitting by a N-Doped Porous g-C$_3$N$_4$ Nanosheet Polymer Photocatalyst Derived from Urea and N,N-Dimethylformamide. *Inorg. Chem. Front.* **2020**, *7*, 1770–1779. [CrossRef]
47. Ângelo, J.; Magalhães, P.; Andrade, L.; Mendes, A. Characterization of TiO$_2$-Based Semiconductors for Photocatalysis by Electrochemical Impedance Spectroscopy. *Appl. Surf. Sci.* **2016**, *387*, 183–189. [CrossRef]
48. Yu, Z.; Yang, K.; Yu, C.; Lu, K.; Huang, W.; Xu, L.; Zou, L.; Wang, S.; Chen, Z.; Hu, J.; et al. Steering Unit Cell Dipole and Internal Electric Field by Highly Dispersed Er Atoms Embedded into NiO for Efficient CO$_2$ Photoreduction. *Adv. Funct. Mater.* **2022**, *32*, 2111999. [CrossRef]
49. Zhu, S.; Liao, W.; Zhang, M.; Liang, S. Design of Spatially Separated Au and CoO Dual Cocatalysts on Hollow TiO$_2$ for Enhanced Photocatalytic Activity towards the Reduction of CO$_2$ to CH$_4$. *Chem. Eng. J.* **2019**, *361*, 461–469. [CrossRef]
50. Jiang, L.; Wang, K.; Wu, X.; Zhang, G. Highly Enhanced Full Solar Spectrum-Driven Photocatalytic CO$_2$ Reduction Performance in Cu$_{2-x}$S/g-C$_3$N$_4$ Composite: Efficient Charge Transfer and Mechanism Insight. *Sol. RRL* **2021**, *5*, 2000326. [CrossRef]
51. Liu, L.; Jiang, Y.; Zhao, H.; Chen, J.; Cheng, J.; Yang, K.; Li, Y. Engineering Coexposed {001} and {101} Facets in Oxygen-Deficient TiO$_2$ Nanocrystals for Enhanced CO$_2$ Photoreduction under Visible Light. *ACS Catal.* **2016**, *6*, 1097–1108. [CrossRef]
52. Katal, R.; Masudy-Panah, S.; Sabbaghan, M.; Hossaini, Z.; Davood Abadi Farahani, M.H. Photocatalytic Degradation of Triclosan by Oxygen Defected CuO Thin Film. *Sep. Purif. Technol.* **2020**, *250*, 117239. [CrossRef]
53. Tian, F.; Liu, Y. Synthesis of p-Type NiO/n-Type ZnO Heterostructure and Its Enhanced Photocatalytic Activity. *Scr. Mater.* **2013**, *69*, 417–419. [CrossRef]
54. Jones, B.M.F.; Maruthamani, D.; Muthuraj, V. Construction of Novel n-Type Semiconductor Anchor on 2D Honey Comb like FeNbO$_4$/RGO for Visible Light Drive Photocatalytic Degradation of Norfloxacin. *J. Photochem. Photobiol. A Chem.* **2020**, *400*, 112712. [CrossRef]
55. Mu, Y.; Zhou, M.; Yang, K.; Zhou, C.; Mi, Y.; Yu, Z.; Lu, K.; Li, Z.; Ouyang, S.; Huang, W.; et al. Sulfur Vacancies Engineered over Cd$_{0.5}$Zn$_{0.5}$S by Yb^{3+}/Er^{3+} Co-Doping for Enhancing Photocatalytic Hydrogen Evolution. *Sustain. Energy Fuels* **2021**, *5*, 5814–5824. [CrossRef]

Disclaimer/Publisher's Note: The statements, opinions and data contained in all publications are solely those of the individual author(s) and contributor(s) and not of MDPI and/or the editor(s). MDPI and/or the editor(s) disclaim responsibility for any injury to people or property resulting from any ideas, methods, instructions or products referred to in the content.

Article

CVD Growth of Hematite Thin Films for Photoelectrochemical Water Splitting: Effect of Precursor-Substrate Distance on Their Final Properties

Leunam Fernandez-Izquierdo [1,2,3], Enzo Luigi Spera [4], Boris Durán [5], Ricardo Enrique Marotti [4], Enrique Ariel Dalchiele [4], Rodrigo del Rio [1,2] and Samuel A. Hevia [1,6,*]

[1] Centro de Investigación en Nanotecnología y Materiales Avanzados (CIEN-UC), Pontificia Universidad Católica de Chile, Casilla 306, Santiago 6904411, Chile
[2] Facultad de Química y de Farmacia, Pontificia Universidad Católica de Chile, Casilla 306, Santiago 6904411, Chile
[3] Department of Material Science and Engineering, The University of Texas at Dallas, 2601 North Floyd Road RL10, Richardson, TX 75080, USA
[4] Instituto de Física, Facultad de Ingeniería, Universidad de la República, Julio Herrera y Reissig 565, C.C. 30, Montevideo 11000, Uruguay
[5] Departamento de Medicina Traslacional, Facultad de Medicina, Universidad Católica del Maule, Talca 3480112, Chile
[6] Instituto de Física, Pontificia Universidad Católica de Chile, Casilla 306, Santiago 6904411, Chile
* Correspondence: sheviaz@uc.cl

Abstract: The development of photoelectrode materials for efficient water splitting using solar energy is a crucial research topic for green hydrogen production. These materials need to be abundant, fabricated on a large scale, and at low cost. In this context, hematite is a promising material that has been widely studied. However, it is a huge challenge to achieve high-efficiency performance as a photoelectrode in water splitting. This paper reports a study of chemical vapor deposition (CVD) growth of hematite nanocrystalline thin films on fluorine-doped tin oxide as a photoanode for photoelectrochemical water splitting, with a particular focus on the effect of the precursor–substrate distance in the CVD system. A full morphological, structural, and optical characterization of hematite nanocrystalline thin films was performed, revealing that no change occurred in the structure of the films as a function of the previously mentioned distance. However, it was found that the thickness of the hematite film, which is a critical parameter in the photoelectrochemical performance, linearly depends on the precursor–substrate distance; however, the electrochemical response exhibits a nonmonotonic behavior. A maximum photocurrent value close to 2.5 mA/cm^2 was obtained for a film with a thickness of around 220 nm under solar irradiation.

Keywords: hematite; chemical vapor deposition; thin film; water splitting

1. Introduction

Hematite is an attractive semiconductor material for photoelectrochemical and photocatalytic purposes due to its stability, abundance, and environmental compatibility, as well as its suitable bandgap and valence band edge position [1–3]. Moreover, this oxide shows a high potential for commercial application in water splitting since its theoretical solar-to-hydrogen efficiency (STH) is as high as 12.9% [1,3,4]. In general, hematite exhibits an n-type semiconducting behavior, which can be due to the tendency of hematite to become oxygen-deficient, irrespective of the preparation method [1]. Unfortunately, because hematite presents a weak optical absorption coefficient, small carrier mobility, and a short hole diffusion length, the illuminated hematite electrodes normally exhibit poor efficiency as photoanodes in water oxidation [5,6]. However, many studies have indicated that nanocrystalline morphology could be one way to overcome those limitations and increase the photon-to-current yield by minimizing the distance that minority carriers must diffuse before reaching the interface and contacts [1,7–10]. In addition, the large surface area exhibited by the nanostructured semiconductors is important in photoelectrochemical devices due to the large number of catalytic sites [3,7,11–13]. Additionally, the nano structuration facilitates the incorporation of other nanomaterials or atoms that improve the catalytic performance [14–18]. Several techniques have been investigated for the synthesis of hematite nanocrystalline thin films: thermal evaporation [19], aqueous chemical growth [20], spray-pyrolysis [10,21,22], sol-gel [23], and electrodeposition methods [1,8,13]. A low-cost and solvent-free method for preparing semiconducting thin films well-suited for the preparation of nanostructures is chemical vapor deposition (CVD) [24,25]. In the last years, several research groups have employed the CVD technique to obtain iron oxide thin films. Thermal CVD synthesis of α-Fe_2O_3 thin films by using different iron precursors such as carbonyls, alkoxides, and β-diketonates has been reported [24–31]. However, it is still necessary to achieve a deep understanding of the hematite thin film growth by CVD in order to develop efficient photoelectrodes for solar photoelectrochemical water splitting.

This work reports a study of the CVD growth process of hematite thin films onto fluorine-doped tin oxide (FTO) substrates by using the ferrocene organometallic compound as an iron precursor. A complete morphological and structural characterization of the hematite films was carried out using scanning electron microscopy (SEM), energy dispersive X-ray spectroscopy (EDS), Raman spectroscopy, X-ray photoelectron spectroscopy (XPS), and X-ray diffraction (XRD). An optical characterization with a thorough analysis was performed due to the importance of this property for the previously mentioned application. Finally, the photoelectrochemical response of the films was evaluated, and the dependence of the distance between the precursor and the substrates in the CVD system was studied in order to find the optimal fabrication condition that maximized the photoelectrochemical response.

2. Results and Discussion

2.1. Morphological and Structural Characterization of the Hematite Thin Films

Nanostructured hematite thin films on FTO substrates were obtained by CVD using ferrocene as precursor in a two-step method. The substrates were located at 10, 14, 18, and 22 cm from the source of the precursor, as shown in Figure 1.

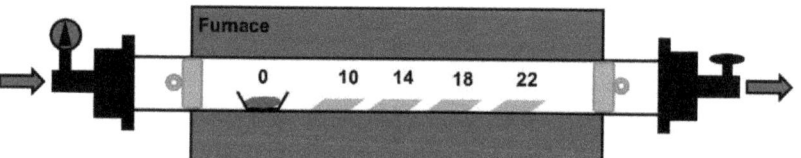

Figure 1. Scheme of the CVD system. Two quartz plugs were placed inside the quartz tube to establish the reaction zone. At the gas entrance (from left to right), was placed a quartz plug, at 10 cm from it was placed the boat with the precursor. The FTO substrates were located 10, 14, 18, and 22 cm from the boat. In the gas exit, a second plug was located (12 cm away from the last substrate). It is important to notice that the plugs do not seal the tube, they only obstruct the flow out from the reaction zone.

It was observed that samples decreased in transparency at the same time as the distance of the precursor, called "x", increased (see Figure 1). Differences in color and transparency of the electrodes can be observed in Figure S1 in the Supplementary Materials. Figure 2 shows SEM images of hematite films grown by CVD on FTO substrates at different values of position X. Figure 2a–d corresponds to 45-degree tilted view micrographs of samples prepared at the positions of 10, 14, 18, and 22 cm, respectively. Figure 2e–h corresponds to cross-section micrographs of the same set of samples. From this set of images, it is observed that the thickness and morphology of the film depend on the position of the substrate in the CVD system.

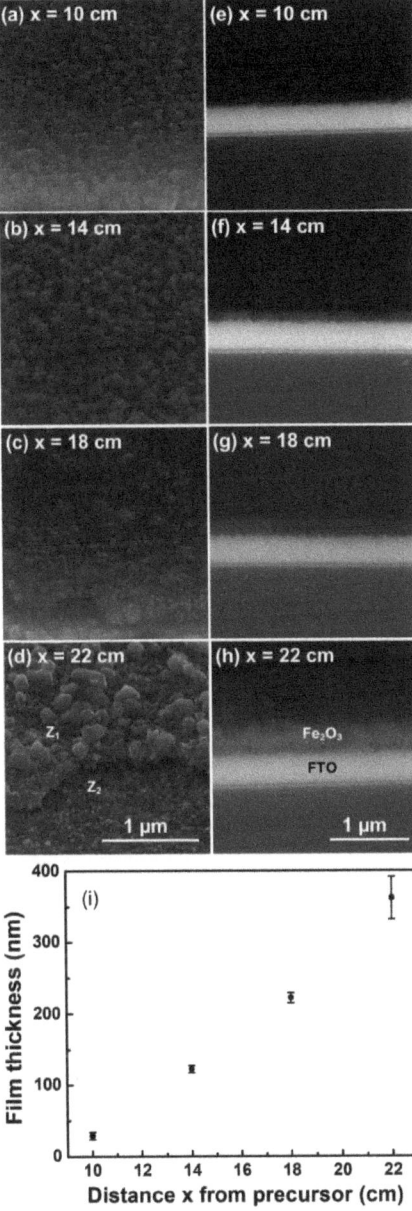

Figure 2. (**a**–**d**) correspond to 45-degree tilted view SEM micrographs of samples prepared at the positions 10, 14, 18, and 22 cm, respectively. In (**d**), Z_1 and Z_2 denote two different zones of the sample, the zone with hematite and the zone of bare FTO, respectively. (**e**–**h**) correspond to cross-section SEM micrographs of the same set of samples. (**i**) Plot of the film thickness (determined by the cross-section micrograph) as a function of their distance to the precursor.

The morphology of the film consists of nanostructures joined to each other, giving the impression that growth in the form of bunches, where the upper ends could have the same origin. So, it could be possible to note that the growth mechanism followed a similar behavior to the one proposed in the Stranski–Krastanov model, where the interaction of the adsorbed atoms among them was similar to that of the adsorbed atoms with the substrate surface [9,32]. In this case, after forming one or more monolayers, subsequent layer growth became unfavorable, and islands were formed [33].

From the cross-section SEM micrographs (Figure 2e–h), it is observed that the thickness of the hematite films rises as the distance between the substrate and the precursor increase. By measuring the different thicknesses in the cross-section micrographs, it was observed that the FTO has a thickness of 315 nm, and the hematite films have thicknesses of 362, 222, 122, and 29 nm for the electrodes prepared at the positions of 10, 14, 18, and 22 cm, respectively. The plot of the film thickness as a function of their distance to the precursor is shown in Figure 2i. The hematite film thickness depends almost linearly on the precursor–substrate distance. The reason for this is that the employed CVD reactor included two quartz plugs, one at the entrance and the other at the exit of the tube reactor (see Figure 1). The purpose of these quartz plugs was to maintain the precursor confined inside the reaction chamber in the zone where the temperature is uniform. However, the presence of these plugs, particularly the outlet plug, jams the gas flow outlet, resulting in an increasing precursor concentration towards the end of the reactor tube. As a consequence of that, the thickness of the grown films follows this concentration gradient.

The dispersive energy X-ray spectroscopy (EDS) analysis, performed under the same conditions for all samples, showed mainly that the percentage of iron in the film rose as their thickness increased (Table S1 in the Supplementary Materials).

Raman spectroscopy was performed in the range of 150–900 cm^{-1}, where the iron oxides and the hematite exhibit their characteristic vibration modes. Figure 3a shows the Raman spectra of the hematite films prepared at 10, 14, 18, and 22 cm, and of the FTO substrate. Seven main resonances can be observed in the hematite film spectra with intensities that increase at the same time that the thickness of the film increases. The peaks located at 227 cm^{-1} and 498 cm^{-1} were assigned to the A_{1g} vibrational modes, and the peaks in 247, 294, 301, 412, and 613 cm^{-1} were assigned to the E_g vibrational modes of the hematite phase of iron oxide [19,26,34]. This analysis confirmed that the annealing treatment for 30 min at 550 °C was suitable to fully transform the initial iron precursor films to hematite. The peak at 660 cm^{-1} has been previously found in hematite. Initially, some researchers attributed this peak to the maghemite or magnetite presence. However, recent publications showed that this resonance has its origin in a Raman-forbidden LO Eu mode, activated through disorder-induced symmetry breaking. This structural disorder can be found in nanostructured systems, or in our case, in nanocrystalline thin films [35].

XPS spectra of the hematite films grown at 14, 18, and 22 cm are shown in Figure 3b. The signals associated with Fe and O are clearly identified in the spectra, and with lower intensity, the adventitious carbon is also present. High-resolution measurements of Fe2p and O1s are shown in Figure 3c,d, respectively. The shape of these signals is very similar between the films. Figure 3e,f presents the Fe2p and O1s signals of the film grown at 18 cm and fitted using the software Multipack (the fits for the other films are shown in Figure S2). The binding energy (BE) of peaks associated with Fe $2p^{1/2}$ and Fe $2p^{3/2}$ are 724.5 and 710.9 eV, respectively. These BE values are characteristic of the Fe(III) in hematite, which is also consistent with the BE of the peak associated with the Fe(III)–O bond in the signal O1s, which is 529.3 eV [3,23,36]. The Raman and XPS results confirm that the films are composed of hematite.

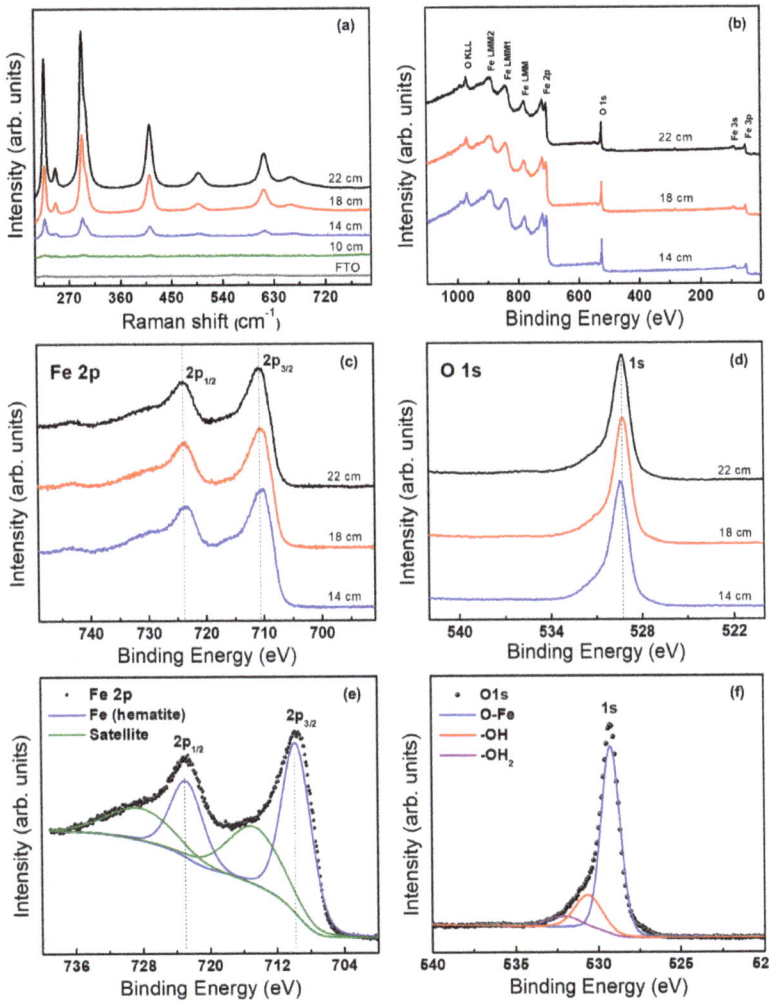

Figure 3. (a) Raman spectra of FTO and electrodes prepared at 10, 14, 18, and 22 cm. XPS spectra of hematite films grown at 14, 18 and 22 cm: (b) survey, (c,d) are the high-resolution measurements of signals Fe2p and O1s, respectively. (e,f) are the Fe2p and O1s signals of the film grown at 18 cm, fitted using the software Multipack.

In order to study the structural properties of the hematite films, X-ray diffraction experiments have been carried out. Figure 4 shows the X-ray diffraction patterns of films prepared at positions 14, 18, and 22 cm. It can be seen that all diffraction peaks (except the peaks of the substrate), can be indexed to the rhombohedrally centered hexagonal structure of Fe$_2$O$_3$ (α-Fe$_2$O$_3$, hematite), which are in agreement with standard reported values [37], see JCPDS pattern at the bottom of each panel in Figure 4. All samples are polycrystalline, and the broadening of the diffraction peaks demonstrates the nanocrystalline character of nanostructured α-Fe$_2$O$_3$ layers. An average crystallite size could be obtained using the Scherrer formula for the crystallite size broadening of diffraction peaks:

$$D = \frac{k\lambda}{\beta cos\theta} \quad (1)$$

where $k = 0.94$, λ is the X-ray wavelength, θ is the Bragg angle, and β is the FWHM of the diffraction peak [38]. By applying the above-mentioned Scherrer equation, typical crystallite size values of ca. 25 nm were estimated from (110) diffraction peak for the hematite films, irrespective of the position in the CVD reactor. Moreover, it can be appreciated in the XRD diffraction patterns depicted in Figure 4, that in contrast to the powder diffractogram α-Fe$_2$O$_3$ JCPDS pattern, the relative intensity of the (110) plane, is anomalous with respect to the other planes, evidencing a preferred crystallographic orientation along the (110) direction axis vertical to the substrate [5]. Therefore, this indicates that the hematite films grow along the (110) crystallographic direction (energetically most favorable [39]), that is, the growth axis is along the c direction [40]. It is worth mentioning that this preferential orientation is the best one for the hematite structure with good electrical conductivity [5], and it is good for the photoelectrochemical process, where the electron can flow through the (001) basal plane (due to anisotropic conductivity of hematite iron oxide [6]), to the back contact, and the hole can still hop laterally between (001) planes to reach the electrolyte interface [2,5].

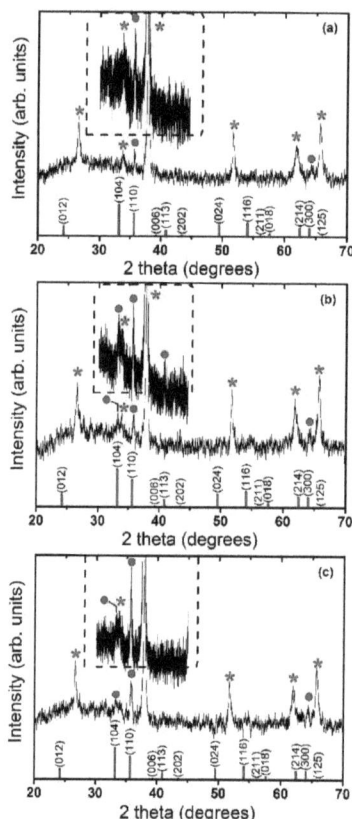

Figure 4. X-ray diffraction patterns of hematite films grown at (**a**) 14 cm, (**b**) 18 cm, and (**c**) 22 cm. JCPDS pattern of rhombohedrally centered hexagonal structure of Fe$_2$O$_3$ (α-Fe$_2$O$_3$, hematite), labeled with the corresponding crystallographic planes is also shown for comparison at the bottom of each panel. A zooming of the XRD pattern in the 2 theta range from 30 to 45 degrees is depicted as an inset in each panel. (•) and (*) symbols indicate the diffraction peaks originated from the hematite phase and from the SnO$_2$:F substrate, respectively.

2.2. Optical Characterization of the Hematite Thin Films

As is shown in Figure S1 in the Supplementary Materials, all samples have a brick-red color, with the sample becoming less translucent as the distance to the precursor increases. The optical transmittance T spectra are shown in Figure 5. For all samples, except x = 10 cm, the transmittance is lower than 25% for wavelengths shorter than 500 nm and has an abrupt increase between 500 and 600 nm. This spectral feature is characteristic of an absorption edge; the position of this edge depends on the sample. A redshift of this absorption edge can be seen at the same time the x value increase. The sample x = 18.0 cm has a decrease in the transmittance after the absorption edge. It may originate in interferences by reflections at the film–air and film–substrate interfaces. The other samples have smaller variations in the transmittance, like oscillations, that could also be because of interferences by reflections.

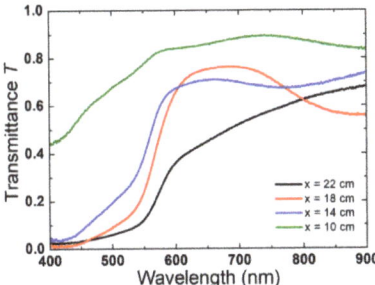

Figure 5. Optical transmittance of hematite films grown by CVD on FTO substrates. The spectra correspond to samples prepared at the positions of 10, 14, 18, and 22 cm.

To study the absorption edge, it is necessary to find the optical band gap E_g. For this optical gap determination, the nature of the absorption edge must be assumed. This is because E_g values are obtained from plots of $(\alpha h\nu)^m$ vs photon energy $h\nu$ where the absorption coefficient α is estimated from the transmittance as $\alpha \sim -\ln T$ [41]. For direct transitions, the plot with $m = 2$ shows a linear region [42]. For indirect transitions, the plot will show a linear region for $m = 1/2$ [43]. In both cases, the corresponding E_g is obtained by extrapolating the linear fitting from the plot and finding the energy value where this linear fitting intersects the zero line. The lowest bandgap of α-Fe$_2$O$_3$ is usually reported to be around 1.9–2.2 eV [7,8,10,13,22,34,36,44–54]. However, while many authors address the corresponding transition to be indirect [13,44,46,47,54], others found direct transitions corresponding to the same edge [7,22,34,49,51,53]. As a way of comparing the results to previous works, several authors report both direct and indirect E_g values from different linear fittings made in the same spectral regions [8,50].

Nevertheless, the absorption coefficient is the sum of all processes occurring in the sample. It has contributions of absorption because of amorphous phases, light dispersion, and reflections on the interfaces. All these processes introduce uncertainty in the zero-absorption line. Because of this, a correction of the absorption coefficient is made. The correction method used is like the one used in nanostructured composite materials [11,12,55,56]. It consists of subtracting an amorphous-like background absorption coefficient (α_{back}) from the measured absorption coefficient (α_{exp}) [43]. This background is obtained from a linear fitting in the transparency region of the $(\alpha_{exp} h\nu)^{1/2}$ vs. $h\nu$ plot, where it should go to zero. After this correction, either a direct or an indirect optical gap can be estimated more precisely from $\alpha_{corr} = \alpha_{exp} - \alpha_{back}$ using the same method described before. The results obtained with this method can be seen in Figure 6a,b. For either case, the same α_{corr} was used. In Figure 6a the value of E_g is estimated, assuming that the gap is direct, and in Figure 6b, is estimated assuming the gap is indirect. The obtained values are summarized in Table 1. All values are within the usually reported range of 1.9–2.2 eV [7,8,10,13,22,34,36,44–54].

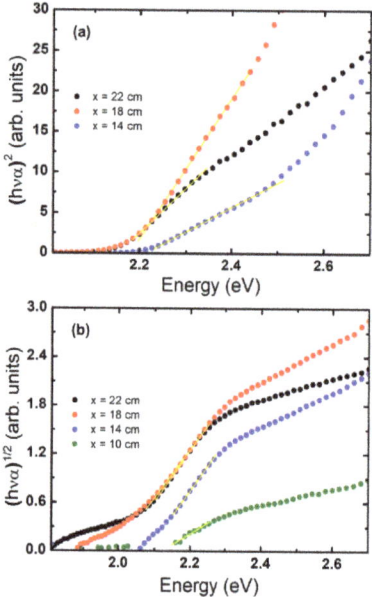

Figure 6. Direct (**a**) and indirect (**b**) optical bandgap of the hematite films. The Tauc plot corresponds to samples prepared at the positions of 10, 14, 18, and 22 cm.

Table 1. Optical bandgap values estimated assuming direct and indirect optical transitions.

Precursor-Substrate Distance Xsp (cm)	Direct Optical Bandgap (eV)	Indirect Optical Bandgap (eV)
22	2.16	2.00
18	2.18	1.99
14	2.21	2.08
10	2.31	2.12

Comparing the values of both determinations, in all cases, the energy gap is lower when it is calculated as an indirect edge. The difference between a direct and indirect energy gap is between 100 and 200 meV. The linear region is clearer and more evident for the case that a direct transition is assumed. Recently, the properties of the hematite were calculated by several ab initio techniques, showing an indirect edge and a direct edge, dozens of meV above, from the calculation of its band structure [57]. Meanwhile, the absorbance obtained from the same calculations shows the behavior of a direct absorption edge like these samples [54]. If both bandgaps are present, the absorption due to direct bandgap dominates because it does not involve interaction with phonons, so it is a transition much more probable and faster. A shift of the E_g to lower energy values from the closest sample to the precursor (x = 10 cm) to the furthest (x = 22 cm) can be observed; the E_g energy decrease and the sample gets darker. It may be related to a decrease in the E_g value with the increase of the film thickness. This tendency was already reported for α-Fe_2O_3 [21,58] and other semiconductor oxides [59].

Finally, the theoretical absorption was compared with the experimental absorption. This comparison can be seen in Figure S3 in the Supplementary Materials. A better agreement is obtained in the direct gap method, so it can be concluded the nature of the gap resembles more a direct transition than an indirect transition, although several works assign Fe_2O_3 an indirect nature [13,44,46,47,54].

2.3. Photoelectrochemical Properties of the Hematite Thin Films

To study the semiconducting and photoelectrochemical properties of hematite films, they were characterized by Mott–Schottky and by linear scan voltammetry in the dark and under illumination. Due to the nanocrystalline morphology of the films, the simple Mott–Schottky equation may not be applicable. However, in a first approximation, this type of analysis can be applied to nanostructured systems. Examples of this approach to characterize nanostructured semiconductor thin films can be encountered in the literature [10]. The Mott–Schottky diagrams for α-Fe_2O_3 electrodes in 1 M NaOH solution measured at 1.0 kHz are shown in Figure 7. All samples exhibit similar behaviors where two regions can be defined, one with a dependence almost parallel to axis X, extended from 0.2 V to 0.6 V, approximately, and the other with a pronounced positive slope from 0.6 to higher potentials. From a linear fit of the data in the second region, the flat band potential and the apparent majority carrier density values, have been obtained from the extrapolation and the slope, respectively. For all samples, an n-type behavior was observed, and a value close to 0.8 V was found for the flat band potential (E_{FB}). The obtained majority carrier density (N_D) values are $4.55 \cdot 10^{21}$, $5.08 \cdot 10^{21}$, $2.27 \cdot 10^{21}$, and $5.13 \cdot 10^{21}$ cm^{-3} for samples fabricated at 10, 14, 18, and, 22 cm, respectively. It can be seen that all the values are in the order of magnitude of 10^{21}, and only the sample prepared at 18 cm has a value close to half of the others. The values of flat band potentials and the density of majority carriers are similar to those reported in the literature [10,60]. In the case of flat band potential, several authors report that same value [1,7,10]; however, it is noteworthy that the density of majority load carriers is between 2 and 3 orders of magnitude higher than those reported by these authors which could be explained by the growth structure obtained by CVD [1].

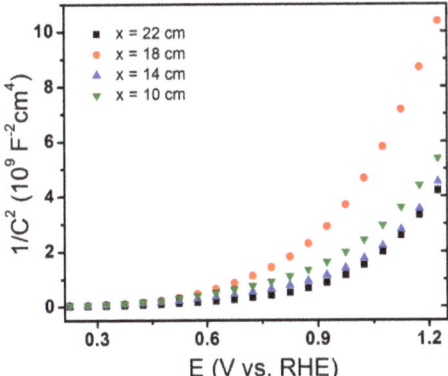

Figure 7. Mott–Schottky diagrams registered in NaOH 1 M, at 1 kHz. The diagrams correspond to samples prepared at the positions of 10, 14, 18, and 22 cm.

In Figure 8, the potentiodynamic j/E profiles of the electrodes with and without white light illumination are shown. In this figure, it is possible to observe that in dark conditions, only capacitive currents are observed in all the potential range studied and from 1.6 V an anodic current is observed due to the water electrooxidation. The increase in the current density of the electrode observed under illumination confirms that the hematite film is photoelectrochemically active. Figure 8 shows the current density versus the potential for samples fabricated at 10, 14, 18, and 22 cm. The start of the photocurrent, that is, the potential at which the current density begins to increase is different for sample 18 being 0.76 V, whereas, for samples 10, 14, and 22, it is 1.20, 1.16, and 1.11 V respectively. Samples were determined by the percentage of photocurrent efficiency being 0.28, 0.50, 1.35, and 0.46% for the sample prepared at 10, 14, 18, and 22 cm, respectively. Differences in sample thicknesses may explain the difference in the percentages of photocurrent efficiency. When the light falls on the hematite film, and the electron–hole pair is generated, the electrons migrate to FTO and the holes at the surface [4,5,25]. When the film is very thin, the formation of a few excitons occurs, and when the thickness is too thick, the separation is limited by the long diffusion distance that the load carriers must travel. A calculation of the effective absorption (A_E) was performed [27], which depends on the absorption of the material and the solar photons flux; therefore, a relationship can be observed between the calculation made and the efficiencies obtained, as shown in Table 2.

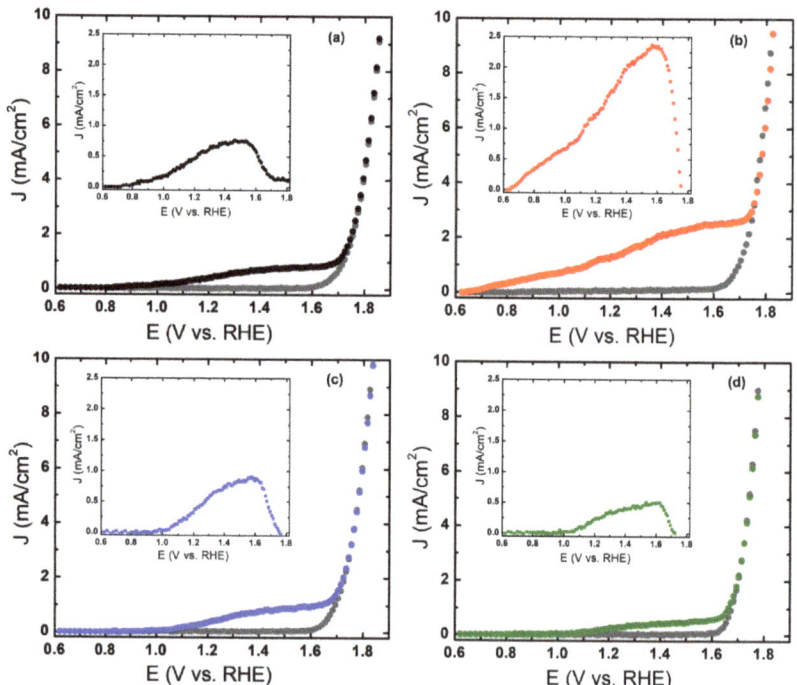

Figure 8. (**a**–**d**) are the current density-potential (J–E) characteristics of α-Fe$_2$O$_3$/FTO electrodes in 1 M NaOH in dark and under illumination for samples prepared at 10, 14, 18, and 22 cm, respectively. (Inset) The photocurrent-potential curve of α-Fe$_2$O$_3$/FTO electrode in 1 M NaOH.

Table 2. Values of Thickness, Effective Absorption (A_E), and Photocurrent Efficiency.

Precursor-Substrate Distance x_{sp} (cm)	Thin-Film Thickness (nm)	A_E (%)	Eff. (%)
22	362	65	0.46
18	222	75	1.35
14	122	56	0.50
10	29	13	0.28

The insets of Figure 8a–d correspond to plots of the difference of the current density under illumination and darkness against the potential. It is clearly observed that sample 18 has a greater photocurrent at all potentials, particularly at 1.23 and 1.58 V, it presents 1.31 and 2.36 mA/cm^2, respectively. The anodic character of the photocurrent indicates that the film exhibits photoactivity with n-type behavior [1,7,9,28]. The oxygen vacancies are the main defects in hematite films, and these are responsible for the n-type behavior, due to the electrons that are donated to the driving band [13].

3. Materials and Methods

The synthesis of the hematite was carried out in two steps, the first being the volatilization and decomposition of the ferrocene (dicyclopentadienyliron from Sigma Aldrich) used as a precursor in an Ar (99.999% purity) atmosphere at 500 °C, and subsequently, the formation of the hematite was obtained in an O_2 (99.999% purity) atmosphere at 550 °C. Both processes were carried out for 30 min with a 200 sccm flow of the respective gas.

Hematite films were characterized by a field-emission SEM (FEI-Quanta 250 FEG). Cross-section, top, and 45 degrees views of each sample were imaged. Raman spectroscopy was performed by using a Witec Alpha 300 equipped with a laser of 785 nm. XPS was performed by using a system from physical electronics (Versa Probe II), and the structural characterization of the hematite films was examined by X-ray diffraction (XRD) by using a PW1840 diffractometer (30 kV, 40 mA, Cu Kα radiation with λ = 1.5406 Å). The transmittance spectra of the samples were measured in equipment consisting of a 1000 W Xe lamp (ORIEL 6271) light source and an ORIEL 77,250 monochromator. The transmitted light was detected with a UDT 11-09-001-1 (100 mm^2 wide-area UV-enhanced unbiased silicon detector). Mott–Schottky diagrams were determined in a three-electrode cell (counter electrode: Pt; reference: Ag/AgCl) using a 1 M NaOH electrolyte with potentiostat 604C from CH Instruments. The potentiodynamic j/E profiles were determined in a three-electrode cell (counter electrode: Pt; reference: Ag/AgCl) using a 1 M NaOH electrolyte with potentiostat EZstat-Pro from NuVant System Inc and a solar simulator 96,000 from Newport-Oriel Instruments equipped with an Air Mass 1.5 G filter.

4. Conclusions

In this work, hematite thin films were grown by the CVD technique by using ferrocene as an iron precursor. The influence of the distance between the precursor and FTO substrate position (in the CVD system), on the different properties of the hematite films, has been exhaustively studied. It was found that the thickness of the hematite films (which is a critical parameter in the photoelectrochemical performance), depends linearly on this precursor–substrate distance. XRD results showed that the obtained films exhibited a single phase of the rhombohedrally centered hexagonal structure of Fe_2O_3 (α-Fe_2O_3, hematite). The XRD study also revealed that the obtained hematite films exhibited a pronounced preferential orientation along the highly conductive (001) basal plane of hematite. All samples are polycrystalline, and the broadening of the diffraction peaks demonstrates the nanocrystalline character of nanostructured α-Fe_2O_3 layers. Raman and XPS characterization confirmed the XRD results, showing that the obtained phase was pure α-Fe_2O_3 hematite without the presence of other impurity phases. The UV–VIS spectra showed an absorption edge in the region between 500 and 600 nm. Both direct and indirect bandgap energies were found close to 2 eV, the indirect one being between 100 and 200 meV lower than the direct one. Both bandgap energies decreased when film thickness increased. All samples exhibited an n-type electronic conductivity type and majority carrier density (N_D) values close to 10^{21} cm^{-3}. It was observed that the thickness of the films has a great influence on the photocurrent efficiency, obtaining the best results for thicknesses of approximately 200 nm. In fact, photoelectrochemical studies give photocurrent efficiencies from 0.28% to 1.35%.

Supplementary Materials: The following supporting information can be downloaded at: https://www.mdpi.com/article/10.3390/molecules28041954/s1, Figure S1: Electrode photography (FTO with hematite). All samples have a reddish bricklike color, getting darker as distance from precursor (x) increases (samples prepared at the positions 10, 14, 18, and 22 cm); Figure S2: XPS spectra of Fe2p and O1s signals fitted by using the software Multipack of hematite films grown at 14, 18, and 22 cm; Figure S3: Comparison between theoretical absorption and the experimental one. From the estimated gap, a theoretical absorption coefficient was constructed using the method that was used for obtaining α^{corr}. Starting from the linear fitting for determining the energy gap, an absorption coefficient was constructed for a direct and an indirect edge. Then the α^{back} was added to make the comparison; Table S1: Values of atomic percentages obtained by EDS of the elements present in the samples.

Author Contributions: Conceptualization, L.F.-I., S.A.H., E.A.D. and R.d.R.; methodology, L.F.-I., S.A.H., R.E.M. and B.D.; formal analysis, R.E.M. and L.F.-I.; investigation, B.D., E.L.S. and L.F.-I.; resources, S.A.H.; data curation, L.F.-I. and E.L.S.; writing original draft preparation, L.F.-I.; writing-review and editing, E.A.D., R.d.R. and S.A.H.; supervision, R.d.R. and S.A.H. All authors have read and agreed to the published version of the manuscript.

Funding: This research was funded by FONDECYT Project No. 1201589, FONDEQUIP EQM150101 and FONDEQUIP EQM170087, CONICYT-ANID National Doctorate Scholarship in Chile, and by Millennium Institute on Green Ammonia as Energy Vector MIGA, ANID/Millennium Science Initiative Program/ICN2021_023.

Institutional Review Board Statement: Not applicable.

Informed Consent Statement: Not applicable.

Data Availability Statement: L. Fernandez-Izquierdo is the depositary of all the data generated by the study.

Acknowledgments: The authors are grateful to Universidad de la República, in Montevideo, Uruguay, PEDECIBA–Física and ANII (Agencia Nacional de Investigación e Innovación).

Conflicts of Interest: The authors declare no conflict of interest.

References

1. Schrebler, R.; Llewelyn, C.; Vera, F.; Cury, P.; Muñoz, E.; Del Rio, R.; Meier, H.G.; Córdova, R.; Dalchiele, E.A. An Electrochemical Deposition Route for Obtaining α-Fe_2O_3 Thin Films. *Electrochem. Solid State Lett.* **2007**, *10*, D95–D99. [CrossRef]
2. Cesar, I.; Kay, A.; Martinez, J.A.G.; Grätzel, M. Translucent Thin Film Fe_2O_3 Photoanodes for Efficient Water Splitting by Sunlight: Nanostructure-Directing Effect of Si-Doping. *J. Am. Chem. Soc.* **2006**, *128*, 4582–4583. [CrossRef] [PubMed]
3. Cai, J.; Chen, H.; Ding, S.; Xie, Q. Promoting photocarrier separation for photoelectrochemical water splitting in α-Fe_2O_3@C. *J. Nanoparticle Res.* **2019**, *21*, 153. [CrossRef]
4. Le Formal, F.; Grätzel, M.; Sivula, K. Controlling Photoactivity in Ultrathin Hematite Films for Solar Water-Splitting. *Adv. Funct. Mater.* **2010**, *20*, 1099–1107. [CrossRef]
5. Mohapatra, S.K.; John, S.E.; Banerjee, S.; Misra, M. Water Photooxidation by Smooth and Ultrathin α-Fe_2O_3 Nanotube Arrays. *Chem. Mater.* **2009**, *21*, 3048–3055. [CrossRef]
6. Kleiman-Shwarstein, A.; Hu, Y.-S.; Forman, A.J.; Stucky, G.D.; McFarland, E.W. Electrodeposition of α-Fe_2O_3 Doped with Mo or Cr as Photoanodes for Photocatalytic Water Splitting. *J. Phys. Chem. C* **2008**, *112*, 15900–15907. [CrossRef]
7. Choi, H.; Hong, Y.; Ryu, H.; Lee, W.-J. Photoelectrochemical properties of hematite thin films grown by MW-CBD. *Surf. Coat. Technol.* **2018**, *333*, 259–266. [CrossRef]
8. Liu, Y.; Yu, Y.-X.; Zhang, W.-D. Photoelectrochemical properties of Ni-doped Fe_2O_3 thin films prepared by electrodeposition. *Electrochim. Acta* **2012**, *59*, 121–127. [CrossRef]
9. Ruan, G.; Wu, S.; Wang, P.; Liu, J.; Cai, Y.; Tian, Z.; Ye, Y.; Liang, C.; Shao, G. Simultaneous doping and growth of Sn-doped hematite nanocrystalline films with improved photoelectrochemical performance. *RSC Adv.* **2014**, *4*, 63408–63413. [CrossRef]
10. Shinde, S.S.; Bansode, R.A.; Bhosale, C.H.; Rajpure, K.Y. Physical properties of hematite α-Fe_2O_3 thin films: Application to photoelectrochemical solar cells. *J. Semicond.* **2011**, *32*, 013001. [CrossRef]
11. Campo, L.; Pereyra, C.J.; Amy, L.; Elhordoy, F.; Marotti, R.E.; Martín, F.; Ramos-Barrado, J.R.; Dalchiele, E.A. Electrochemically Grown ZnO Nanorod Arrays Decorated with CdS Quantum Dots by Using a Spin-Coating Assisted Successive-Ionic-Layer-Adsorption and Reaction Method for Solar Cell Applications. *ECS J. Solid State Sci. Technol.* **2013**, *2*, 26–28. [CrossRef]
12. Guerguerian, G.; Elhordoy, F.; Pereyra, C.J.; Marotti, R.E.; Martin, F.; Leinen, D.; Ramos-Barrado, J.R.; Dalchiele, E.A. ZnO nanorod/CdS nanocrystal core/shell-type heterostructures for solar cell applications. *Nanotechnology* **2011**, *22*, 505401. [CrossRef]
13. Enache, C.; Liang, Y.; van de Krol, R. Characterization of structured α-Fe_2O_3 photoanodes prepared via electrodeposition and thermal oxidation of iron. *Thin Solid Films* **2011**, *520*, 1034–1040. [CrossRef]
14. Pina, J.; Dias, P.; Serpa, J.; Azevedo, J.; Mendes, A.; de Melo, J.S.S. Phenomenological Understanding of Hematite Photoanode Performance. *J. Phys. Chem. C* **2021**, *125*, 8274–8284. [CrossRef]
15. Berardi, S.; Cristino, V.; Bignozzi, C.A.; Grandi, S.; Caramori, S. Hematite-based photoelectrochemical interfaces for solar fuel production. *Inorganica Chim. Acta* **2022**, *535*, 120862. [CrossRef]
16. Baldovi, H. Optimization of α-Fe_2O_3 Nanopillars Diameters for Photoelectrochemical Enhancement of α-Fe_2O_3-TiO_2 Heterojunction. *Nanomaterials* **2021**, *11*, 2019. [CrossRef] [PubMed]
17. Cai, J.; Liu, H.; Liu, C.; Xie, Q.; Xu, L.; Li, H.; Wang, J.; Li, S. Enhanced photoelectrochemical water oxidation in Hematite: Accelerated charge separation with Co doping. *Appl. Surf. Sci.* **2021**, *568*, 150606. [CrossRef]
18. Talibawo, J.; Kyesmen, P.I.; Cyulinyana, M.C.; Diale, M. Facile Zn and Ni Co-doped hematite nanorods for efficient photocatalytic water oxidation. *Nanomaterials* **2022**, *12*, 2961. [CrossRef]

19. Jubb, A.; Allen, H.C. Vibrational Spectroscopic Characterization of Hematite, Maghemite, and Magnetite Thin Films Produced by Vapor Deposition. *ACS Appl. Mater. Interfaces* **2010**, *2*, 2804–2812. [CrossRef]
20. Chen, X.; Fu, Y.; Kong, T.; Shang, Y.; Niu, F.; Diao, Z.; Shen, S. Protected Hematite Nanorod Arrays with Molecular Complex Co-Catalyst for Efficient and Stable Photoelectrochemical Water Oxidation. *Eur. J. Inorg. Chem.* **2018**, *2019*, 2078–2085. [CrossRef]
21. Kumar, A.; Yadav, K. Optical properties of nanocrystallite films of α-Fe$_2$O$_3$ and α-Fe$_{2-x}$Cr$_x$O$_3$ ($0.0 \leq x \leq 0.9$) deposited on glass substrates. *Mater. Res. Express* **2017**, *4*, 075003. [CrossRef]
22. Akl, A.A. Optical properties of crystalline and non-crystalline iron oxide thin films deposited by spray pyrolysis. *Appl. Surf. Sci.* **2004**, *233*, 307–319. [CrossRef]
23. Mirzaei, A.; Janghorban, K.; Hashemi, B.; Bonyani, M.; Leonardi, S.; Neri, G. Highly stable and selective ethanol sensor based on α-Fe$_2$O$_3$ nanoparticles prepared by Pechini sol–gel method. *Ceram. Int.* **2016**, *42*, 6136–6144. [CrossRef]
24. Leduc, J.; Goenuellue, Y.; Ghamgosar, P.; You, S.; Mouzon, J.; Choi, H.; Vomiero, A.; Grosch, M.; Mathur, S. Electronically-Coupled Phase Boundaries in α-Fe$_2$O$_3$/Fe$_3$O$_4$ Nanocomposite Photoanodes for Enhanced Water Oxidation. *ACS Appl. Nano Mater.* **2019**, *2*, 334–342. [CrossRef]
25. Stadler, D.; Brede, T.; Schwarzbach, D.; Maccari, F.; Fischer, T.; Gutfleisch, O.; Volkert, C.A.; Mathur, S. Anisotropy control in magnetic nanostructures through field-assisted chemical vapor deposition. *Nanoscale Adv.* **2019**, *1*, 4290–4295. [CrossRef]
26. Liardet, L.; Katz, J.E.; Luo, J.; Grätzel, M.; Hu, X. An ultrathin cobalt–iron oxide catalyst for water oxidation on nanostructured hematite photoanodes. *J. Mater. Chem. A* **2019**, *7*, 6012–6020. [CrossRef]
27. Lévy-Clément, C.; Elias, J. Optimization of the Design of Extremely Thin Absorber Solar Cells Based on Electrodeposited ZnO Nanowires. *ChemPhysChem* **2013**, *14*, 2321–2330. [CrossRef]
28. Li, S.; Zhang, P.; Song, X.; Gao, L. Ultrathin Ti-doped hematite photoanode by pyrolysis of ferrocene. *Int. J. Hydrogen Energy* **2014**, *39*, 14596–14603. [CrossRef]
29. Tamirat, A.G.; Rick, J.; Dubale, A.A.; Su, W.-N.; Hwang, B.-J. Using hematite for photoelectrochemical water splitting: A review of current progress and challenges. *Nanoscale Horiz.* **2016**, *1*, 243–267. [CrossRef]
30. Wang, H.; He, X.; Li, W.; Chen, H.; Fang, W.; Tian, P.; Xiao, F.; Zhao, L. Hematite nanorod arrays top-decorated with an MIL-101 layer for photoelectrochemical water oxidation. *Chem. Commun.* **2019**, *55*, 11382–11385. [CrossRef]
31. Gasparotto, A.; Carraro, G.; Maccato, C.; Barreca, D. On the use of Fe(dpm)$_3$ as precursor for the thermal-CVD growth of *hematite* nanostructures. *Phys. Status Solidi (A)* **2017**, *214*, 1600779. [CrossRef]
32. Fornari, C.I.; Fornari, G.; Rappl, P.H.D.O.; Abramof, E.; Travelho, J.D.S. Monte Carlo Simulation of Epitaxial Growth. In *Epitaxy*; IntechOpen: Rijeka, Croatia, 2018; pp. 113–129. [CrossRef]
33. Ohring, M. *The Materials Science of thin Films*; Academic Press: San Diego, CA, USA, 1992.
34. Mansour, H.; Letifi, H.; Bargougui, R.; De Almeida-Didry, S.; Negulescu, B.; Autret-Lambert, C.; Gadri, A.; Ammar, S. Structural, optical, magnetic and electrical properties of hematite (α-Fe$_2$O$_3$) nanoparticles synthesized by two methods: Polyol and precipitation. *Appl. Phys. A* **2017**, *123*, 787. [CrossRef]
35. Xu, Y.; Zhao, D.; Zhang, X.; Jin, W.; Kashkarov, P.; Zhang, H. Synthesis and characterization of single-crystalline α-Fe$_2$O$_3$ nanoleaves. *Phys. E Low Dimens. Syst. Nanostructures* **2009**, *41*, 806–811. [CrossRef]
36. Ingler, W.B.; Khan, S.U. Photoresponse of spray pyrolytically synthesized copper-doped p-Fe$_2$O$_3$ thin film electrodes in water splitting. *Int. J. Hydrogen Energy* **2005**, *30*, 821–827. [CrossRef]
37. Powder Diffraction File, Joint Committee for Powder Diffraction Studies (JCPDS) File No. JCPDS 86-0550 (Rhombohedrally Centered Hexagonal Structure of Fe$_2$O$_3$, Hematite), n.d.
38. Cullity, B.D. *Elements of X-ray Diffraction*, 2nd ed.; Addison-Wesley: Reading, MA, USA, 1956.
39. Nasibulin, A.G.; Rackauskas, S.; Jiang, H.; Tian, Y.; Mudimela, P.R.; Shandakov, S.D.; Nasibulina, L.I.; Jani, S.; Kauppinen, E.I. Simple and rapid synthesis of α-Fe$_2$O$_3$ nanowires under ambient conditions. *Nano Res.* **2009**, *2*, 373–379. [CrossRef]
40. Liao, L.; Zheng, Z.; Yan, B.; Zhang, J.X.; Gong, H.; Li, J.C.; Liu, C.; Shen, Z.X.; Yu, T. Morphology Controllable Synthesis of α-Fe$_2$O$_3$ 1D Nanostructures: Growth Mechanism and Nanodevice Based on Single Nanowire. *J. Phys. Chem. C* **2008**, *112*, 10784–10788. [CrossRef]
41. Pankove, J.I. *Optical Processes in Semiconductors*, 2nd ed.; Dover Publications: New York, NY, USA, 1976.
42. Sapoval, B.; Hermann, C. *Physics of Semiconductors*; Springer: New York, NY, USA, 1995.
43. Tauc, J. Absorption edge and internal electric fields in amorphous semiconductors. *Mater. Res. Bull.* **1970**, *5*, 721–729. [CrossRef]
44. Mohanty, S.; Ghose, J. Studies on some α-Fe$_2$O$_3$ photoelectrodes. *J. Phys. Chem. Solids* **1992**, *53*, 81–91. [CrossRef]
45. Cherepy, N.J.; Liston, D.B.; Lovejoy, J.A.; Deng, H.; Zhang, J.Z. Ultrafast Studies of Photoexcited Electron Dynamics in γ- and α-Fe$_2$O$_3$ Semiconductor Nanoparticles. *J. Phys. Chem. B* **1998**, *102*, 770–776. [CrossRef]
46. Duret, A.; Grätzel, M. Visible Light-Induced Water Oxidation on Mesoscopic α-Fe$_2$O$_3$ Films Made by Ultrasonic Spray Pyrolysis. *J. Phys. Chem. B* **2005**, *109*, 17184–17191. [CrossRef]
47. Glasscock, J.; Barnes, P.; Plumb, I.; Bendavid, A.; Martin, P. Structural, optical and electrical properties of undoped polycrystalline hematite thin films produced using filtered arc deposition. *Thin Solid Films* **2008**, *516*, 1716–1724. [CrossRef]
48. Gilbert, B.; Frandsen, C.; Maxey, E.R.; Sherman, D.M. Band-gap measurements of bulk and nanoscale hematite by soft X-ray spectroscopy. *Phys. Rev. B* **2009**, *79*, 035108. [CrossRef]
49. Rühle, S.; Anderson, A.Y.; Barad, H.-N.; Kupfer, B.; Bouhadana, Y.; Rosh-Hodesh, E.; Zaban, A. All-Oxide Photovoltaics. *J. Phys. Chem. Lett.* **2012**, *3*, 3755–3764. [CrossRef] [PubMed]

50. Al-Kuhaili, M.; Saleem, M.; Durrani, S. Optical properties of iron oxide (α-Fe$_2$O$_3$) thin films deposited by the reactive evaporation of iron. *J. Alloy. Compd.* **2012**, *521*, 178–182. [CrossRef]
51. Lassoued, A.; Lassoued, M.S.; Dkhil, B.; Ammar, S.; Gadri, A. Synthesis, structural, morphological, optical and magnetic characterization of iron oxide (α-Fe$_2$O$_3$) nanoparticles by precipitation method: Effect of varying the nature of precursor. *Phys. E Low Dimens. Syst. Nanostructures* **2018**, *97*, 328–334. [CrossRef]
52. Kung, H.H.; Jarrett, H.S.; Sleight, A.W.; Ferretti, A. Semiconducting oxide anodes in photoassisted electrolysis of water. *J. Appl. Phys.* **1977**, *48*, 2463–2469. [CrossRef]
53. Merchant, P.; Collins, R.; Kershaw, R.; Dwight, K.; Wold, A. The electrical, optical and photoconducting properties of Fe$_{2-x}$Cr$_x$O$_3$ ($0 \leq x \leq 0.47$). *J. Solid State Chem.* **1979**, *27*, 307–315. [CrossRef]
54. Shinar, R.; Kennedy, J.H. Photoactivity of doped αFe$_2$O$_3$ electrodes. *Sol. Energy Mater.* **1982**, *6*, 323–335. [CrossRef]
55. Bojorge, C.D.; Kent, V.R.; Teliz, E.; Cánepa, H.R.; Henríquez, R.; Gómez, H.; Marotti, R.E.; Dalchiele, E.A. Zinc-oxide nanowires electrochemically grown onto sol-gel spin-coated seed layers. *Phys. Status Solidi (A)* **2011**, *208*, 1662–1669. [CrossRef]
56. Guerguerian, G.; Elhordoy, F.; Pereyra, C.J.; Marotti, R.E.; Martín, F.; Leinen, D.; Ramos-Barrado, J.R.; Dalchiele, E.A. ZnO/Cu$_2$O heterostructure nanopillar arrays: Synthesis, structural and optical properties. *J. Phys. D Appl. Phys.* **2012**, *45*, 245301. [CrossRef]
57. Piccinin, S. The band structure and optical absorption of hematite (α-Fe$_2$O$_3$): A first-principles GW-BSE study. *Phys. Chem. Chem. Phys.* **2019**, *21*, 2957–2967. [CrossRef] [PubMed]
58. Kumar, P.; Rawat, N.; Hang, D.-R.; Lee, H.-N.; Kumar, R. Controlling band gap and refractive index in dopant-free α-Fe$_2$O$_3$ films. *Electron. Mater. Lett.* **2015**, *11*, 13–23. [CrossRef]
59. Marotti, R.; Guerra, D.; Bello, C.; Machado, G.; Dalchiele, E. Bandgap energy tuning of electrochemically grown ZnO thin films by thickness and electrodeposition potential. *Sol. Energy Mater. Sol. Cells* **2004**, *82*, 85–103. [CrossRef]
60. Zhu, Q.; Yu, C.; Zhang, X. Ti, Zn co-doped hematite photoanode for solar driven photoelectrochemical water oxidation. *J. Energy Chem.* **2019**, *35*, 30–36. [CrossRef]

Disclaimer/Publisher's Note: The statements, opinions and data contained in all publications are solely those of the individual author(s) and contributor(s) and not of MDPI and/or the editor(s). MDPI and/or the editor(s) disclaim responsibility for any injury to people or property resulting from any ideas, methods, instructions or products referred to in the content.

Article

Surface Coordination of Pd/ZnIn₂S₄ toward Enhanced Photocatalytic Activity for Pyridine Denitrification

Deling Wang [1], Erda Zhan [1], Shihui Wang [1], Xiyao Liu [2,3], Guiyang Yan [2,3,*], Lu Chen [2,3,*] and Xuxu Wang [1,*]

[1] State Key Laboratory of Photocatalysis on Energy and Environment, Fuzhou University, Fuzhou 350002, China
[2] Province University Key Laboratory of Green Energy and Environment Catalysis, Ningde Normal University, Ningde 352100, China
[3] Fujian Provincial Key Laboratory of Featured Materials in Biochemical Industry, Ningde Normal University, Ningde 352100, China
* Correspondence: ygyfjnu@163.com (G.Y.); chenlu199104@163.com (L.C.); xwang@fzu.edu.cn (X.W.); Tel.: +86-13809566652 (G.Y.); +86-156959097359 (L.C.); +86-13600887951 (X.W.)

Abstract: New surface coordination photocatalytic systems that are inspired by natural photosynthesis have significant potential to boost fuel denitrification. Despite this, the direct synthesis of efficient surface coordination photocatalysts remains a major challenge. Herein, it is verified that a coordination photocatalyst can be constructed by coupling Pd and CTAB-modified ZnIn₂S₄ semiconductors. The optimized Pd/ZnIn₂S₄ showed a superior degradation rate of 81% for fuel denitrification within 240 min, which was 2.25 times higher than that of ZnIn₂S₄. From the in situ FTIR and XPS spectra of 1% Pd/ZnIn₂S₄ before and after pyridine adsorption, we find that pyridine can be selectively adsorbed and form Zn···C-N or In···C-N on the surface of Pd/ZnIn₂S₄. Meanwhile, the superior electrical conductivity of Pd can be combined with ZnIn₂S₄ to promote photocatalytic denitrification. This work also explains the surface/interface coordination effect of metal/nanosheets at the molecular level, playing an important role in photocatalytic fuel denitrification.

Keywords: ZnIn₂S₄; surface coordination; denitrification; pyridine; photocatalysis

Citation: Wang, D.; Zhan, E.; Wang, S.; Liu, X.; Yan, G.; Chen, L.; Wang, X. Surface Coordination of Pd/ZnIn₂S₄ toward Enhanced Photocatalytic Activity for Pyridine Denitrification. *Molecules* **2023**, *28*, 282. https://doi.org/10.3390/molecules28010282

Academic Editor: Yuanfu Chen

Received: 16 November 2022
Revised: 20 December 2022
Accepted: 27 December 2022
Published: 29 December 2022

Copyright: © 2022 by the authors. Licensee MDPI, Basel, Switzerland. This article is an open access article distributed under the terms and conditions of the Creative Commons Attribution (CC BY) license (https://creativecommons.org/licenses/by/4.0/).

1. Introduction

Nitrogen-containing compounds (NCCs) have been regarded as some of the most significant atmospheric pollutants, whose combustion will cause serious environmental pollution [1–3] and the excess release of NOx into the air, causing acid rain [4]. Petroleum contains high amounts of NCCs [5–8], such as pyridine. Pyridine, as a representative nitrogen heterocycle, is reported to exert toxic, teratogenic, and carcinogenic effects, and it is classified as a priority pollutant [9,10]. Meanwhile, the stable structures of pyridine are difficult to degrade and they persist for long periods in the environment [11,12]. Therefore, the removal of pyridine from crude fuels has become a global research hotspot [13–15]. The question of how to deal with these problems has attracted the attention of researchers.

Thus far, the main denitrification method is HDN, which is expensive and inefficient because of its strict temperature and pressure requirements [16,17]. Alternative technologies have been studied, such as absorption [18], ODN [19], and photocatalysis [20]. Among them, photocatalytic technology has been regarded as a prospective technology for converting NCCs into fuels because of its good selectivity, the direct utilization of sunlight at room temperature, and high efficiency [20–22], which is consistent with the concepts of environmental protection and sustainability. Moreover, oxidation processes play an important role in photocatalytic pyridine denitrification. It was reported that pyridine was attacked by free radicals or adsorbed photons, leading to ring opening and causing a range of oxygen-containing products in the process of photocatalysis [15,23]. Meanwhile, a study found that the pyridine denitrification efficiency has been restricted by the rapid recombination of photogenerated carriers, insufficient reactive sites, and sluggish kinetics of the electron transfer process [24]. Therefore, it is desirable to design a photocatalyst with superior electron transfer and separation.

Two-dimensional nanosheets not only possess a high specific surface area and abundant catalytically active sites but also shorten the diffusion distance of charge carriers [25]. Notably, $ZnIn_2S_4$, with a 2D structure, serves as visible light response photocatalyst with a distinctive and tunable electronic structure for improved photocatalytic performance [26,27]. Consequently, $ZnIn_2S_4$ photocatalysts have aroused attention in photocatalytic hydrogen production [28], CO_2 reduction [29], and contaminant degradation [30–33]. However, for prospective uses in the actual world, $ZnIn_2S_4$'s photocatalytic activity needs to be substantially enhanced [34,35]. It is well known that semiconductor photocatalysts' photocatalytic activity can be significantly increased by coupling cocatalysts [36–42]. In current photocatalytic applications, noble metals such as Pt, Au, and Pd are commonly used because of their ability to form Schottky barriers, which can accelerate the separation of photogenerated carriers [43]. Moreover, compared with other noble metals, Pd has a relatively high Fermi energy level, low electronic affinity for effective electron transfer to protons, and a high state density close to the semiconductor Fermi energy level [44]. Meanwhile, numerous studies have shown that the use of surfactants can significantly improve the dispersibility and contact area of semiconductors, further improving the reaction efficiency [45–49].

Herein, $Pd/ZnIn_2S_4$ heterostructures were successfully synthesized for the selective transformation of pyridine under visible light. Systematic studies showed that the 1% $Pd/ZnIn_2S_4$ heterojunction exhibited the optimal degradation rate of 81% in the photocatalytic denitrification of pyridine, being nearly 2.25 times higher than that of pristine $ZnIn_2S_4$. Such remarkably enhanced photocatalytic performance was mainly ascribed to the Zn···C-N or In···C-N coordination for pyridine molecules' selective adsorption and activation. Meanwhile, the electron separation and transfer ability of Pd, as well as SPR, affect Pd. Under visible light irradiation, the $Pd/ZnIn_2S_4$ composites outperform pure $ZnIn_2S_4$ in photocatalytic performance and cycle stability, confirming that the material has excellent fuel denitrification performance. Finally, we suggest a potential photocatalytic mechanism for the degradation of pyridine by $Pd/ZnIn_2S_4$ at the molecular level.

2. Results

2.1. Characterization

The crystal structure of the synthesized photocatalyst was characterized by XRD. As shown in Figure 1, all characteristic peaks of the pure $ZnIn_2S_4$ (ZIS) were consistent with the $ZnIn_2S_4$ hexagonal. The diffraction peaks located at 27.7°, 30.5°, and 47.2° were ascribed to the (102), (104), and (110) crystal planes of hexagonal $ZnIn_2S_4$ (PDF: 65-2023) [29]. However, due to the low quantity and homogeneous dispersion of Pd, the distinctive diffraction peaks of Pd were not observed.

The morphologies and microstructures of ZIS and Pd/ZIS were characterized by SEM and TEM. As displayed in Figure 2a, the synthesized ZIS exhibited a flower-like structure with a diameter of 4–5 μm, which was composed of abundant 2D nanosheets. After Pd was loaded on ZIS, the flower-like structure was well maintained (Figure 2b). As shown in Figure 2c,d, the Pd nanoparticles had a uniform distribution on the ZIS ultrathin nanosheet surface. In addition, the magnified HRTEM images shown in Figure 2d display obvious lattice fringes of ca. to 0.32 and 0.22 nm, which correspond to the (102) crystal plane of hexagonal ZIS and the (111) plane of face-centered cubic Pd [36]. It is further confirmed that the development of a 2D nano-junction between Pd and ZIS indicates intimate contact between them. Meanwhile, the EDX spectrum of the Pd/ZIS heterojunction demonstrates the presence of Zn, In, S, and Pd elements in Pd/ZIS composites (Figure 2e). In addition, the actual proportion of Pd in the 1% Pd/ZIS photocatalyst is shown in Table S1. According to the above results, it is indicated that the Pd nanoparticles were successfully uniformly dispersed on the ZIS surface.

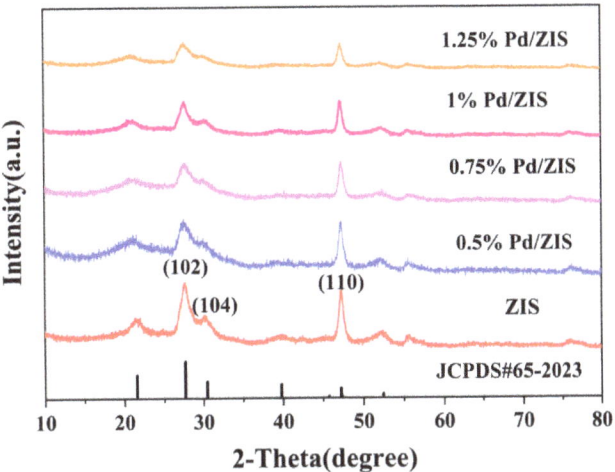

Figure 1. XRD patterns of ZIS and X wt % Pd/ZIS (X = 0.5, 0.75, 1, 1.25).

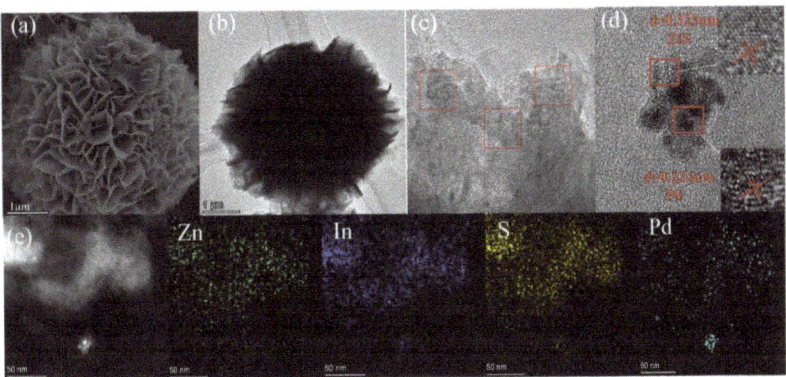

Figure 2. (**a**) SEM of ZIS; (**b**,**c**) TEM of 1% Pd/ZIS; (**d**) HRTEM of 1% Pd/ZIS; (**e**) HAADF-TEM and corresponding elemental mapping images Zn, In, S, Pd.

The surface chemical states of the as-prepared samples were investigated by XPS analysis. The binding energies in the XPS spectra were calibrated by the adventitious carbon C 1s peak at 284.8 eV. As shown in Figure 3, four elements, Zn, In, S, and Pd, can be observed, which are consistent with the EDS results. The binding energy of Zn 2p can be resolved into two peaks located at 1044.48 eV and 1021.45 eV, which correspond to Zn $2p_{1/2}$ and Zn $2p_{3/2}$, respectively [32]. As shown in Figure 3b, the two peaks exhibited at 451.99 and 444.45 eV could be assigned to In $3d_{3/2}$ and In $3d_{5/2}$ [39]. Meanwhile, the binding energies of S 2p can be divided into two peaks located at 162.54 and 161.25 eV, which were attributed to the S $2p_{1/2}$ and S $2p_{3/2}$ [43]. Compared with the ZIS, the Zn 2p, In 3d, and S 2p of 1% Pd/ZIS exhibited a 0.4 eV blue shift, which could be ascribed to the electron transfer between the two components [40]. In addition, the peaks at 341.10 and 335.42 eV correspond to the Pd $3d_{3/2}$ and Pd $3d_{5/2}$, which are associated with the metallic Pd [36]. The above results confirm that Pd was successfully deposited on the ZIS surface and existed in strong interactions.

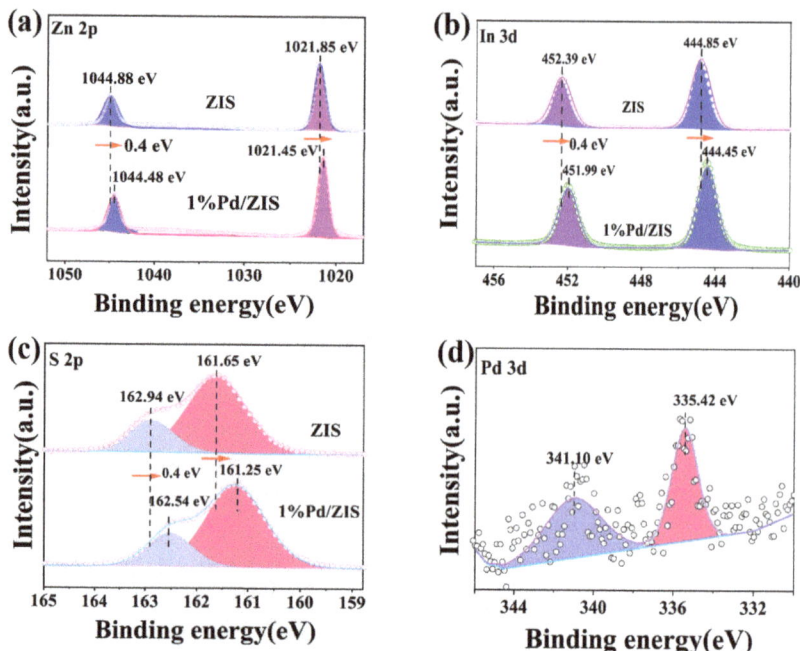

Figure 3. XPS spectra of ZIS and 1% Pd/ZIS: (**a**) Zn 2p; (**b**) In 3d; (**c**) S 2p; (**d**) Pd 3d.

2.2. Photocatalytic Performance

UV–vis DRS was applied to investigate the optical properties and band gaps of the synthesized samples. The absorption band edge of ZIS, as illustrated in Figure 4a, was approximately 536 nm, which corresponded to a band gap of 2.53 eV. The light absorption of Pd/ZIS in the visible area was improved when Pd was added. With an increase in the mass ratio of Pd in the composite, the adsorption intensity in the whole visible region gradually became stronger. The band gaps of ZIS and 1% Pd/ZIS were obtained from the Kubelka–Munk plots [50]. As displayed in Figure 4b, the band gaps of pure ZIS and 1% Pd/ZIS were calculated to be 2.53 and 2.48 eV, respectively. Moreover, the Mott–Schottky curves of ZIS exhibited positive slopes, suggesting that this semiconductor is an n-type one [51]. The flat-band potential of the sample was determined by extrapolating the lines to $1/C^2 = 0$ and was found to be −1.11 eV (vs. Ag/AgCl, pH = 7). According to the conversion relation between the Ag/AgCl electrode and the standard hydrogen electrode, $E_{NHE} = E_{Ag/AgCl} + 0.197$ [29], the CB position of the sample was finally calculated to be −0.91 eV (vs. NHE, pH = 6.8). Consequently, VB-XPS was able to identify the VB potential of ZIS. According to the VB-XPS plots of ZIS, the VB potential of ZIS was calculated to be 1.51 eV. The contact potential difference between the sample and the XPS analyzer was used to compute the VB potential of the normal hydrogen electrode (EVB-NHE vs. NHE, pH = 6.8) using the following calculation [24]:

$$EVB\text{-}NHE = \varphi + EVB - XPS - 4.44 \quad (1)$$

φ was the electron work function (4.55 eV) of the XPS analyzer, and EVB-XPS was the VB measured from the VB-XPS plots. Therefore, the EVB-NHE value of ZIS was calculated to be 1.62 eV. The band gap of ZIS could be computed using Mott–Schottky and VB-XPS spectral analyses, and it was matched with the bandgap values obtained from the UV–vis DRS spectra to obtain a value of 2.53 eV.

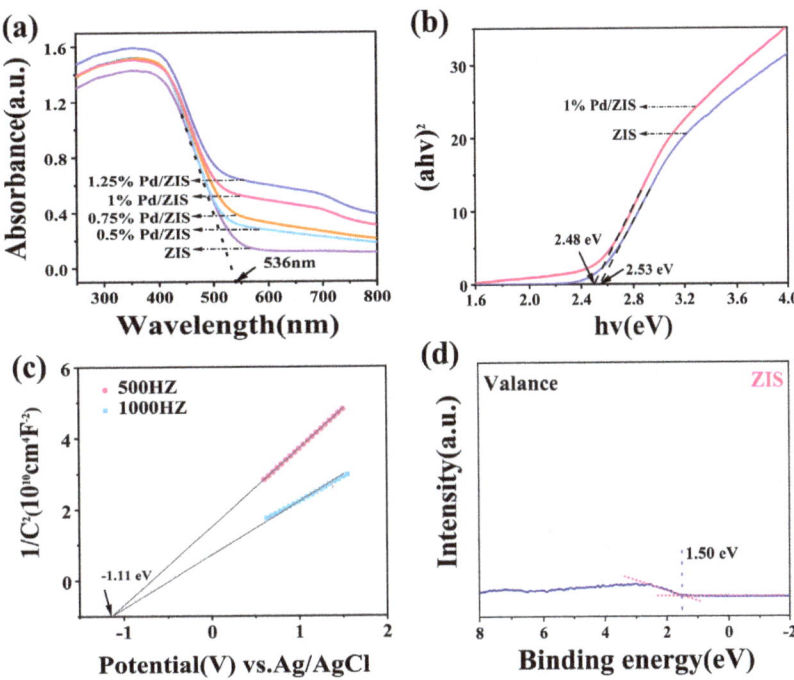

Figure 4. UV–vis diffuse reflectance (**a**) and optical band gap spectra (**b**) of ZIS and Pd/ZIS composites; (**c**) Mott–Schottky plots of ZIS; (**d**) valence band curves of ZIS.

The photocatalytic activity of the samples was evaluated based on the photocatalytic degradation of pyridine under visible light (λ > 420 nm). As shown in Figure 5a, when the reaction system did not contain catalysts, the concentration of the pyridine solution did not decrease remarkably. Moreover, pure ZIS demonstrated the lowest activity, with only 36% pyridine degradation, which was ascribed to the rapid recombination of photogenerated carriers. However, after the Pd and CTAB loading, the activity of pyridine degradation significantly improved (Figure S8). Furthermore, the 1% Pd/ZIS composite exhibited the most effective photocatalytic activity and fuel denitrification efficiency of up to 81%. With the increase in the Pd loading ratio, the degradation rate of the Pd/ZIS decreased significantly, which was caused by the massive Pd load, resulting in the blocking of active sites. Moreover, we investigated the kinetics of the denitrification process in photocatalytic fuel using the pseudo-first-order model. As displayed in Figure 5d, the 1% Pd/ZIS photocatalyst exhibited the maximal rate constant (0.45 h^{-1}), which was nearly 3.6 times greater than that of pristine ZIS (0.12 h^{-1}). It was confirmed that there was a strong interaction between ZIS and Pd. Meanwhile, compared with the other catalysts, it was lower than that of the Pd/ZIS composite sample (Table S2).

To explore the catalytic stability, we sought to detect the stability of 1% Pd/ZIS. As shown in Figure 6a, the fuel denitrification of the composite remained at a relatively high level. The degradation rate of pyridine still reached 73.5% after five cycles. Further research on the stability of 1% Pd/ZIS, XRD, and TEM was conducted to characterize the 5-cycle using 1% Pd/ZIS. From the XRD patterns (Figure 6b) and TEM images (Figure S1), no obvious change could be observed when comparing the used and fresh 1% Pd/ZIS, indicating that the composition and morphology of 1% Pd/ZIS were maintained after the reaction.

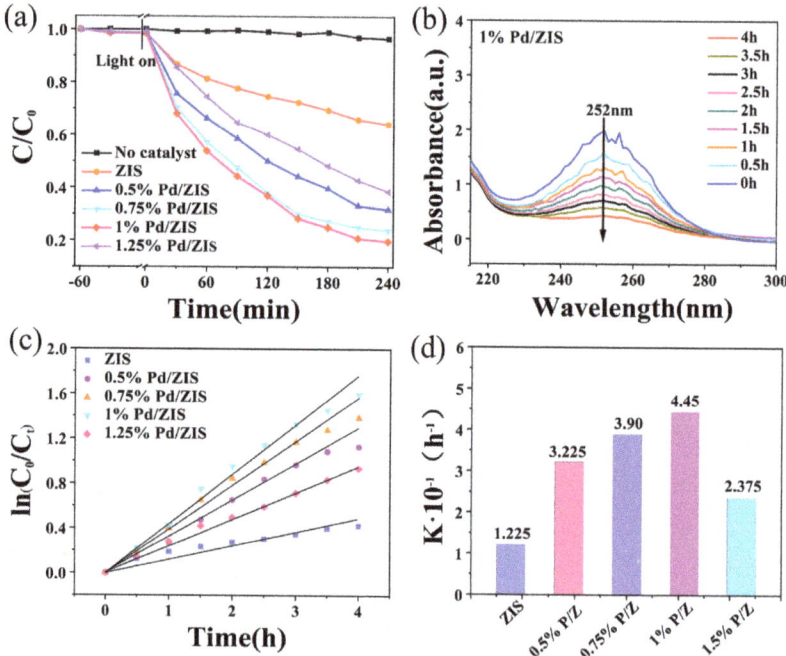

Figure 5. (**a**) Photocatalytic denitrification of pyridine under visible light irradiation; (**b**) UV–vis absorption spectral changes of 1% Pd/ZIS composites; (**c**) corresponding pseudo-first-order kinetics fitting curves of Pd/ZIS composites with different compositions; (**d**) apparent rate constants over different catalysts.

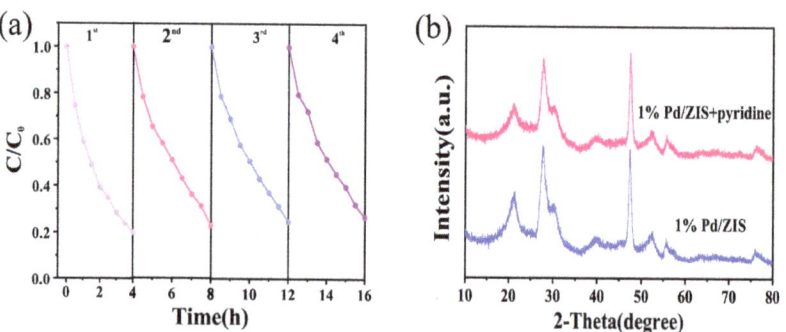

Figure 6. (**a**) Cycling experiments of photocatalytic denitrification of 1% Pd/ZIS; (**b**) XRD patterns of 1% Pd/ZIS after five cycles of photocatalysis.

2.3. Adsorption Performance

According to the above experimental results for photocatalytic fuel denitrification, it is found that ZIS, Pd, and pyridine play vital roles in the photocatalytic performance. To further reveal the adsorption phenomenon of pyridine molecules on the Pd/ZIS surface, the adsorption behavior of pyridine molecules over the catalyst was investigated. As shown in Figure 7a (2), several characteristic peaks ascribed to pyridine (3080, 3037, 1582, and 1442 cm^{-1}) could be observed after absorbing pyridine for 60 min. Meanwhile, these characteristic peaks could still be observed after vacuum degassing (Figure 7a (3)), indicating the strong chemisorption interaction between Pd/ZIS and pyridine molecules. The characteristic peaks at 3080 and 3037 cm^{-1} were attributed to the C-H stretching vibration of pyridine, while the characteristic peaks at 1582 and 1442 cm^{-1} were assigned to the stretching vibration of C-N and C-C [52]. Furthermore, with the evacuation at 150 °C, the characteristic peak of the stretching vibration of C-N at 1582 cm^{-1} showed an obvious shift, while the characteristic peak of C-H and C-C remained unchanged. It is speculated that the C-N groups of pyridine could be selectively chemisorbed on Pd/ZIS via the formation of surface coordination species to promote activity for fuel denitrification [53]. In addition, as shown in Figure 7b, the strong peaks at 1450 cm^{-1} and 1540 cm^{-1} corresponded to Lewis acids and Bronsted acids, revealing that the presence of abundant acid sites on Pd/ZIS and combined with pyridine (a Lewis base) [24]. Based on the above analysis, it can be concluded that Pd/ZIS has a strong binding affinity for pyridine.

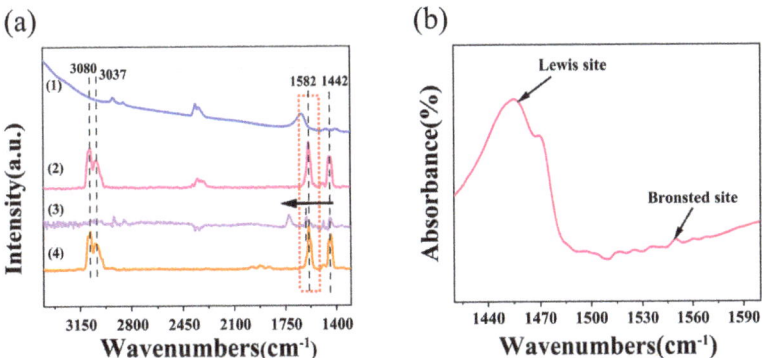

Figure 7. (a) In situ FTIR spectra of Pd/ZIS: (1) after degassing in vacuum at 180 °C for 2h, (2) adsorbing pyridine for 1 h at room temperature (physisorption and chemisorption), (3) further degassing in vacuum at 150 °C for 30 min (chemisorption), and (4) FTIR spectra of pyridine; (b) Fourier transform infrared spectrum of 1% Pd/ZIS during pyridine adsorption.

To further study the interaction mechanisms between the representative pyridine molecules and the Pd/ZIS surface, we investigated the changes in elements' (Zn, In, S, Pd) binding energy by XPS characterization of the ZIS and Pd/ZIS samples before and after pyridine adsorption. As demonstrated in Figure 8 and Figure S2, the peaks of Zn 2p were shifted to a low binding energy following pyridine adsorption in the XPS spectra of ZIS and Pd/ZIS samples, indicating that the electron cloud density of Zn was increased. It validated the production of Zn···C-N coordination between pyridine molecules and Pd/ZIS Zn^{2+} sites, leading to electron transfer from the C-N group to Zn^{2+} via Zn···C-N coordination [53]. Meanwhile, the binding energy shift of In 3d following pyridine adsorption was comparable to Zn 2p. These findings revealed that surface Zn^{2+} and In^{3+} would combine with C-N groups via Zn···C-N and In···C-N coordination, respectively. Notably, peaks of Zn 2p and In 3d of Pd/ZIS were higher than the shift of ZIS following pyridine adsorption, while 1% Pd/ZIS was the highest. It is proven that Pd loaded on the ZIS can lead to more electron transfer from the C-N group to Zn^{2+} and In^{3+} via Zn···C-N coordination and In···C-N coordination. It was further shown that the abundant surface Zn^{2+} and In^{3+} sites of Pd/ZIS as Lewis acid sites make a substantial contribution to the selective adsorption and activate C-N groups, significantly enhancing the pyridine selectivity. In addition, the UV–vis DRS spectra of Pd/ZIS adsorbed pyridine are displayed in Figure S3. The UV–vis DRS spectra of Pd/ZIS showed a redshift after adsorbing the pyridine molecule, which also indicated that surface coordination species were formed between the produced materials and pyridine [54].

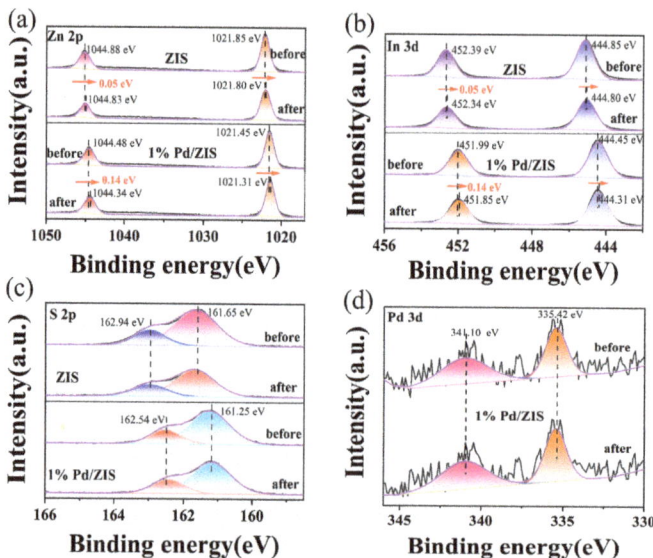

Figure 8. XPS spectra of the ZIS and 1% Pd/ZIS samples before and after pyridine adsorption: (**a**) Zn 2p; (**b**) In 3d; (**c**) S 2p; (**d**) Pd 3d.

2.4. Photocatalytic Mechanism

To explain further the efficiency of charge carrier transfer and separation, it was demonstrated by photoelectrochemical and photoluminescence (PL) studies. As shown in Figure 9a, the photocurrent intensity of Pd/ZIS was significantly higher than that of pure ZIS, while that of 1% Pd/ZIS was the highest, implying that 1% Pd/ZIS can more efficiently promote photogenerated carriers separation and transfer. Meanwhile, Figure 9b shows that the radius of 1% Pd/ZIS was the smallest, indicating that 1% Pd/ZIS can maximally reduce the interfacial charge transfer resistance. At the same time, according to the PL spectra of ZIS and Pd/ZIS, the PL intensity of 1% Pd/ZIS was the lowest, indicating that the photogenerated electrons of 1% Pd/ZIS can be more effectively separated [55]. From the above results, it is concluded that the Pd loaded on the ZIS can accelerate the separation and transfer of photogenerated carriers.

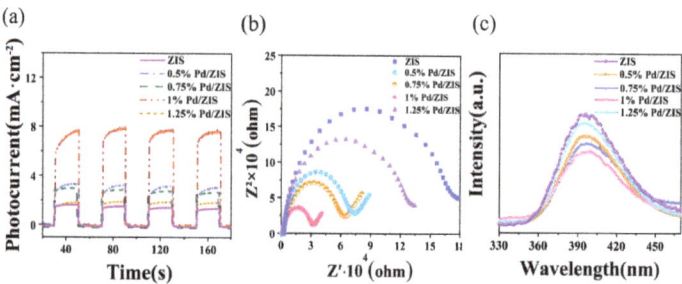

Figure 9. (**a**) Transient photocurrent responses; (**b**) electrochemical impedance spectroscopy analysis; (**c**) PL spectra of ZIS and Pd/ZIS.

To provide a direct representation of the enhanced photocatalytic fuel denitrification after Pd loading, it was measured by HPLC-MS and ESR spectroscopy. The peak intensity of pyridine at around $m/z = 80.1$ was gradually decreased during the reaction process (Figure S4), indicating that the pyridine was successfully denitrogenated. Meanwhile, new peaks formed at m/z = 110.1, 116, and 61.1, revealing the conversion of pyridine to protonated intermediates such as $C_4H_4O_3$, $C_4H_4O_4$, and NH_2COOH. Meanwhile, the pyridine adsorption value was obtained by conversion in Table S3. The reactive species in the reaction system that the ESR detected should be further explored. As illustrated in Figure 10, TEMPO molecules were regarded as hole probes during the reaction process, because their free radicals could be oxidized by holes [56]. In the presence of 1% Pd/ZIS, the signal of TEMPO gradually decreases, revealing the generation of photogenerated holes (h^+). With the extension of the illumination time, the signal of DMPO-•O_2^- was enhanced step by step, which indicated that the •O_2^- was the main active species during the reaction process. Meanwhile, the signal of DMPO-•OH was relatively weak, which was ascribed to the VB position of ZIS, which was lower than that of E (-OH/•OH; 1.99 eV) [57]. However, the generation of •OH radicals may originate from the •O_2^- combined with H^+ to form •OH [58].

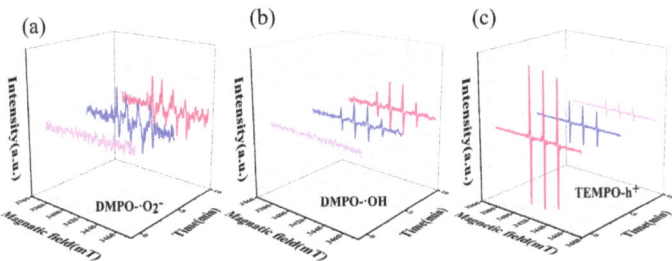

Figure 10. (a–c) Electron spin resonance spectra of various radical adducts.

Based on the above results, the photocatalytic mechanism (Figure 11, Figure S5) for the denitrification of pyridine over the Pd/ZIS composite was proposed [11]. When 1% Pd/ZIS composites are excited by visible light, the photogenerated electrons of Pd nanoparticles could migrate to the surface of the ZIS; once electrons reach ZIS, they react with pyridine and oxygen in the solution to yield a range of oxygen-containing products and •O_2^-, respectively. Meanwhile, the generated •O_2^- reacts with H^+ to form •OH radicals. The e^-, •O_2^-, and •OH of reactive species could participate in the degradation of pyridine. The possible mechanism of denitrification can be found in the Supplementary Materials. In the denitrification process, pyridine combines with the H^+ and the •O_2^- to generate a ring and a ring-opening intermediate product successively [11,59]. The intermediate product is converted into mineralization products including NO_3^-, CO_2, and H_2O in the presence of e^-, •OH, and •O_2^-.

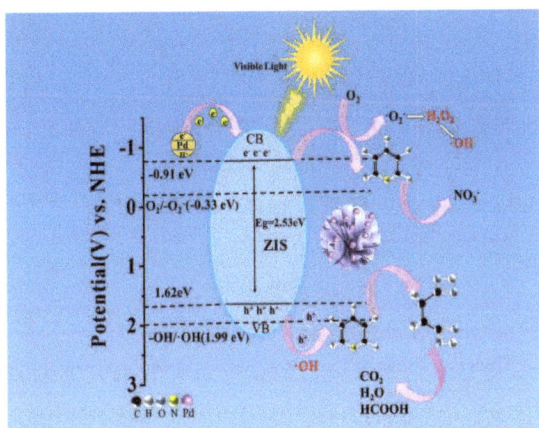

Figure 11. Possible photocatalytic denitrification mechanism of Pd-modified ZIS.

3. Materials and Methods

3.1. Materials

Zinc chloride ($ZnCl_2$), indium chloride tetrahydrate ($InCl_3 \cdot 4H_2O$), thioacetamide (TAA), cetyltrimethylammonium bromide (CTAB), palladium chloride ($PdCl_2$), methanol, and ethanol were purchased from Sinopharm Chemical Reagent Co. Ltd.(Shanghai, China). and Macklin Reagent Co. Ltd. (Shanghai, China).

3.2. Synthesis of $ZnIn_2S_4$ Nanosheets (ZIS)

The ZIS was synthesized via a typical hydrothermal method [21]. First, 1 mmol $ZnCl_2$, 2 mmol $InCl_3 \cdot 4H_2O$, and 8 mmol TAA were dissolved in 30 mL of ethanol aqueous (ethanol:water = 1:2). Then, the mixture was transferred into a 50 mL Teflon-lined autoclave and heated at 180 °C for 2 h. After the reaction, the solid products were washed with deionized water and ethanol several times. Finally, the obtained products were dried at 60 °C for 12 h.

3.3. Synthesis of Pd/$ZnIn_2S_4$ Nanosheets (Pd/ZIS)

First, 100 mg ZIS was dispersed in a 20 mL aqueous solution containing 25 % methanol as a sacrificial agent. Pd (1%) was loaded on the ZIS surface using an in-situ photodeposition method with $PdCl_2$ as a precursor (Scheme 1). While under the N_2 atmosphere, the mixture solution was stirred to remove O_2. Afterward, the above solution was irradiated by a 300 W Xenon lamp for 1 h. Then, 70 mg CTAB was added to the above solution and stirred overnight. The final product was washed with ethanol and deionized water and collected by centrifugation and dried at 60 °C for 12 h.

Scheme 1. Flow diagram for the fabrication of Pd/ZIS.

3.4. Dispersion Experiment

A 5 mg sample was dissolved in 5 mL pyridine/octane solution (Figure S6).

3.5. Characterization Methods

The crystalline phase of the photocatalysts was obtained using a Bruker D8 Advance X-ray diffractometer (Salbruken, Germany). The surface morphology and microstructures were observed with SEM (Czech TESCAN MIRA LMS, Brno, Czech Republic) and TEM (FEI Talos F200s, Massachusetts, American) instruments. UV–vis absorption spectra were characterized by a Shimadzu UV-2700 (Kyoto, Japan), using BaSO4 as a reflectance standard. X-ray photoelectron spectroscopy (XPS) measurements were conducted on a Thermo Scientific K-Alpha+ spectrometer (Massachusetts, American). Photoluminescence (PL) spectra were characterized with a fluorescence spectrophotometer (Edinburgh FLS1000, Livingston, Scotland, UK) with an excitation wavelength of 225 nm. The in situ FTIR spectra of pyridine absorbed over the photocatalysts were collected on a Tensor 27 Fourier transform infrared (FT-IR) spectrometer (Salbruken, Germany) with a resolution of 4 cm^{-1} for 64 scans. EPR spectra of the prepared samples were obtained using a Germany Bruker EMX nano-spectrometer (Salbruken, Germany). Photoelectrochemical measurements were performed using an electrochemical workstation (CHI-660, Shanghai, China) with a conventional three-electrode cell, including the reference electrode (Pt wire), the counter electrode (Ag/AgCl), and the working electrode (FTO as a support). Mott–Schottky plots were generated with different frequencies of 500and 1000 Hz.

3.6. Performance Testing

The fuel denitrification of the photocatalyst was performed in a quartz reactor. First, 50 mL of pyridine/n-octane solution (70 mg/mL) was used to distribute 50 mg of photocatalyst. The solution was stirred for 1 h in the dark to achieve adsorption equilibrium. Then, using a 420 nm cutoff filter, the suspension was performed under 300 W Xe lamp irradiation (Figure S7). At given time intervals, 1.5 mL aliquots were sampled. A Varian Cary 50 spectrometer (Palo Alto, American) was used to monitor the residual concentration of pyridine in the supernatant at a 252 nm peak position.

4. Conclusions

In summary, we have successfully prepared a Pd/ZIS photocatalyst for the efficient photocatalytic denitrification of pyridine. When exposed to visible light, 1% Pd/ZIS produces a 2.25-fold greater reaction rate for the photocatalytic denitrification of pyridine than ZIS. From all the experimental results, the superior photocatalytic performance of 1% Pd/ZIS was ascribed to the enhanced visible light absorption, as well as the Zn···C-N or In···C-N coordination for pyridine molecules' selective adsorption and activation. Meanwhile, the electron separation and transfer ability of Pd can be combined with ZIS to promote photocatalytic denitrification. This work provides a pathway for the photocatalytic denitrification of pyridine and explores organic molecules' selective adsorption, activation, and photocatalytic conversion on metal/nanosheet surfaces.

Supplementary Materials: The following supporting information can be downloaded at: https://www.mdpi.com/article/10.3390/molecules28010282/s1, Table S1: Actual mass content of Pd from EDS analysis of the prepared samples; Table S2: Comparison between the photocatalytic activity of 1% Pd/ZnIn$_2$S$_4$ and that of other reported catalysts for pyridine denitrogenation; Table S3: Pyridine peak area from LC-MS analysis of the prepared samples; Figure S1: (a,b) HRTEM of Pd/ZIS after 5-times recycling and element mapping images; Figure S2: XPS spectra of the other ratio Pd/ZIS sample before and after pyridine adsorption. Figure S3: UV–vis DRS spectra of the Pd/ZIS before and after pyridine adsorption; Figure S4: High-performance liquid chromatography profiles of pyridine after different irradiation times: (a) 0 and (b) 4 h; Figure S5: Possible denitrification pathway of pyridine; Figure S6: Dispersion experiment before and after CTAB absorption; Figure S7: Schematic diagram of photocatalytic fuel denitrification reaction; Figure S8: Photocatalytic denitrification of pyridine under visible light irradiation over ZIS, Pd/ZIS (no CTAB), CTAB/ZIS, and 1% Pd/ZIS. Refs. [60–63] are cited in the Supplementary Materials.

Author Contributions: Conceptualization: X.W. and G.Y.; investigation: D.W., E.Z., L.C., X.L. and S.W.; writing—review and editing: D.W. All authors have read and agreed to the published version of the manuscript.

Funding: This work is funded by the National Nature Science Foundation of China (22108129), Natural Science Foundation of Fujian Province (2021J05253, 2020J05225), and Scientific Research Fund project of Ningde Normal University (2021Y05, 2021Q102, 2021ZDK09, 2019T03, 2019Y15).

Institutional Review Board Statement: Not applicable.

Informed Consent Statement: Not applicable.

Data Availability Statement: The data presented in the study are available from the corresponding author.

Conflicts of Interest: The authors declare no conflict of interest.

Sample Availability: Not applicable.

References

1. Yao, Z.; Miras, H.N.; Song, Y.F. Efficient concurrent removal of sulfur and nitrogen contents from complex oil mixtures by using polyoxometalate-based composite materials. *Inorg. Chem. Front.* **2016**, *3*, 1007–1013. [CrossRef]
2. Padoley, K.V.; Mudliar, S.N.; Banerjee, S.K.; Deshmukh, S.C.; Pandey, R.A. Fenton oxidation: A pretreatment option for improved biological treatment of pyridine and 3-cyanopyridine plant wastewater. *Chem. Eng. J.* **2011**, *166*, 1–9. [CrossRef]
3. Jiirgen, M.L.; Dieter, H.S.; Wei, M.H.; Paul, J.C. Importance of biomass burning in the atmospheric budgets of nitrogen-containing gases. *Nature* **1990**, *346*, 552–553. [CrossRef]
4. Ahmed, I.; Khan, N.A.; Jhung, S.H. Adsorptive denitrogenation of model fuel by functionalized UiO-66 with acidic and basic moieties. *Chem. Eng. J.* **2017**, *321*, 40–47. [CrossRef]
5. Bello, S.S.; Wang, C.; Zhang, M.; Gao, H.; Han, Z.; Shi, L.; Su, F.; Xu, G. A Review on the Reaction Mechanism of Hydrodesulfurization and Hydrodenitrogenation in Heavy Oil Upgrading. *Energy Fuels* **2021**, *35*, 10998–11016. [CrossRef]
6. Misra, P.; Badoga, S.; Dalai, A.K.; Adjaye, J. Enhancement of sulfur and nitrogen removal from heavy gas oil by using polymeric adsorbent followed by hydrotreatment. *Fuel* **2018**, *226*, 127–136. [CrossRef]
7. Shu, C.; Wang, L.; Kong, P.; Gao, M. Adsorptive removal of nitrogen compounds from fuel oil by a poly ionic liquid PVIm-PXDC. *Sep. Sci. Technol.* **2018**, *54*, 2567–2576. [CrossRef]
8. Han, D.Y.; Li, G.X.; Cao, Z.B.; Zhai, X.Y.; Yuan, M.M. A Study on the Denitrogenation of Fushun Shale Oil. *Energy Sources Part A Recovery Util. Environ. Eff.* **2013**, *35*, 622–628. [CrossRef]
9. Li, J.; Cai, W.; Cai, J. The characteristics and mechanisms of pyridine biodegradation by Streptomyces sp. *J. Hazard. Mater.* **2009**, *165*, 950–954. [CrossRef]
10. Sun, J.Q.; Xu, L.; Tang, Y.Q.; Chen, F.M.; Liu, W.Q.; Wu, X.L. Degradation of pyridine by one Rhodococcus strain in the presence of chromium (VI) or phenol. *J. Hazard. Mater.* **2011**, *191*, 62–68. [CrossRef]
11. Li, Y.; Yi, R.; Yi, C.; Zhou, B.; Wang, H. Research on the degradation mechanism of pyridine in drinking water by dielectric barrier discharge. *J. Environ. Sci.* **2017**, *53*, 238–247. [CrossRef]
12. Liu, T.; Ding, Y.; Liu, C.; Han, J.; Wang, A. UV activation of the pi bond in pyridine for efficient pyridine degradation and mineralization by UV/H_2O_2 treatment. *Chemosphere* **2020**, *258*, 127208. [CrossRef]
13. Zhang, C.; Li, M.; Liu, G.; Luo, H.; Zhang, R. Pyridine degradation in the microbial fuel cells. *J. Hazard. Mater.* **2009**, *172*, 465–471. [CrossRef]
14. Yao, H.; Ren, Y.; Deng, X.; Wei, C. Dual substrates biodegradation kinetics of m-cresol and pyridine by Lysinibacillus cresolivorans. *J. Hazard. Mater.* **2011**, *186*, 1136–1140. [CrossRef]
15. Zhang, Y.; Chang, L.; Yan, N.; Tang, Y.; Liu, R.; Rittmann, B.E. UV photolysis for accelerating pyridine biodegradation. *Environ. Sci. Technol.* **2014**, *48*, 649–655. [CrossRef]
16. Luana, V.; Wallace, C.; Ricardo, J.; de Claudia, O.; Sandra, S.; Marco, A.G. Removal of sulfur and nitrogen compounds from diesel oil by adsorption using clays as adsorbents. *Energy Fuels* **2017**, *31*, 11731–11742. [CrossRef]
17. Qu, D.; Feng, X.; Li, N.; Ma, X.; Shang, C.; Chen, X.D. Adsorption of heterocyclic sulfur and nitrogen compounds in liquid hydrocarbons on activated carbons modified by oxidation: Capacity, selectivity and mechanism. *RSC Adv.* **2016**, *6*, 41982–41990. [CrossRef]
18. Laredo, G.C.; Vega-Merino, P.M.; Trejo-Zárraga, F.; Castillo, J. Denitrogenation of middle distillates using adsorbent materials towards ULSD production: A review. *Fuel Process. Technol.* **2013**, *106*, 21–32. [CrossRef]
19. Roman, F.F.; Diaz de Tuesta, J.L.; Silva, A.M.T.; Faria, J.L.; Gomes, H.T. Carbon-Based Materials for Oxidative Desulfurization and Denitrogenation of Fuels: A Review. *Catalysts* **2021**, *11*, 1239. [CrossRef]
20. Li, Y.; Lv, K.; Ho, W.; Zhao, Z.; Huang, Y. Enhanced visible-light photo-oxidation of nitric oxide using bismuth-coupled graphitic carbon nitride composite heterostructures. *Chin. J. Catal.* **2017**, *38*, 321–329. [CrossRef]
21. Zhang, W.; Liu, X.; Dong, X.a.; Dong, F.; Zhang, Y. Facile synthesis of $Bi_{12}O_{17}Br_2$ and $Bi_4O_5Br_2$ nanosheets: In situ DRIFTS investigation of photocatalytic NO oxidation conversion pathway. *Chin. J. Catal.* **2017**, *38*, 2030–2038. [CrossRef]
22. Zhao, X.; Du, Y.; Zhang, C.; Tian, L.; Li, X.; Deng, K.; Chen, L.; Duan, Y.; Lv, K. Enhanced visible photocatalytic activity of TiO_2 hollow boxes modified by methionine for RhB degradation and NO oxidation. *Chin. J. Catal.* **2018**, *39*, 736–746. [CrossRef]
23. Low, G.K.C.; McEvoy, S.R.; Mathews, R.W. Formation of nitrate and ammonium ions in titanium dioxide mediated photocatalytic degradation of organic compounds containing nitrogenatoms. *Environ. Sci. Technol.* **1991**, *25*, 460–467. [CrossRef]
24. Liang, R.; Wang, S.; Lu, Y.; Yan, G.; He, Z.; Xia, Y.; Liang, Z.; Wu, L. Assembling Ultrafine SnO_2 Nanoparticles on MIL-101(Cr) Octahedrons for Efficient Fuel Photocatalytic Denitrification. *Molecules* **2021**, *26*, 7566. [CrossRef]

25. Xu, B.; He, P.; Liu, H.; Wang, P.; Zhou, G.; Wang, X. A 1D/2D helical CdS/ZnIn$_2$S$_4$ nano-heterostructure. *Angew. Chem. Int. Ed. Engl.* **2014**, *53*, 2339–2343. [CrossRef]
26. Yang, W.; Zhang, L.; Xie, J.; Zhang, X.; Liu, Q.; Yao, T.; Wei, S.; Zhang, Q.; Xie, Y. Enhanced Photoexcited Carrier Separation in Oxygen-Doped ZnIn$_2$S$_4$ Nanosheets for Hydrogen Evolution. *Angew. Chem. Int. Ed. Engl.* **2016**, *55*, 6716–6720. [CrossRef]
27. Chen, Y.; Tian, G.; Ren, Z.; Pan, K.; Shi, Y.; Wang, J.; Fu, H. Hierarchical core-shell carbon nanofiber@ ZnIn$_2$S$_4$ composites for enhanced hydrogen evolution performance. *ACS Appl. Mater. Interfaces* **2014**, *6*, 13841–13849. [CrossRef]
28. Du, C.; Zhang, Q.; Lin, Z.; Yan, B.; Xia, C.; Yang, C. Half-unit-cell ZnIn$_2$S$_4$ monolayer with sulfur vacancies for photocatalytic hydrogen evolution. *Appl. Catal. B Environ.* **2019**, *248*, 193–201. [CrossRef]
29. Yang, C.; Li, Q.; Xia, Y.; Lv, K.; Li, M. Enhanced visible-light photocatalytic CO$_2$ reduction performance of ZnIn$_2$S$_4$ microspheres by using CeO$_2$ as cocatalyst. *Appl. Surf. Sci.* **2019**, *464*, 388–395. [CrossRef]
30. Gao, B.; Liu, L.; Liu, J.; Yang, F. Photocatalytic degradation of 2,4,6-tribromophenol on Fe$_2$O$_3$ or FeOOH doped ZnIn$_2$S$_4$ heterostructure: Insight into degradation mechanism. *Appl. Catal. B Environ.* **2014**, *147*, 929–939. [CrossRef]
31. Xiao, Y.; Wang, H.; Jiang, Y.; Zhang, W.; Zhang, J.; Wu, X.; Liu, Z.; Deng, W. Hierarchical Sb$_2$S$_3$/ZnIn$_2$S$_4$ core-shell heterostructure for highly efficient photocatalytic hydrogen production and pollutant degradation. *J. Colloid. Interface Sci.* **2022**, *623*, 109–123. [CrossRef]
32. Tang, M.; Ao, Y.; Wang, P.; Wang, C. All-solid-state Z-scheme WO$_3$ nanorod/ZnIn$_2$S$_4$ composite photocatalysts for the effective degradation of nitenpyram under visible light irradiation. *J. Hazard. Mater.* **2020**, *387*, 121713. [CrossRef]
33. Bo, L.; He, K.; Tan, N.; Gao, B.; Feng, Q.; Liu, J.; Wang, L. Photocatalytic oxidation of trace carbamazepine in aqueous solution by visible-light-driven ZnIn$_2$S$_4$: Performance and mechanism. *J. Environ. Manag.* **2017**, *190*, 259–265. [CrossRef]
34. Zuo, G.; Wang, Y.; Teo, W.L.; Xie, A.; Guo, Y.; Dai, Y.; Zhou, W.; Jana, D.; Xian, Q.; Dong, W.; et al. Ultrathin ZnIn$_2$S$_4$ Nanosheets Anchored on Ti$_3$C$_2$ TX MXene for Photocatalytic H$_2$ Evolution. *Angew. Chem. Int. Ed. Engl.* **2020**, *59*, 11287–11292. [CrossRef]
35. Song, Y.; Zhang, J.; Dong, X.; Li, H. A Review and Recent Developments in Full-Spectrum Photocatalysis using ZnIn$_2$S$_4$-Based Photocatalysts. *Energy Technol.* **2021**, *9*, 2100033. [CrossRef]
36. Liu, C.; Zhang, Y.; Wu, J.; Dai, H.; Ma, C.; Zhang, Q.; Zou, Z. Ag-Pd alloy decorated ZnIn$_2$S$_4$ microspheres with optimal Schottky barrier height for boosting visible-light-driven hydrogen evolution. *J. Mater. Sci. Technol.* **2022**, *114*, 81–89. [CrossRef]
37. Jia, T.; Liu, M.; Zheng, C.; Long, F.; Min, Z.; Fu, F.; Yu, D.; Li, J.; Lee, J.H.; Kim, N.H. One-Pot Hydrothermal Synthesis of La-Doped ZnIn$_2$S$_4$ Microspheres with Improved Visible-Light Photocatalytic Performance. *Nanomaterials* **2020**, *10*, 2026. [CrossRef]
38. Mohamed, R.M.; Shawky, A.; Aljahdali, M.S. Palladium/zinc indium sulfide microspheres: Enhanced photocatalysts prepare methanol under visible light conditions. *J. Taiwan Inst. Chem. Eng.* **2016**, *65*, 498–504. [CrossRef]
39. Wang, B.; Deng, Z.; Fu, X.; Xu, C.; Li, Z. Photodeposition of Pd nanoparticles on ZnIn$_2$S$_4$ for efficient alkylation of amines and ketones' α-H with alcohols under visible light. *Appl. Catal. B Environ.* **2018**, *237*, 970–975. [CrossRef]
40. Shi, X.; Dai, C.; Wang, X.; Hu, J.; Zhang, J.; Zheng, L.; Mao, L.; Zheng, H.; Zhu, M. Protruding Pt single-sites on hexagonal ZnIn$_2$S$_4$ to accelerate photocatalytic hydrogen evolution. *Nat. Commun.* **2022**, *13*, 1287. [CrossRef]
41. Mandal, S.; Adhikari, S.; Shengyan, P.; Hui, M.; Kim, D.-H. Constructing gold-sensitized ZnIn$_2$S$_4$ microarchitectures for efficient visible light-driven photochemical oxidation and sensing of micropollutants. *Appl. Surf. Sci.* **2019**, *498*, 143840. [CrossRef]
42. Zhu, T.; Ye, X.; Zhang, Q.; Hui, Z.; Wang, X.; Chen, S. Efficient utilization of photogenerated electrons and holes for photocatalytic redox reactions using visible light-driven Au/ZnIn$_2$S$_4$ hybrid. *J. Hazard. Mater.* **2019**, *367*, 277–285. [CrossRef] [PubMed]
43. An, H.; Li, M.; Liu, R.; Gao, Z.; Yin, Z. Design of Ag$_x$Au$_{1−x}$ alloy/ZnIn$_2$S$_4$ system with tunable spectral response and Schottky barrier height for visible-light-driven hydrogen evolution. *Chem. Eng. J.* **2020**, *382*, 122593. [CrossRef]
44. Liao, G.; Gong, Y.; Zhang, L.; Gao, H.; Yang, G.; Fang, B. Semiconductor polymeric graphitic carbon nitride photocatalysts: The "holy grail" for the photocatalytic hydrogen evolution reaction under visible light. *Energy Environ. Sci.* **2019**, *12*, 2080–2147. [CrossRef]
45. Bowker, M.; Morton, C.; Kennedy, J.; Bahruji, H.; Greves, J.; Jones, W.; Davies, P.R.; Brookes, C.; Wells, P.P.; Dimitratos, N. Hydrogen production by photoreforming of biofuels using Au, Pd and Au–Pd/TiO$_2$ photocatalysts. *J. Catal.* **2014**, *310*, 10–15. [CrossRef]
46. Mohamed, M.M.; Bayoumy, W.A.; Khairy, M.; Mousa, M.A. Synthesis of micro–mesoporous TiO$_2$ materials assembled via cationic surfactants: Morphology, thermal stability and surface acidity characteristics. *Microporous Mesoporous Mater.* **2007**, *103*, 174–183. [CrossRef]
47. Zhang, S.; Liu, Q.; Cheng, H.; Gao, F.; Liu, C.; Teppen, B.J. Thermodynamic Mechanism and Interfacial Structure of Kaolinite Intercalation and Surface Modification by Alkane Surfactants with Neutral and Ionic Head Groups. *J. Phys. Chem. C Nanomater. Interfaces* **2017**, *121*, 8824–8831. [CrossRef]
48. Xu, L.; Dong, S.; Hao, J.; Cui, J.; Hoffmann, H. Surfactant-Modified Ultrafine Gold Nanoparticles with Magnetic Responsiveness for Reversible Convergence and Release of Biomacromolecules. *Langmuir* **2017**, *33*, 3047–3055. [CrossRef]
49. Yang, G.; Guo, Q.; Yang, D.; Peng, P.; Li, J. Disperse ultrafine amorphous SiO$_2$ nanoparticles synthesized via precipitation and calcination. *Colloids Surf. A Physicochem. Eng. Asp.* **2019**, *568*, 445–454. [CrossRef]
50. Lv, J.; Zhang, J.; Liu, J.; Li, Z.; Dai, K.; Liang, C. Bi SPR-Promoted Z-Scheme Bi$_2$MoO$_6$/CdS-Diethylenetriamine Composite with Effectively Enhanced Visible Light Photocatalytic Hydrogen Evolution Activity and Stability. *ACS Sustain. Chem. Eng.* **2017**, *6*, 696–706. [CrossRef]

51. Xia, Y.; Li, Q.; Lv, K.; Tang, D.; Li, M. Superiority of graphene over carbon analogs for enhanced photocatalytic H_2-production activity of $ZnIn_2S_4$. *Appl. Catal. B Environ.* **2017**, *206*, 344–352. [CrossRef]
52. Manhas, F.M.; Fatima, A.; Verma, I.; Siddiqui, N.; Muthu, S.; AlSalem, H.S.; Savita, S.; Singh, M.; Javed, S. Quantum computational, spectroscopic (FT-IR, NMR and UV–Vis) profiling, Hirshfeld surface, molecular docking and dynamics simulation studies on pyridine-2,6-dicarbonyl dichloride. *J. Mol. Struct.* **2022**, *1265*, 133374. [CrossRef]
53. Shi, Y.; Wang, Z.; Liu, C.; Wu, T.k.; Liu, R.; Wu, L. Surface synergetic effects of Pt clusters/monolayer Bi_2MoO_6 nanosheet for promoting the photocatalytic selective reduction of 4-nitrostyrene to 4-vinylaniline. *Appl. Catal. B Environ.* **2022**, *304*, 121010. [CrossRef]
54. Liang, S.; Wen, L.; Lin, S.; Bi, J.; Feng, P.; Fu, X.; Wu, L. Monolayer HNb_3O_8 for selective photocatalytic oxidation of benzylic alcohols with visible light response. *Angew. Chem. Int. Ed. Engl.* **2014**, *53*, 2951–2955. [CrossRef]
55. Hu, T.; Dai, K.; Zhang, J.; Chen, S. Noble-metal-free Ni_2P modified step-scheme $SnNb_2O_6$/CdS-diethylenetriamine for photocatalytic hydrogen production under broadband light irradiation. *Appl. Catal. B Environ.* **2020**, *269*, 118844. [CrossRef]
56. Wang, H.; Wu, Y.; Xiao, T.; Yuan, X.; Zeng, G.; Tu, W.; Wu, S.; Lee, H.Y.; Tan, Y.Z.; Chew, J.W. Formation of quasi-core-shell In_2S_3/anatase TiO_2@metallic Ti_3C_2Tx hybrids with favorable charge transfer channels for excellent visible-light-photocatalytic performance. *Appl. Catal. B Environ.* **2018**, *233*, 213–225. [CrossRef]
57. Zhu, C.; He, Q.; Yao, H.; Le, S.; Chen, W.; Chen, C.; Wang, S.; Duan, X. Amino-functionalized NH_2-MIL-125(Ti)-decorated hierarchical flowerlike $ZnIn_2S_4$ for boosted visible-light photocatalytic degradation. *Environ. Res.* **2022**, *204*, 112368. [CrossRef]
58. He, Z.; Liang, R.; Zhou, C.; Yan, G.; Wu, L. Carbon quantum dots (CQDs)/noble metal co-decorated MIL-53(Fe) as difunctional photocatalysts for the simultaneous removal of Cr(VI) and dyes. *Sep. Purif. Technol.* **2021**, *255*, 117725. [CrossRef]
59. Zhang, X.; Song, H.; Sun, C.; Chen, C.; Han, F.; Li, X. Photocatalytic oxidative desulfurization and denitrogenation of fuels over sodium doped graphitic carbon nitride nanosheets under visible light irradiation. *Mater. Chem. Phys.* **2019**, *226*, 34–43. [CrossRef]
60. Lu, Y.; Liang, R.; Yan, G.; Liang, Z.; Hu, W.; Xia, Y.; Huang, R. Solvothermal synthesis of TiO_2@MIL-101(Cr) for efficient photocatalytic fuel denitrification. *J. Fuel Chem. Technol.* **2021**, *50*, 1–8. Available online: https://kns.cnki.net/kcms/detail/14.1140.tq.20211025.1659.002.html (accessed on 21 February 2022). [CrossRef]
61. Hu, W.; Jiang, M.; Liang, R.; Huang, R.; Xia, Y.; Liang, Z.; Yan, G. Construction of Bi_2MoO_6/CdS heterostructures with enhanced visible light photocatalytic activity for fuel denitrification. *Dalton Trans.* **2021**, *50*, 2596–2605. [CrossRef] [PubMed]
62. Zheng, L.; Yan, G.; Huang, Y.; Wang, X.; Long, J.; Li, L.; Xu, T. Visible-light photocatalytic denitrogenation of nitrogen-containing compound in petroleum by metastable $Bi_{20}TiO_{32}$. *Int. J. Hydrogen Energy* **2014**, *39*, 13401–13407. [CrossRef]
63. Lu, Y.; Pan, H.; Lai, J.; Xia, Y.; Chen, L.; Liang, R.; Yan, G.; Huang, R. Affiliation Bimetallic CoCu-ZIF material for efficient visible light photocatalytic fuel denitrification. *RSC Adv.* **2022**, *12*, 12702–12709. [CrossRef] [PubMed]

Disclaimer/Publisher's Note: The statements, opinions and data contained in all publications are solely those of the individual author(s) and contributor(s) and not of MDPI and/or the editor(s). MDPI and/or the editor(s) disclaim responsibility for any injury to people or property resulting from any ideas, methods, instructions or products referred to in the content.

Review

Advances in Bi$_2$WO$_6$-Based Photocatalysts for Degradation of Organic Pollutants

Haiyan Jiang [1], Jiahua He [2], Changyi Deng [2], Xiaodong Hong [2,*] and Bing Liang [3]

1. Basic Department, Liaoning Institute of Science and Technology, Benxi 117004, China
2. School of Materials Science and Hydrogen Energy, Foshan University, Foshan 528000, China
3. College of Materials Science and Engineering, Shenyang University of Chemical Technology, Shenyang 110142, China
* Correspondence: hongxiaodong@lntu.edu.cn

Abstract: With the rapid development of modern industries, water pollution has become an urgent problem that endangers the health of human and wild animals. The photocatalysis technique is considered an environmentally friendly strategy for removing organic pollutants in wastewater. As an important member of Bi-series semiconductors, Bi$_2$WO$_6$ is widely used for fabricating high-performance photocatalysts. In this review, the recent advances of Bi$_2$WO$_6$-based photocatalysts are summarized. First, the controllable synthesis, surface modification and heteroatom doping of Bi$_2$WO$_6$ are introduced. In the respect of Bi$_2$WO$_6$-based composites, existing Bi$_2$WO$_6$-containing binary composites are classified into six types, including Bi$_2$WO$_6$/carbon or MOF composite, Bi$_2$WO$_6$/g-C$_3$N$_4$ composite, Bi$_2$WO$_6$/metal oxides composite, Bi$_2$WO$_6$/metal sulfides composite, Bi$_2$WO$_6$/Bi-series composite, and Bi$_2$WO$_6$/metal tungstates composite. Bi$_2$WO$_6$-based ternary composites are classified into four types, including Bi$_2$WO$_6$/g-C$_3$N$_4$/X, Bi$_2$WO$_6$/carbon/X, Bi$_2$WO$_6$/Au or Ag-based materials/X, and Bi$_2$WO$_6$/Bi-series semiconductors/X. The design, microstructure, and photocatalytic performance of Bi$_2$WO$_6$-based binary and ternary composites are highlighted. Finally, aimed at the existing problems in Bi$_2$WO$_6$-based photocatalysts, some solutions and promising research trends are proposed that would provide theoretical and practical guidelines for developing high-performance Bi$_2$WO$_6$-based photocatalysts.

Keywords: Bi$_2$WO$_6$; photocatalysis; degradation performance; composite

1. Introduction

With the rapid development of various industries, pollution problems, including soil pollution, air pollution, and water pollution, are becoming more and more serious. The problem of water pollution is endangering the health of humans and wild animals; thus, wastewater treatment has become an important task of scientists and technicians. The photocatalysis technique is regarded as an environmentally friendly route for solving water pollution [1]. Therefore, the development of highly active photocatalysts is regarded as the major task of photocatalysis [2]. Lots of semiconductors are adopted as photocatalysts for the degradation of organic pollutants under the irradiation of visible light or UV light [3]. However, the general photocatalytic mechanism of different photocatalysts can be classified into the following three steps: The first step is the generation of photogenerated electrons and holes. When the light irradiation energy surpasses the energy gap (Eg) of the semiconductor, the electrons jump from valence band (VB) to conduction band (CB) and leave holes in the VB [4]. The second step is the transfer of charge carriers. The electrons transfer to the surface of catalysts or recombine with holes. In this process, preventing the recombination of electrons and holes increases the lifetime of photogenerated electrons and holes and enhances the photocatalytic activity by prolonging the redox reaction time [5]. The last step is the surface redox reaction between electrons or holes with O$_2$ in H$_2$O.

Consequently, highly active oxidants are produced and degrade organic pollutants into H_2O and CO_2, without secondary pollution. Based on the photocatalytic mechanism, the design principles of high-performance photocatalysts include, but are not limited to, enhancing the light harvesting capability and suppressing the recombination of electrons and holes [6]. Bi_2WO_6 is a commonly-used Bi-series semiconductor [7], which has widely served as a visible light photocatalyst for the degradation of organic pollutants due to its advantages of a perovskite-type layered structure, light harvesting ability (band gap of 2.8 eV), low cost, chemical stability, and non-toxicity [8]. In the respect of controllable synthesis, hydrothermal/solvothermal, sol-gel process, calcination, and electrodeposition can be used to prepare Bi_2WO_6 nanoplates, nanosheets, nanorods, and nanoflowers or spheres. Besides microstructure control, surface modification and heteroatom doping are adopted for enhancing the photocatalytic activity through generating surface vacancies or defects. In addition, the construction of Bi_2WO_6-based composite photocatalysts has become the most popular research topic for the diversity of alternative semiconductive materials.

In view of the important role of Bi_2WO_6 in Bi-series semiconductor photocatalysts, herein, we summarize the recent advances of Bi_2WO_6-based photocatalysts. Figure 1 shows that the whole review involves three sections. The first section is the controllable synthesis, surface modification, and heteroatom doping of Bi_2WO_6. In the second section, we majorly introduce the progress of Bi_2WO_6-based binary composite photocatalysts, including Bi_2WO_6/carbon or MOF composite, Bi_2WO_6/g-C_3N_4 composite, Bi_2WO_6/metal oxides composite, Bi_2WO_6/metal sulfides composite, Bi_2WO_6/Bi-series composite, and Bi_2WO_6/metal tungstates composite. In the last section, Bi_2WO_6-based ternary composites are reviewed, including Bi_2WO_6/g-C_3N_4/X, Bi_2WO_6/carbon/X, Bi_2WO_6/Au or Ag-based materials/X, and Bi_2WO_6/Bi-series semiconductors/X. The design, microstructure, and photocatalytic performance of Bi_2WO_6-based binary and ternary composites are highlighted in detail. Lastly, we summarize the research highlights and existing problems in Bi_2WO_6-based photocatalysts. Based on the existing problems, some solutions and promising research trends are put forward, finally, that would provide theoretical and practical guidelines for developing novel Bi_2WO_6-based photocatalysts.

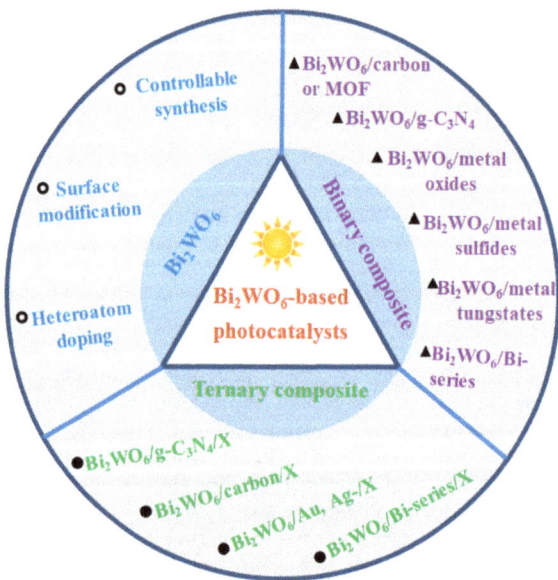

Figure 1. The related research directions of Bi_2WO_6-based photocatalysts.

2. Morphology Control, Surface Modification, and Heteroatom Doping of Bi_2WO_6

2.1. Morphology Control

The microstructure of Bi_2WO_6 seriously affects the specific surface area and photocatalytic performance. Therefore, the controllable synthesis and surface modification of Bi_2WO_6 have become the basic topic in preparing high-performance Bi_2WO_6-based photocatalysts. The synthetic methods of Bi_2WO_6 include the hydrothermal/solvothermal method, sol-gel process and calcination, and the electrodeposition method. However, the hydrothermal and solvothermal methods have been widely adopted for fabricating Bi_2WO_6, due to the easy operation and controllable microstructure. In this respect, Lai et al. [9] synthesized various Bi_2WO_6 photocatalysts via the solvothermal route and discussed the influence of reaction temperature on photocatalytic activity. The BWO-140 sample prepared at 140 °C delivered the best activity for the removal of Erichrome Black T (EBT) dye due to the oxygen vacancies, small size, and large surface area. Selvi et al. [10] discussed the influence of reaction time on the performance of Bi_2WO_6 photocatalysts. Under the same hydrothermal conditions, the nanoplates of Bi_2WO_6-24 h delivered the optimum photocatalytic activity for the degradation of MB due to the narrow band, smaller crystallite size, and hierarchical structure.

2.2. Surface Modification

Besides the effect of reaction conditions, hexadecyl trimethyl ammonium bromide (CTAB) and polyvinylpyrrolidone (PVP) are adopted to adjust the microstructure of Bi_2WO_6. By using CTAB and PVP surfactants, Guo et al. [11] prepared nanosheet-assembled Bi_2WO_6 microspheres and investigated the growth mechanism of the assembled microspheres. When used for the degradation of Rh B under visible light, the degradation efficiency was 98% within 50 min. By adopting CTAB surfactant and the mixed solvent of ethyl alcohol and ethylene glycol, Bai et al. [12] synthesized Bi_2WO_6 photocatalysts with abundant oxygen vacancies. The generation of oxygen vacancy enhanced the photogenerated carrier separation efficiency and visible light absorption ability, which resulted in superior photocatalytic activity for the degradation of ciprofloxacin, and the degradation rate reached 90% within 6 h. In addition, the CTAB-capped Bi_2WO_6 photocatalyst was synthesized with flower-like structures [13]. The CTAB surfactant affected the microstructure, which facilitated the physical adsorption of Rh B dye and enhanced the photoactivity of the resulting product. Consequently, the 0.20CTAB-Bi_2WO_6 sample degraded 100% Rh B within 120 min. By using the CTAB surfactant and an ethylene glycol-water mixed solvent, Zhou et al. [14] synthesized Bi_2WO_6 and Au-decorated Bi_2WO_6 hollow microspheres. When used for the degradation of phenol under visible light, Au nanoparticle-decorated Bi_2WO_6 delivered enhanced photocatalytic activity due to the cooperative electron trapping abilities and the SPR effect of Au nanoparticles.

2.3. Heteroatom Doping

Heteroatom doping can be adopted to adjust the crystal plane structure of Bi_2WO_6 and broaden the light absorption range, which is widely reported to enhance the photoactivity of Bi_2WO_6. In this section, we summarize the advances of non-metal-doped Bi_2WO_6 and metal-doped Bi_2WO_6.

In the existing non-metal dopants, N, F, Cl, and I serve as dopants for fabricating non-metal-doped Bi_2WO_6 photocatalysts, and the major topics mostly involve the influence of the doping amount on the microstructures and photodegradation performance. In this respect, Hoang et al. [15] synthesized N-doped Bi_2WO_6 nanoparticles and discussed the influence of the N-doping amount on photocatalytic activity. Through a comparison, the doped sample with a N/Bi atomic ratio of 0.5% presented the best photodegradation performance for removing Rh B under the irradiation of visible light. Chen et al. [16] prepared F-doped Bi_2WO_6 via a hydrothermal route and investigated the effect of the F-doping amount on the sample morphology and degradation performance. With an increase of the F-doping amount, ultrathin Bi_2WO_6 nanosheets were transformed into hierarchical nanoflowers. When used for the degradation of tetracycline (TC), the degradation rate constant of the optimized F-BWO4 sample was about 4.5 times higher than that of pristine Bi_2WO_6 due to the hierarchical structure and strong electronegativity. Phuruangrat et al. [17] reported the preparation of I-doped Bi_2WO_6 photocatalysts and their degradation performance. The results showed that 3 wt% I-doped Bi_2WO_6 presented the best performance, which degraded 100% Rh B in 100 min under the radiation of visible light, with a degradation rate of 0.044 min^{-1}.

Compared to non-metal doping, there are abundant works about metal-doped Bi_2WO_6, including Fe, Ti, Sr, Er, La, Au, Ag, and Mo dopants. Besides mono-metal doping, dual-metal doping was reported to further improve the photocatalytic activity of Bi_2WO_6. Among various metal dopants, the incorporation of Fe dopants would accelerate the electron-hole separation and improve the photocatalytic activity. In respect of the Fe-doping mechanism, Hu et al. [18] confirmed that Fe doping narrowed the energy band gap and induced abundant oxygen vacancies, which enhanced the separation efficiency of photogenerated carriers and the light absorption capability. When used for the degradation of Rh B and salicylic acid (SA), the optimized BW-Fe-0.10 sample showed 11.9 and 8.0 times higher than that of pristine Bi_2WO_6, respectively. Arif et al. [19] prepared Ti-doped Bi_2WO_6 photocatalysts and confirmed that the presence of Ti^{3+}/Ti^{4+} in Ti-doped Bi_2WO_6 promoted the generation of reactive oxygen species, which greatly enhanced the photocatalytic activity of Bi_2WO_6. Furthermore, the layered 3D hierarchical structure adjusted the band structure of Bi_2WO_6, further facilitating the enhancement of the photocatalytic performance. Maniyazagan et al. [20] synthesized hierarchical Sr-Bi_2WO_6 photocatalysts for the degradation of 4-NP and MB. As shown in Figure 2a, by optimizing the content of Sr^{2+} ions, the composite of 15% Sr-Bi_2WO_6 delivered the highest photocatalytic activity. Under the irradiation of UV light with $NaBH_4$, the optimized sample degraded 99.5% MB in 25 min and 99.4% 4-NP reduction in 15 min, respectively. The major reason was the enhanced charge carrier separation and the generation of oxygen vacancies. Qiu et al. [21] fabricated an Er^{3+}-mixed Bi_2WO_6 photocatalyst by a one-step hydrothermal route (Figure 2b). After adding Er^{3+} ions, Bi_2WO_6 was transformed into a layered nanosheet with a high specific surface area. The sample of 16% Er^{3+}-Bi_2WO_6 presented a high degradation rate of 94.58% TC within 60 min. The enhanced activity was attributed to the porous structure and enhanced separation efficiency of photogenerated electrons. Ning et al. [22] synthesized La^{3+}-doped Bi_2WO_6 nanoplates for the degradation of Rh B. Compared to Bi_2WO_6, the doped sample showed a higher specific surface area, and the band gap reduced to 2.81 eV from 2.89 eV. Furthermore, the La^{3+}-doping enhanced the separation efficiency of electron and hole pairs. Therefore, the La^{3+}-doped Bi_2WO_6 presented a higher degradation rate constant than pure Bi_2WO_6.

In respect of noble metal dopants, Phuruangrat et al. [23] synthesized Au-doped Bi_2WO_6 and incorporated Au^{3+} ions into Bi_2WO_6 lattice. By adjusting the doping amount of Au, 3% Au-doped Bi_2WO_6 nanoplates presented the highest Rh B degradation rate of 96.25% within 240 min, which was 2.15 times higher than that of pure Bi_2WO_6. The superior performance was ascribed to the enhanced separation efficiency of photogenerated electrons and holes. In another work, Phu et al. [24] discussed the photocatalytic activities of Ag-doped Bi_2WO_6 and Ag nanoparticle-decorated Bi_2WO_6. For Ag-doped sample, Ag ions substituted the lattice of Bi_2WO_6, while, for the decorated sample, abundant Ag nanoparticles were dispersed on the surface of Bi_2WO_6 nanoparticles with no lattice change. When used for the degradation of Rh B by visible light, the activity of the Ag nanoparticles-modified sample was more than two times higher than that of the Ag-doped sample due to the enhanced surface plasmon resonance caused by Ag nanoparticles. Besides mono-metal doping, (La, Mo) co-doped Bi_2WO_6 was reported [25]. The introduction of La and Mo adjusted the particle size and lattice spacing of Bi_2WO_6. Moreover, the La and Mo co-dopants inhibited the charge recombination. Consequently, the (0.25La, 0.25Mo)-Bi_2WO_6 sample containing 0.25 mol% La and 0.25 mol% Mo showed the highest activity for the photodegradation of MB.

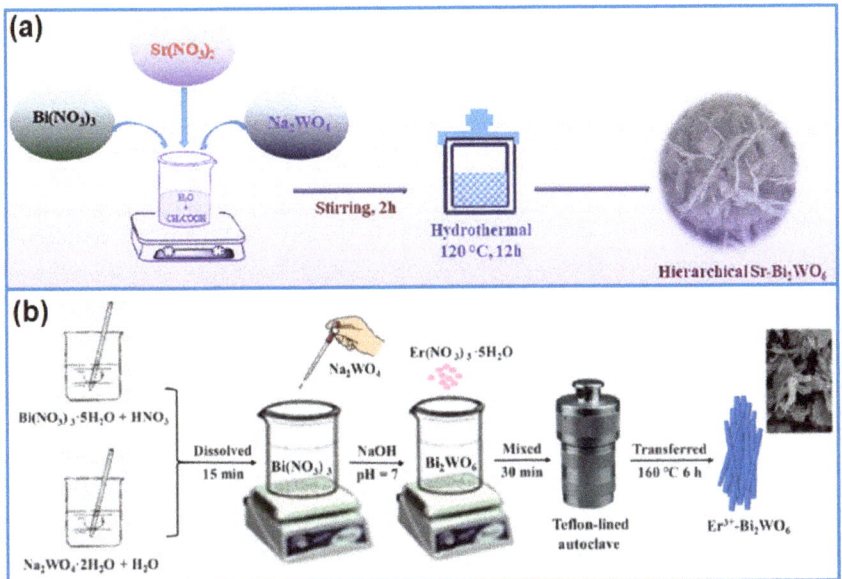

Figure 2. Fabrication process of (**a**) hierarchical Sr-Bi$_2$WO$_6$ [20]. Copyright (2022) Elsevier. (**b**) Er^{3+}-Bi$_2$WO$_6$ photocatalyst [21]. Copyright (2022) Elsevier.

3. Bi$_2$WO$_6$-Based Binary Composite

Besides the decoration or doping of Bi$_2$WO$_6$, the construction of Bi$_2$WO$_6$-based composite is widely reported for enhancing the photoactivity of Bi$_2$WO$_6$. According to the type of candidate materials, we classified the existing Bi$_2$WO$_6$-based binary composites into six types: Bi$_2$WO$_6$/carbon or MOF composite, Bi$_2$WO$_6$/g-C$_3$N$_4$ composite, Bi$_2$WO$_6$/metal oxides composite, Bi$_2$WO$_6$/metal sulfides composite, Bi$_2$WO$_6$/Bi-series composite, and Bi$_2$WO$_6$/metal tungstates composite. The design idea, microstructure, and photocatalytic performance of these binary composite photocatalysts are summarized in detail.

3.1. Bi$_2$WO$_6$/Carbon or MOF Composite

Carbon materials exhibit good conductivity and a large specific surface area, which are more suitable for loading Bi$_2$WO$_6$ nanostructures. In this section, various carbon materials, including graphene, carbon nanotube, carbon dots, and biomass-derived carbon are used for hybridizing with Bi$_2$WO$_6$. Among carbon materials, graphene oxide (GO) is a typical two-dimension template with abundant oxygen-containing groups, which can be served as ideal 2D substrates for loading semiconductors. Compared to GO, reduced GO (rGO) exhibits a superior electronic conductivity, which was adopted to hybridize with Bi$_2$WO$_6$ to enhance the photodegradation efficiency. For example, Zhao et al. [26] fabricated an rGO/Bi$_2$WO$_6$ composite photocatalyst via the hydrothermal method, and the composite degraded 87.49% norfloxacin within 180 min. The photocatalytic activity was much higher than that of pure Bi$_2$WO$_6$ due to the efficient charge separation and enhanced light-harvesting capacity. Arya et al. [27] also prepared an rGO-Bi$_2$WO$_6$ heterostructure via hydrothermal route and investigated their photocatalytic activity for the removal of levofloxacin. When kept in visible light at room temperature, the rGO-Bi$_2$WO$_6$ composite achieved a high degradation rate of 74.3% within 120 min due to the inhibition of charge carrier recombination. Furthermore, multi-walled carbon nanotubes (MWNTs) were coupled with Bi$_2$WO$_6$ to fabricate 3D mesoporous MWNTs-Bi$_2$WO$_6$ microspheres [28]. The MWNTs promoted the transfer and separation of hole and electron pairs, which enhanced the light absorption capability of Bi$_2$WO$_6$. The composite containing 3% MWNTs showed the optimum photoactivity, and the degradation efficiency was 1.35 times higher than that of pure Bi_2WO_6. In addition, carbon dots (CDs) were used for decorating a 3D Cl-doped Bi_2WO_6 hollow microsphere to construct $CDs/Cl-Bi_2WO_6$ composite photocatalysts [29]. The introduced CDs and Cl doping enhanced the visible light absorption capability and inhibited the recombination of electron–hole pairs. The optimized 0.5% $CDs/Cl-Bi_2WO_6$ composite degraded 85.1% TCH within 60 min, which was much better than Bi_2WO_6 and $Cl-Bi_2WO_6$.

In addition, biomass-derived carbon materials were coupled with Bi_2WO_6 to fabricate Bi_2WO_6/C hybrid photocatalysts. In this respect, Liang et al. [30] firstly prepared bamboo leave-derived carbon and then fabricated 3D flower-like Bi_2WO_6/C composites by hydrothermal route. The large specific surface area of biomass carbon enhanced the adsorption capacity; meanwhile, the good conductivity promoted the separation of charge carriers. The optimized Bi_2WO_6/C (6: 1) sample had a high degradation rate of 85.4% for the removal of TC within 90 min. Wang et al. [31] prepared a Bi_2WO_6/N-modified biochar (BW/N-B) composite for the degradation of Rh B pollutant. The loading of N-B improved the photocatalytic activity due to the enhanced separation and transfer of electron–hole pairs. Under the irradiation of visible light, the BW/N1-B sample with a urea/biochar ratio of 2:1 presented the best photocatalytic activity, which degraded 99.1% Rh B within 45 min. In addition, N and S co-doped corn straw biochar (NSBC) was used to hybridize with Bi_2WO_6 to form $Bi_2WO_6/NSBC$ composite photocatalysts (Figure 3a) for the degradation of ciprofloxacin (CIP) [32]. The N and S co-doped biochar exhibited a high specific surface area and interconnected fiber structure and high catalytic property, which effectively prevented the agglomeration of Bi_2WO_6. The combination of NSBC and Bi_2WO_6 extended the visible light response, adjusted the band gap, and promoted the separation and transfer of photoinduced carriers, which achieved the fast degradation of CIP (5 mg/L) within 75 min.

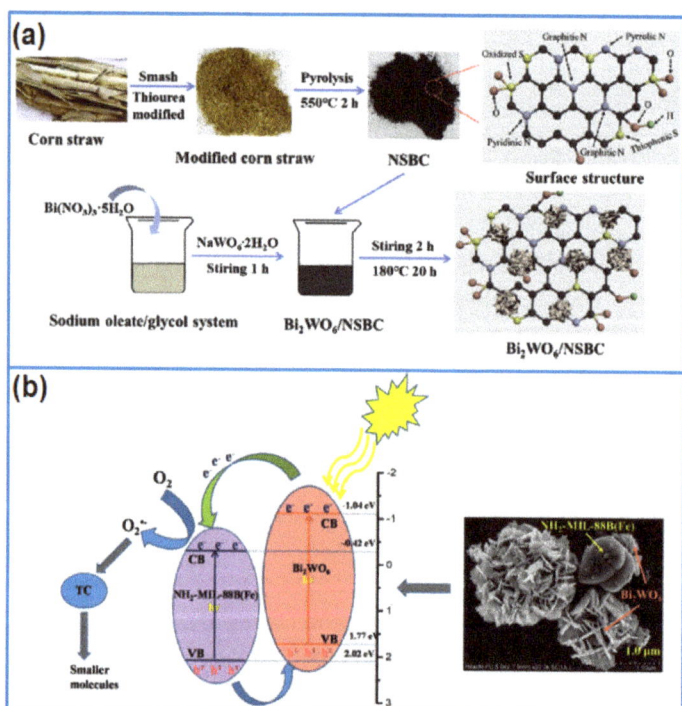

Figure 3. Preparation of (**a**) $Bi_2WO_6/NSBC$ composite [32]. Copyright (2021) Elsevier. (**b**) $Bi_2WO_6/NH_2-MIL-88B$ (Fe) heterostructure [33]. Copyright (2021) Elsevier.

Metal-organic frameworks (MOFs) consist of metal ions or metal clusters and organic ligands, which show some advantages in permanent porosity, tunable pore size, high specific surface area, and active surface chemistry. Various MOFs are widely utilized in adsorption, gas storage and separation, energy storage, and photocatalysis. In the field of photocatalysis, various MOFs, including Fe-based MOFs and Zn-based MOFs, are coupled with Bi_2WO_6 to fabricate hybrid photocatalysts. In this respect, MIL-100(Fe) nanoparticles were hybridized with Bi_2WO_6 nanosheets to construct MIL-100(Fe)/Bi_2WO_6 Z-scheme heterojunction [34]. When tested for the degradation of TC under sunlight, the optimized 12%MIL/BWO delivered the highest photocatalytic activity due to the large specific surface area, enhanced light adsorption range, and high charge transfer and separation efficiency. In addition, the MIL-88A(Fe)/Bi_2WO_6 heterojunction was synthesized for the degradation of Rh B and TC [35]. The combination of MIL-88A(Fe) and Bi_2WO_6 effectively inhibited the recombination of photogenerated carriers. Under the irradiation of visible light, the heterojunction degraded 96% Rh B and 71% TC within 50 min and 80 min, respectively. Kaur et al. [33] prepared a Bi_2WO_6/NH_2-MIL-88B(Fe) heterostructure for the degradation of TC (Figure 3b). Due to the promoted separation and transfer of photoexcited charges caused by the interfacial contact, the heterostructure presented a high degradation efficiency of 89.4% within 130 min under solar illumination. Tu et al. [36] synthesized MIL-53(Fe)/Bi_2WO_6 heterostructure photocatalysts. The formation of heterojunction extended the visible light absorption capability and accelerated the transfer of photogenerated electrons. When used for the degradation of Rh B and phenol Rh B under visible light irradiation, the degradation rate constants of the heterostructure containing 5 wt% MIL-53(Fe) were 3.75-fold and 3.27-fold higher than that of pristine Bi_2WO_6.

Besides various Fe-based MOFs, Zhang et al. [37] prepared hydrangea-like Bi_2WO_6/ZIF-8 (BWOZ) hybrid photocatalysts by using a flower-like Bi_2WO_6 template. The generation of heterojunction induced the fast separation of photogenerated carriers. Meanwhile, the optimized BWOZ containing 7.0 wt% Bi_2WO_6 showed a large specific surface area, which degraded 85.7% MB within 240 min, and the reaction kinetic constant was 23-fold and 1.61-fold higher than that of pure Bi_2WO_6 and ZIF-8, respectively. In addition, Dai et al. [38] also fabricated Bi_2WO_6/ZIF-8 composite photocatalysts for the degradation of TC. Under the irradiation of UV light, the optimized sample achieved the fast degradation of 97.8% TC (20 mg/L) within 80 min, and the degradation rate constant was about 3-fold higher than that of pure Bi_2WO_6.

3.2. Bi_2WO_6/g-C_3N_4 Composite

Serving as a non-metal semiconductor, graphitic carbon nitride (g-C_3N_4) presents two-dimensional graphite-like structures with a high specific surface area, which are widely adopted as active templates for loading various semiconductor photocatalysts. In this section, we introduce the research progress of Bi_2WO_6/g-C_3N_4 composite photocatalysts. Qi et al. [39] prepared Bi_2WO_6/g-C_3N_4 heterojunction by hydrothermal reaction and discussed the effect of g-C_3N_4/Bi_2WO_6 ratio on photodegradation performance. When the Bi_2WO_6 ratio was 10 wt%, the composite presented the highest photocatalytic activity for the degradation of MB, and the reaction rate constant was 4 and 1.94 times higher than that of pristine g-C_3N_4 and Bi_2WO_6, respectively. In addition, Zhao et al. [40] also prepared Bi_2WO_6/g-C_3N_4 (BW/CNNs, Figure 4a) composite photocatalysts for the degradation of Ceftriaxone sodium in an aquatic environment. The combination of the two components enhanced the absorption capability of visible light and accelerated the separation of photogenerated electron–hole pairs. As a result, the 40%-BW/CNNs delivered the best photocatalytic activity, which degraded 94.50% Ceftriaxone sodium within 120 min under the irradiation of visible light. Chen et al. [41] synthesized Bi_2WO_6 on g-C_3N_4 to fabricate Bi_2WO_6/g-C_3N_4 heterojunctions by a hydrothermal route. When used for the degradation of Rh B and phenol solution, the composite with a Bi/g-C_3N_4 molar ratio of 4% presented the highest photodegradation activity for the highest separation efficiency of photogenerated electron–hole pairs. Moreover, the separation and transfer of electron–hole pairs were proven as a direct Z-scheme mechanism.

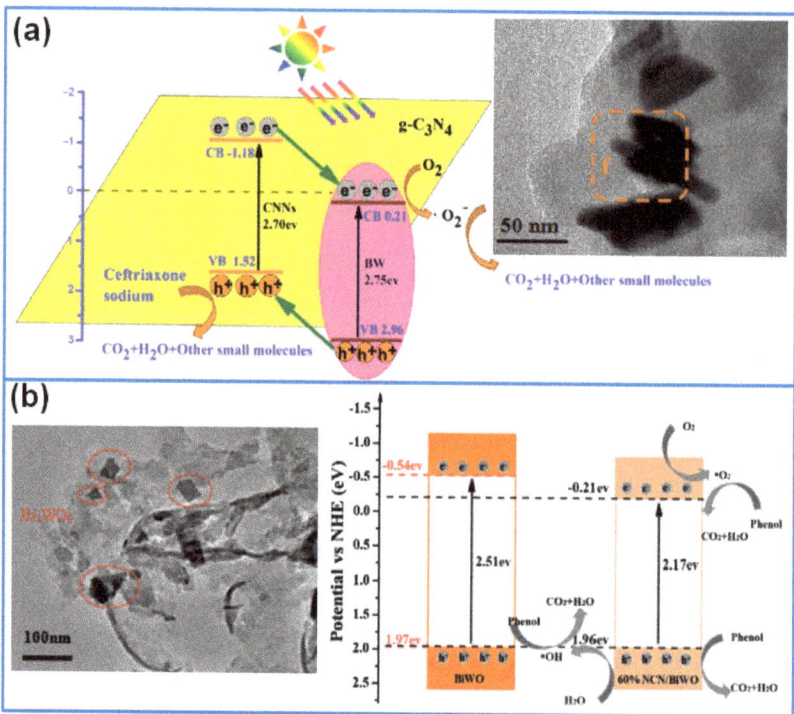

Figure 4. Photocatalytic mechanism of (**a**) $Bi_2WO_6/g\text{-}C_3N_4$ [40]. Copyright (2018) Elsevier. (**b**) N-doped $g\text{-}C_3N_4$ (NCN)/Bi_2WO_6 [42]. Copyright (2020) Elsevier.

In respect of the microstructure design, Zhu et al. [42] prepared various N-doped $g\text{-}C_3N_4$ (NCN)/Bi_2WO_6 (BWO) composites by using NCN as the template, as shown in Figure 4b. The ratio of NCN/BWO was adjusted to reduce the band gap and increase the surface area. The 60% NCN/BWO presented the best photocatalytic activity, which achieved the fast degradation of 93.1% phenol within 5 h, and the degradation rate constant was 18.5 times higher than that of Bi_2WO_6. The major reason was ascribed to the enhanced visible light absorption capability caused by NCN. Wang et al. [43] prepared core-shell structured $g\text{-}C_3N_4@Bi_2WO_6$ composite by in situ, forming an ultrathin $g\text{-}C_3N_4$ shell layer on Bi_2WO_6 nanosheets. By adjusting the thickness of the $g\text{-}C_3N_4$ shell layer, the interface of $g\text{-}C_3N_4@Bi_2WO_6$ was optimized to promote the separation efficiency of photogenerated electron–hole pairs. As a result, the composite photocatalyst with a 1 nm-thick shell layer delivered the optimum degradation phenol activity under visible light, which was about 1.9 times higher than that of Bi_2WO_6 and 5.7 times higher than that of $g\text{-}C_3N_4$. Zhang et al. [44] prepared $g\text{-}C_3N_4$ quantum dot (CNQD)-decorated ultrathin Bi_2WO_6 nanosheets for the degradation of Rh B and tetracycline (TC) under the irradiation of visible light. By optimizing the ratio, the composite of 5% CNQDs/BWO delivered the highest photocatalytic performance, which degraded 87% TC and 92.51% Rh B within 60 min. The superior activity can be ascribed to the Z-scheme charge transfer mechanism, the up-conversion behavior of CNQDs, and the enhanced separation and transfer rates of photo-generated charges. Regarding the performance of different S-scheme heterojunctions, Gordanshekan et al. [45] compared the photocatalytic activity of $Bi_2WO_6/g\text{-}C_3N_4$ with Bi_2WO_6/TiO_2. When serving as photocatalysts for removing cefixime (CFX) in polluted water, $Bi_2WO_6/g-C_3N_4$ showed a removal efficiency of 94%, which was slightly higher than that of Bi_2WO_6/TiO_2 (91%).

3.3. Bi_2WO_6/Metal Oxides Composite

Metal oxides have a wide range, and most of them can serve as semiconductor photocatalysts. To further enhance the photoactivity of Bi_2WO_6, various metal oxide semiconductors, including TiO_2, ZnO, and other metal oxides, are reported to hybridize with Bi_2WO_6, and the research progress of Bi_2WO_6/metal oxides composites is introduced in this section.

3.3.1. Bi_2WO_6/TiO_2 Composite

Compared to other metal oxides, TiO_2 exhibits some advantages of low cost, low toxicity, strong redox ability, and high catalytic activity, and it is regarded as the most popular photocatalyst in the past few decades. However, the wide band gap of 3.0~3.2 eV limits the visible light adsorption, only allowing a UV light response. Moreover, the fast recombination of photogenerated electron–hole pairs further weakens the photocatalytic activity of TiO_2. The combination of TiO_2 with Bi_2WO_6 overcomes the shortage of TiO_2, and various TiO_2/Bi_2WO_6 composite photocatalysts were developed in recent years. For example, Li et al. [46] fabricated TiO_2/Bi_2WO_6 microflowers via a one-step hydrothermal reaction, in which TiO_2 nanoparticles (10 nm) were dispersed on Bi_2WO_6 microflowers. The TiO_2/Bi_2WO_6 composite degraded 100% Rh B within 60 min under visible light or 30 min under UV-vis light. The enhanced photocatalytic activity was ascribed to the synergistic effect of two components. Furthermore, the improved light adsorption capacity and carrier separation efficiency also facilitated the photoactivity. In addition, an electrospinning technique was adopted to fabricate Bi_2WO_6/TiO_2 nanofibers (BTNF) by decorating Bi_2WO_6 nanosheets on the TiO_2 fiber surface [47]. In this composite, the Bi_2WO_6 extended the light absorption range, and the Bi_2WO_6/TiO_2 heterojunction promoted the charge separation. Therefore, BTNF presented a superior visible light activity for degrading Rh B, which was much better than pure Bi_2WO_6, TiO_2, and their mixture. Lu et al. [48] deposited Bi_2WO_6 nanosheets onto TiO_2 nanotube arrays (TNTAs) to fabricate BWO/TNTAs composite photocatalysts. The 0.2BWO/TNTAs sample achieved the fast degradation of 92.2% TC within 180 min due to the promoted charge separation and extended light absorption range.

Besides the direct combination of Bi_2WO_6 and TiO_2, Wang et al. [49] prepared Sb^{3+} doped Bi_2WO_6/TiO_2 nanotube photocatalysts. A rose-like BWO-10 sample showed superior photocatalytic activity, which achieved the fast degradation of 80.58% Rh B, 77.23% MO, and 99.06% MB under the irradiation of visible light, and the best performance resulted from the uniform rose-like structure and adjusted energy level. Sun et al. [50] prepared N/Ti^{3+} co-doped TiO_2/Bi_2WO_6 heterojunctions (NT-TBWx) and proved that the degradation rate order of organic pollutants was photocatalysis < sonocatalysis < sonophotocatalysis. The superior performance was attributed to the doping level, heterophase junction, and heterojunction.

3.3.2. Bi_2WO_6/ZnO Composite

To achieve the combination of narrow-band gap Bi_2WO_6 and wide-band gap ZnO, various ZnO/Bi_2WO_6 (ZBW) heterostructures were developed to improve the photocatalytic activity of Bi_2WO_6. In this respect, Liu et al. [51] synthesized the ZBW heterostructure via the hydrothermal method and investigated the photocatalytic activity for degrading MB under ultraviolet light. ZBW degraded 95.48% MB within 120 min, and the excellent photocatalytic performance was due to the promoted separation of electrons and holes caused by the heterojunction. Koutavarapu et al. [52] synthesized a hetero-structured Bi_2WO_6/ZnO composite via a hydrothermal route for the degradation of Rh B under solar irradiation. The formation of a Bi_2WO_6/ZnO interface reduced the charge transfer resistance and inhibited the recombination of charge carriers. By adjusting the additional amount of ZnO, the optimized BWZ-20 composite showed the best photocatalytic activity, which degraded 99% Rh B within 50 min. Chen et al. [53] synthesized a ZnO/Bi_2WO_6 heterostructure on flexible carbon cloth (CC) substrate. The optimized Z3B-CC sample containing 3 wt% Bi_2WO_6 degraded 96.9% MB within 100 min, which also facilitated the reuse of the photocatalyst. The enhanced activity was related to the enhanced light absorption range and the formation of a type-II energy band structure. To further enhance the activity, Zhao et al. [54] prepared a Z-scheme C and N-co-doped ZnO/Bi_2WO_6 (CZB) hybrid photocatalyst, and the influence of the C and N-co-doped ZnO content on the photodegradation performance of CZB composites was investigated. In this complicated structure, C and N co-doping adjusted the energy level and enhanced light absorption. Furthermore, residual C accelerated the separation and transfer of photogenerated carries. Through a comparison, CZB containing 5 wt% C/N-ZnO presented the best activity for the removal of tetracycline, enrofloxacin, and norfloxacin under visible light, and the photodegradation mechanism was confirmed as the formation of a Z-scheme heterojunction.

3.3.3. Bi_2WO_6/Other Metal Oxides Composite

Besides commonly used TiO_2 and ZnO, other metal oxides, including SnO_2, MnO_2, Co_3O_4, Fe_3O_4, CuO, WO_3, Bi_2O_4, and In_2O_3, were hybridized with Bi_2WO_6 to enhance photocatalytic activity. For example, Salari et al. [55] prepared Z-scheme flower-like Bi_2WO_6/MnO_2 composite photocatalysts, in which MnO_2 nanoparticles that were dispersed on 3D Bi_2WO_6 flowers enhanced the transfer and separation of charge carriers. As a result, the optimized Bi_2WO_6/MnO_2 (1:10) degraded 100% MB within 100 min. Mallikarjuna et al. [56] deposited small SnO_2 nanoparticles onto Bi_2WO_6 nanoplates to fabricate SnO_2/2D-Bi_2WO_6 photocatalysts (Figure 5a). The loading of SnO_2 nanoparticles adjusted the visible light absorption region and promoted charge separation and transfer efficiency. When used for degrading the Rh B pollutant, the photocatalytic activity of the composite was more than 2.7 times higher than that of 2D-Bi_2WO_6 nanoplates.

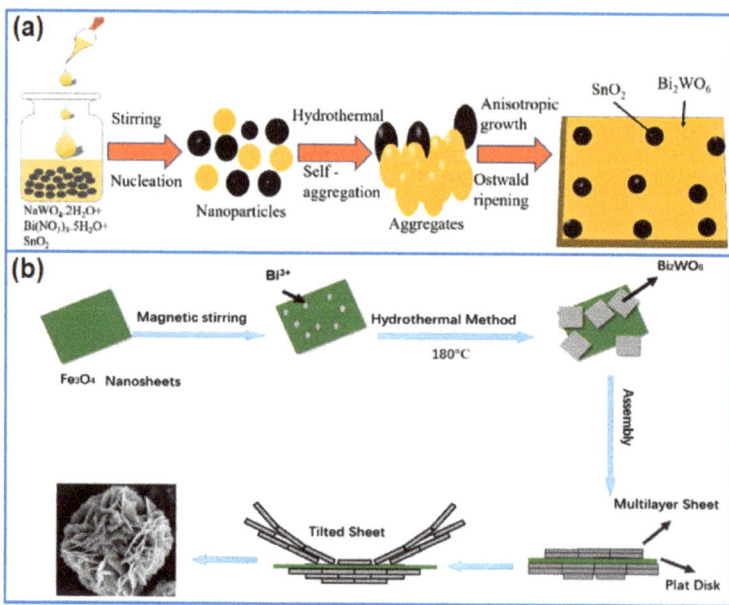

Figure 5. Fabrication process of (**a**) SnO_2/2D-Bi_2WO_6 [56]. Copyright (2021) Elsevier. (**b**) Fe_3O_4/Bi_2WO_6 heterojunction [57]. Copyright (2021) American Chemical Society.

Zhang et al. [58] prepared flower-like Co_3O_4 QDs/Bi_2WO_6 composite photocatalysts to achieve the uniform dispersion of Co_3O_4 QDs on flower-like nanosheets. By adjusting the ratio of Co_3O_4 QDs, the sample of 10%-Co_3O_4 QDs/Bi_2WO_6 presented the optimum performance for the removal of TC, and the degradation rate constant was about 1.55 and 3.40 times higher than that of Bi_2WO_6 and Co_3O_4 QDs, respectively. The superior performance was ascribed to the formation of a p–n heterojunction and enhanced the visible light absorption capacity. Zhu et al. [57] prepared Z-scheme Fe_3O_4/Bi_2WO_6 heterojunctions as photocatalysts for degrading ciprofloxacin (CIP), in which the flower-like composite was assembled by abundant nanosheets (Figure 5b). The formation of Z-scheme heterojunctions facilitated the light-harvesting capacity and suppressed the recombination of photogenerated carriers. Under the irradiation of visible light, the FB-180 sample prepared at 180 °C with 4% Fe delivered optimum photoactivity, which degraded about 99.7% CIP within 15 min. Moreover, the sample showed superior reusability and stability. Koutavarapu et al. [59] fabricated CuO/Bi_2WO_6 (CuBW) composite photocatalysts for degrading TC and MB. In this composite, Bi_2WO_6 provided transfer pathways for photogenerated electrons, while CuO was used to receive carriers from Bi_2WO_6 and inhibited the recombination of charge carriers; thus, the formation of the heterostructure improved the photocatalytic activity. The optimized CuBW sample containing 10 mg Bi_2WO_6 showed the highest degradation efficiency, which degraded 97.72% TC within 75 min and 99.43% MB within 45 min.

In respect of interface engineering design, Chen et al. [60] achieved the in-situ growth of (001)- and (110)-exposed WO_3 on (010)-exposed Bi_2WO_6 to form Z-scheme heterojunction photocatalysts. The facet control produced some dislocation defects for promoting the carriers transfer. Furthermore, Z-scheme transfer mode optimized the transfer of photogenerated electrons and improved the oxidization ability of photogenerated holes. Consequently, the WO_3 (001) and (110)/Bi_2WO_6 achieved a high removal rate of 74.5% for salicylic acid within 6 h, and the kinetic constant was 2.4 times higher than that of WO_3 (001)/Bi_2WO_6.

In addition, Bi_2O_4 micro-rods were in situ grown on Bi_2WO_6 microspheres to form a Bi_2O_4/Bi_2WO_6 heterojunction [61]. The heterojunction facilitated the separation and transfer of charge carriers. When used for the degradation of Rh B under visible light, the degradation rate constant of the composite was 5 times higher than that of pure Bi_2WO_6. While degrading MO, the enhanced factor reached 90-fold. Besides Bi_2O_4, a flower-like Bi_2WO_6/Bi_2O_3 photocatalyst was also synthesized by the ionic liquid solvothermal method and calcination [62], and it presented a higher photocatalytic H_2 production activity than pure Bi_2WO_6. Qin et al. [63] prepared a rich oxygen vacancy (OVs) Bi_2WO_6/In_2O_3 hybrid photocatalyst for the degradation of Rh B. The formation of a Bi_2WO_6/In_2O_3 heterostructure extended the lifetime of photogenerated charge carriers. Furthermore, the OVs in Bi_2WO_6/In_2O_3 accelerated the separation of photogenerated electron–hole pairs. Through comparison, the BiIn80 sample containing 80 wt% Bi_2WO_6 exhibited the best photocatalytic activity, and the reaction rate constant was about 4.17-fold and 15-fold higher than that of Bi_2WO_6 and In_2O_3, respectively.

3.4. Bi_2WO_6/Metal Sulfides Composite

To effectively optimize the band edge of Bi_2WO_6, except for metal oxides, various metal sulfides were used to construct a Bi_2WO_6-containing Z-scheme heterojunction for enhancing visible light harvesting capability, such as, Cu_2S, MoS_2 or $MoSe_2$, SnS_2, WS_2, CdS, Bi_2S_3, In_2S_3, $CuInS_2$, and $FeIn_2S_4$.

In this field, Tang et al. [64] prepared hierarchical flower-like Cu_2S/Bi_2WO_6 photocatalysts via a three-step method, in which Cu_2S particles were distributed on the surface of Bi_2WO_6 nanosheets (Figure 6a). Attributed to the hierarchical structure, the enhanced visible light absorption capacity, and the Z-scheme transfer mechanism, the sample of 1%Cu_2S/Bi_2WO_6 presented the highest photocatalytic activity for the removal of glyphosate. Based on the exfoliated MoS_2 nanosheets as substrates, Zhang et al. [65] synthesized Z-scheme hetero-structured MoS_2/Bi_2WO_6 hierarchical flower-like microspheres, in which the MoS_2 substrate greatly affected the morphology and photocatalytic activity of the heterostructure (Figure 6b). The optimized composite degraded 100% Rh B within 90 min and killed almost all of Pseudomonas aeruginosa within 60 min. Similar to MoS_2, layered $MoSe_2/Bi_2WO_6$ composite photocatalysts were reported for the photocatalytic oxidation of gaseous toluene [66]. The formation of a p–n heterojunction provided a strong interlayer interaction, which effectively inhibited the recombination of photoinduced electron–hole pairs. The optimized 1.5%-$MoSe_2/Bi_2WO_6$ presented the highest activity, which degraded 80% gaseous toluene within 3 h under the irradiation of visible light, and the rate constant was about 7 times and 6 times higher than that of pure $MoSe_2$ and Bi_2WO_6, respectively.

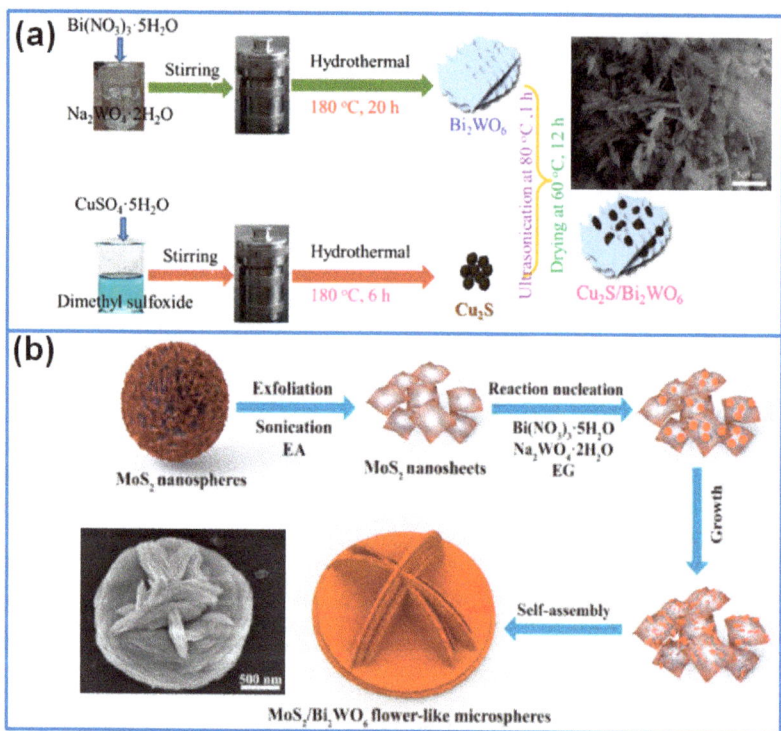

Figure 6. Fabrication process of (**a**) flower-like Cu_2S/Bi_2WO_6 [64]. Copyright (2020) Elsevier. (**b**) MoS_2/Bi_2WO_6 microspheres [65]. Copyright (2021) Elsevier.

Kumar et al. [67] synthesized a SnS_2/Bi_2WO_6 plate-on-plate composite via a two-step hydrothermal route for the degradation of tetracycline (TC) and ciprofloxacin (CIP). Due to the formation of a Z-scheme heterojunction, the optimized $0.10SnS_2/Bi_2WO_6$ exhibited superior photocatalytic activity, which degraded 97% TC and 93% CIP within 90 min under sunlight exposure, and the degradation rate constant was three-fold higher than that of pure Bi_2WO_6. Similar to the plate-on-plate structure, Su et al. [68] synthesized an sTable 2D/2D WS_2/Bi_2WO_6 heterostructure photocatalyst via a hydrothermal reaction. The generated Z-scheme heterostructure promoted the separation and transfer of photogenerated carriers, which showed much higher photocatalytic activity for the degradation of Rh B and OTC than pure WS_2 or Bi_2WO_6.

Su et al. [69] synthesized CdS quantum dots (QDs) on a Bi_2WO_6 monolayer via an in situ hydrothermal method to construct a S-scheme heterojunction. The Bi–S coordination at the junction interface enhanced the charge separation and interfacial charge migration. The optimized composite containing 7% CdS exhibited the best photocatalytic activity, which completely decomposed 100 ppm C_2H_4 within 15 min, and the degradation rate constant was 88 times and 194 times higher than that of pure CdS and Bi_2WO_6. In addition,

CdS nanocrystals were decorated on the surface of Bi_2WO_6 clusters to form $CdS@Bi_2WO_6$ photocatalysts [70]. By adjusting the microstructure of CdS, the CdS nanorod-decorated Bi_2WO_6 showed a higher charge separation capability than that of CdS cluster-decorated Bi_2WO_6, which degraded 96.1% Rh B within 120 min. Xu et al. [71] fabricated Bi_2S_3/2D-Bi_2WO_6 composite photocatalysts by using the ion exchange method, in which the Bi_2S_3 nanoparticle loading on Bi_2WO_6 nanosheets enhanced the light absorption ability and promoted the transfer and separation of photogenerated carriers. The optimized BWS-2 sample achieved the complete degradation of Rh B much better than that of Bi_2WO_6 nanosheets. He et al. [72] fabricated a core-shell structured In_2S_3/Bi_2WO_6 composite by using In_2S_3 microspheres as templates. The combination of the core and Bi_2WO_6 shell extended the visible-light absorption range and enabled the Z-scheme transfer pathway. Therefore, the core-shell composite achieved the fast degradation of TCH, and the activity was 2.1 and 2.4 times higher than that of pure Bi_2WO_6 and In_2S_3, respectively.

As important semiconductors, $CuInS_2$ and $FeIn_2S_4$ exhibit a strong light absorption capability, with suitable energy band edges, which serve as high-performance visible light photocatalysts in the photocatalytic field. Lu et al. [73] synthesized Bi_2WO_6 on the surface of network-like $CuInS_2$ microspheres to fabricate a Z-scheme $CuInS_2$/Bi_2WO_6 heterojunction. The heterojunction interface enhanced the charge transfer capability, further promoting the separation of charge carriers. When used for the degradation of tetracycline hydrochloride (TCH), the activity of 15% $CuInS_2$/Bi_2WO_6 was three-fold higher than that of pure $CuInS_2$ and 17% higher than that of pure Bi_2WO_6. Shangguan et al. [74] prepared Z-scheme $FeIn_2S_4$/Bi_2WO_6 composite photocatalysts for the degradation of TCH. The formation of a direct Z-scheme heterojunction promoted the separation of photogenerated holes and electrons, which presented enhanced activity for the removal of TCH, much better than pure Bi_2WO_6 and $FeIn_2S_4$.

3.5. Bi_2WO_6/Bi-Series Composite

Except for Bi_2WO_6, other Bi-containing semiconductors, including BiOCl, BiOBr, BiOI, $Bi_2O_2CO_3$, $BiPO_4$, $Bi_2Sn_2O_7$, $BiFeO_3$, and $CuBi_2O_4$, also exhibit excellent visible light adsorption capability and photocatalytic activity. The combination of Bi_2WO_6 and other Bi-containing semiconductors can be regarded as an effective strategy for enhancing the photoactivity of Bi_2WO_6.

In this field, Guo et al. [75] synthesized Bi_2WO_6 nanoparticles on layered BiOCl nanosheets to fabricate 0D/2D Bi_2WO_6/BiOCl composite photocatalysts. The formed heterojunction interface promoted the separation of photogenerated charge carriers. As a result, the optimized 1%Bi_2WO_6/BiOCl sample showed a superior degradation performance for removing OTC and phenol, and the degradation rate of OTC and phenol was 2.7-fold and 6.1-fold higher than that of pure BiOCl. Liang et al. [76] prepared a Bi_2WO_6/BiOCl heterojunction via the one-step hydrothermal method for the degradation of Rh B and TC. The formed heterojunction at the Bi_2WO_6/BiOCl interface promoted the separation of photogenerated electron–hole pairs, further improving the photocatalytic activity. Liu et al. [77] synthesized a 2D-3D BiOBr/Bi_2WO_6 composite with 2D Bi_2WO_6 nanosheets inserted in BiOBr microspheres. 3D BiOBr microspheres reduced the aggregation of Bi_2WO_6 nanosheets and enhanced the visible light absorption capability by providing interfacial contact. Through optimization, the BiOBr/Bi_2WO_6 (8:1) delivered the highest degradation efficiency for removing Rh B, TC, CIP, and MB (100%, 96%, 90% and 94%). Ren et al. [78] synthesized Bi_2WO_6/BiOBr composites via a one-step solvothermal route by using [C16mim] Br ionic liquid as the Br source. In this composite, Bi_2WO_6 nanoparticles wrapped on flower-like BiOBr and formed a type II heterojunction, which promoted the transfer and separation of charge carriers and enhanced visible light harvesting. The optimized composite with a W/Br ratio of 1:2 delivered the highest photocatalytic activity for the gradation of MB, Rh B, and TC. Chen et al. [79] synthesized a BiOBr/Bi/Bi_2WO_6 composite via the hydrothermal method to construct a Z-scheme heterojunction for enhancing photocatalytic activity. Due to the synergistic effect of the Z-scheme BiOBr/Bi_2WO_6 heterojunction and the surface plasmon resonance (SPR) effect of Bi, the optimized 20%BiOBr/7%Bi/Bi$_2$WO$_6$ achieved the fast degradation of Rh B under visible light, and the degradation rate of Rh B reached 98.02% within 60 min. He et al. [80] prepared a hydrangea-like BiOBr/Bi$_2$WO$_6$ composite via an ionic liquid-assisted hydrothermal route. The core-shell structured 3D/2D BiOBr/Bi$_2$WO$_6$ (Figure 7a) displayed an enhanced degradation performance for removing organic dye and drugs due to the formation of the Z-scheme heterojunction.

Besides BiOCl and BiOBr, flower-like BiOI/Bi$_2$WO$_6$ microspheres were prepared via the hydrothermal route for the degradation of phenol [81]. The formed heterojunction between Bi$_2$WO$_6$ and BiOI enhanced the separation efficiency of the electron and hole, further improving the photocatalytic activity. To further improve the photoactivity of the Bi$_2$WO$_6$/BiOI composite, Zheng et al. [82] deposited Ag nanoparticles onto the surface of Bi$_2$WO$_6$/BiOI to fabricate a Bi$_2$WO$_6$/BiOI/Ag heterojunction (Figure 7b). Besides the function of the heterojunction structure, Ag particles contributed to the SPR effect, which extended the visible-light absorption and accelerated the separation/transfer of photogenerated carriers. The sample of Bi$_2$WO$_6$/BiOI/Ag-8 displayed the highest activity for the degradation of tetracycline and a superior recycling performance.

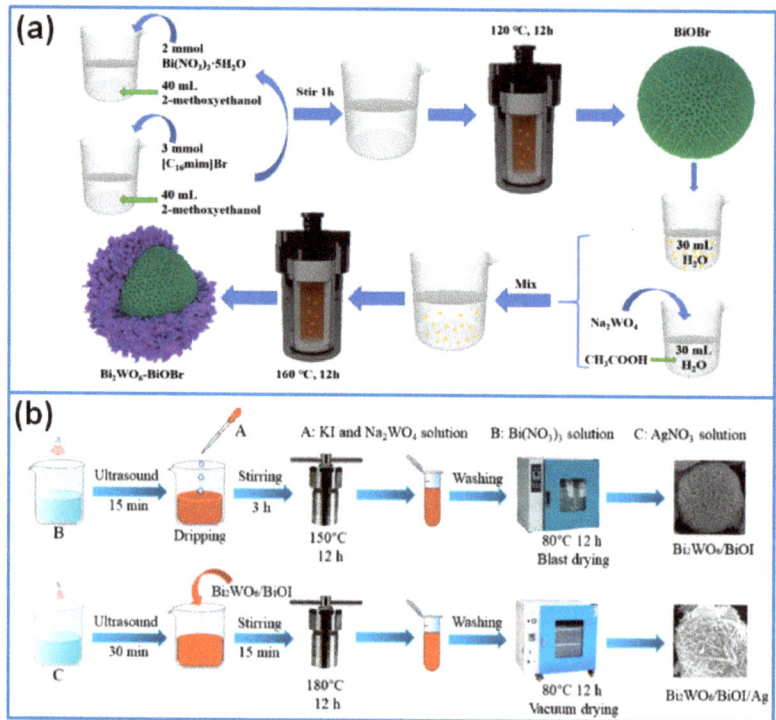

Figure 7. Fabrication process of (**a**) 3D/2D BiOBr/Bi$_2$WO$_6$ [80]. Copyright (2021) Elsevier. (**b**) Bi$_2$WO$_6$/BiOI/Ag heterojunction [82]. Copyright (2022) Elsevier.

Qiang et al. [83] synthesized I-doped Bi$_2$O$_2$CO$_3$/Bi$_2$WO$_6$ heterojunction microspheres via the ionic liquid-assisted solvothermal method. The I-doped heterojunction adjusted the energy band structure and enhanced visible light adsorption, charge separation, and proton reduction. Therefore, the doped composite presented outstanding photocatalytic performance for the degradation of TC and Rh B. Wu et al. [84] prepared 3D flower-like BiPO$_4$/Bi$_2$WO$_6$ composites via the hydrothermal method for the degradation of Rh B under visible light irradiation. The hybridization of two components accelerated the separation efficiency of charge carriers and inhibited their recombination. The composite containing

15% BiPO$_4$ displayed the highest photocatalytic activity, which degraded 92% Rh B within 100 min, about 3.7 and 1.4 times higher than that of BiPO$_4$ and Bi$_2$WO$_6$, respectively. Zhang et al. [85] deposited Bi$_2$Sn$_2$O$_7$ (BSO) nanoparticles onto Bi$_2$WO$_6$ (BWO) nanosheets to fabricate flower-like Bi$_2$Sn$_2$O$_7$/Bi$_2$WO$_6$ hierarchical composite photocatalysts. When used for the degradation of Rh B under visible light, the 7% BSO/BWO composite displayed the best photocatalytic activity, much better than that of pure BWO or BSO, due to the promoted separation of the photogenerated electron-hole pairs.

Tao et al. [86] synthesized Bi$_2$WO$_6$ nanosheets (NSs) on electrospun BiFeO$_3$ nanofibers (NFs) to fabricate 1D discrete heterojunction nanofibers. The ferromagnetic feature of BiFeO$_3$ facilitated the recycling treatment, and the high surface area facilitated the photocatalytic reaction by providing abundant active sites. Moreover, the 1D heterojunction promoted the separation/transport of photogenerated charges. As a result, the reaction rate constant of the nanofiber-like photocatalyst for Rh B degradation was 36.7 times and 8.7 times higher than that of pure BiFeO$_3$ and Bi$_2$WO$_6$, respectively. Integrating the solvothermal reaction with the electrospinning technique, Teng et al. [87] prepared a one-dimensional CuBi$_2$O$_4$/Bi$_2$WO$_6$ fiber composite. Due to the formation of a Z-type heterojunction, a fiber-like photocatalyst achieved the fast degradation of more than 90% TCH within 120 min. In addition, Wang et al. [88] created flower-flake-like CuBi$_2$O$_4$/BWO composite via a hydrothermal route. Under the irradiation of visible light, the composite containing 60 wt% CuBi$_2$O$_4$ delivered the highest photocatalytic activity, which degraded 93% TC (20 mg/L) within 1 h. The superior performance was ascribed to the improved visible light absorption, interfacial charge transfer and separation, and the prolonged lifetime of photogenerated carriers.

3.6. Bi$_2$WO$_6$/Metal Tungstates Composite

Metal tungstates (MWO$_4$) show the wolframite-type monoclinic structure and scheelite-type tetragonal structure, which received more attention in the photocatalytic field. The combination of Bi$_2$WO$_6$ with metal tungstates integrates the advantage of each component and presents enhanced photocatalytic activity for the degradation of organic pollutants. In this field, Kumar et al. [89] synthesized a Bi$_2$WO$_6$/ZnWO$_4$ composite photocatalyst via the modified hydrothermal method. The bi-crystalline framework of Bi$_2$WO$_6$ and ZnWO$_4$ played a synergistic effect, which reduced the crystallite size and band gap and effectively separated and transferred the photo-generated electron–hole pairs. Under UV irradiation, the optimized 30% Bi$_2$WO$_6$/ZnWO$_4$ delivered the maximum degradation performance for the removal of Plasmocorinth B dye. Miao et al. [90] prepared Sb$_2$WO$_6$/Bi$_2$WO$_6$ composite photocatalysts and evaluated their photocatalytic activity for the degradation of Rh B and MO. The composite showed an increased specific surface area and an enhanced visible-light absorption capability, and it suppressed the recombination of electron–hole pairs. The composite containing 6% Sb achieved the fast degradation of 100% Rh B and 70% MO within 90 min, much better than pure single phase Bi$_2$WO$_6$ and Sb$_2$WO$_6$. Ni et al. [91] synthesized a flower-like Ag$_2$WO$_4$/Bi$_2$WO$_6$ (AWO/BWO) composite for the degradation of Rh B. AWO and BWO formed a direct Z-scheme heterojunction, which promoted the migration of interface charges, enhanced the light absorption capability, and inhibited the recombination of the electron-hole pairs. The composite containing 3 wt% AWO exhibited the highest activity, which degraded nearly 100% Rh B within 150 min, 11.5-fold and 1.5-fold higher than that of pristine AWO and BWO.

4. Bi$_2$WO$_6$-Based Ternary Composite

Besides Bi$_2$WO$_6$-based binary heterojunction composites, lots of Bi$_2$WO$_6$-containing ternary composites were developed as high-performance photocatalysts. From the reported Bi$_2$WO$_6$-based ternary composites, commonly used components involve carbon materials, g-C$_3$N$_4$, BiOX, AgBr, Ag$_2$CO$_3$, Ag$_2$O, Cu$_2$O, TiO$_2$, ZnO, Ti$_3$C$_2$, Bi$_2$MoO$_6$, BiPO$_4$, and Au/Ag nanoparticles. Compared to binary composites, the advantage of ternary composites exhibits optimized light harvesting capability and photocatalytic activity due to the constructed double heterostructure interfaces and the synergistic effect derived from three components. In view of the broad selectivity of semiconductor candidates, it is hard to summarize the design principle of Bi_2WO_6-based ternary composites. However, the fixed binary combination of Bi_2WO_6/g-C_3N_4, Bi_2WO_6/carbon materials, Bi_2WO_6/Au or Ag-based materials, and Bi_2WO_6/Bi-series semiconductors is reported to hybridize with the third component.

4.1. The Composite of Bi_2WO_6/g-C_3N_4/Other Materials

In respect of the Bi_2WO_6/g-C_3N_4 combination, Zhang et al. [92] prepared a dual Z-scheme BiSI/Bi_2WO_6/g-C_3N_4 photocatalyst via hydrothermal method. The generation of dual Z-scheme heterojunction promoted the transfer and separation of photogenerated electron–hole pairs. When used for the degradation of TC, Rh B, and chlortetracycline (CTC), the optimized BiSI/Bi_2WO_6/20%g-C_3N_4 exhibited the highest photocatalytic activity, much better than that of the single and binary systems. Sun et al. [93] fabricated a double Z-scheme g-C_3N_4/Bi_2MoO_6/Bi_2WO_6 (CN/MO/WO) composite for the degradation of TC. In this ternary system, g-C_3N_4 enhanced the specific surface area and accelerated the carrier transfer. Bi_2WO_6 and Bi_2MoO_6 extended the light absorption range and inhibited the recombination of photogenerated electron–hole pairs. As a result, the optimized 15% CN/MO/WO composite achieved the fast photodegradation of 98% TC within 30 min under the irradiation of visible light. Zhou et al. [94] synthesized dual Z-scheme BiOBr/g-C_3N_4/Bi_2WO_6 photocatalysts via one-pot hydrothermal reaction. The dual heterojunction effectively suppressed the recombination of photogenerated carriers and presented superior photocatalytic activity, which degraded 90% TC within 40 min under the irradiation of visible light. In addition, a Bi_2WO_6/BiOI/g-C_3N_4 ternary composite photocatalyst was prepared for the degradation of TC [95], and the optimized composite degraded over 90% TC within 120 min. Moreover, the ternary photocatalyst also exhibited a superior performance for the degradation of municipal waste transfer station leachate.

Hu et al. [96] fabricated a ternary heterojunction g-C_3N_4/$BiVO_4$-Bi_2WO_6 photocatalyst by the intercalation of a $BiVO_4$-Bi_2WO_6 composite into compressed layered g-C_3N_4 nanosheets. The compressed layer structure accelerated the transfer of electrons and the generation of superoxide radicals, which enhanced photocatalytic activity, and the degradation efficiency of Rh B and TC was 96.7% and 94.8% within 60 min, respectively. In another work, a 2D/2D/2D Bi_2WO_6/g-C_3N_4/Ti_3C_2 composite (Figure 8a) was prepared via a one-step hydrothermal reaction [97]. In this system, seamless interfacial contact of the 2D heterojunction facilitated the separation and transfer of photogenerated electron–hole pairs. Moreover, Ti_3C_2 also promoted charge separation. As a result, the composite achieved the fast photodegradation of CIP, and the reaction rate constant was 4.78 times higher than that of Bi_2WO_6. Li et al. [98] prepared a 2D/2D Z-scheme g-C_3N_4/Au/Bi_2WO_6 (CN/Au/BWO) composite for the photodegradation of Rh B. Serving as a redox mediator, Au nanoparticles accelerated the transmission and separation of photogenerated carriers. Moreover, the 2D/2D Z-scheme structure provided abundant active sites for enhancing the photocatalytic activity. The CN/Au(1)/BWO sample degraded 88.7% Rh B within 30 min, and the rate constant was 1.48-fold and 1.62-fold higher than that of pure BWO and CN, respectively. To further enhance the photocatalytic activity of the g-C_3N_4/Bi_2WO_6 Z-scheme heterojunction, Jia et al. [99] introduced nitrogen-doped carbon quantum dots (NCQs) onto a g-C_3N_4/Bi_2WO_6 interface to form g-C_3N_4/Bi_2WO_6/NCQs ternary composites (Figure 8b). The NCQs extended the light absorption range and promoted the transfer and separation of photogenerated electron–hole pairs. Compared to single or binary composites, the ternary composite showed the highest degradation efficiency for the removal of Rh B and TC under visible light irradiation.

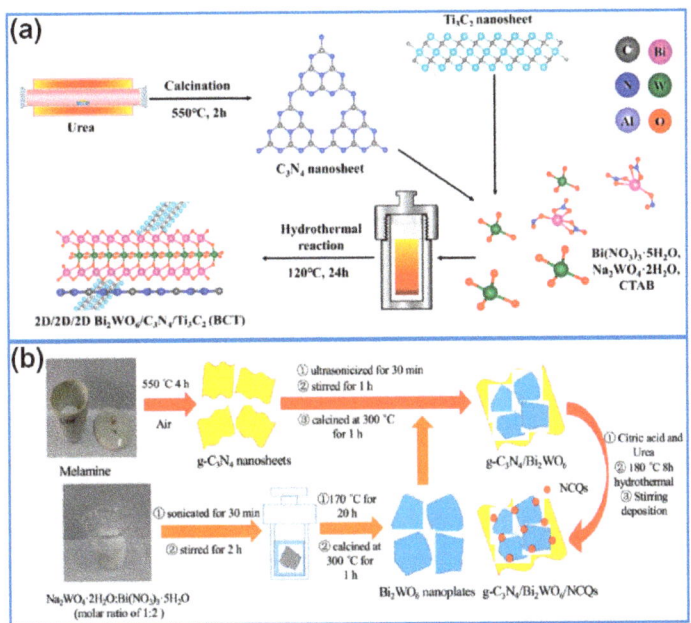

Figure 8. Preparation of (**a**) Bi$_2$WO$_6$/g-C$_3$N$_4$/Ti$_3$C$_2$ [97]. Copyright (2020) Elsevier. (**b**) g-C$_3$N$_4$/Bi$_2$WO$_6$/NCQs ternary composite [99]. Copyright (2020) Elsevier.

4.2. The Composite of Bi$_2$WO$_6$/Carbon/Other Materials

In the field of Bi$_2$WO$_6$/carbon materials, Guan et al. [100] synthesized a ternary AgBr/GO/Bi$_2$WO$_6$ Z-scheme photocatalyst and discussed the effect of AgBr and GO fractions on photocatalytic activity. The optimized 15%AgBr/5GO/Bi$_2$WO$_6$ delivered the highest degradation efficiency for the removal of 84% TC under visible light, and the reaction kinetic constant was about 3.16-fold and 4.60-fold higher than that of pure Bi$_2$WO$_6$ and AgBr, respectively. The superior performance was due to an extended visible light adsorption range and an enhanced charge separation and transfer. Zhu et al. [101] prepared a GO@BiOI/Bi$_2$WO$_6$ composite for the removal of Bisphenol A (BPA). In this ternary system, GO effectively modified the surface of BiOI/Bi$_2$WO$_6$ and improved the physico-chemical property. The optimized composite degraded 81% BPA within 5 h under the irradiation of UV-vis light. Tian et al. [102] prepared a Z-scheme flower-like Bi$_2$MoO$_6$/Bi$_2$WO$_6$/MWCNTs photocatalyst via hydrothermal route. Under visible light irradiation, the ternary composite degraded 96% reactive blue 19 (RB-19) within 4 h, the photocatalytic efficiency was much higher than that of Bi$_2$MoO$_6$/MWCNTs, and pure Bi$_2$MoO$_6$ and Bi$_2$WO$_6$. Niu et al. [103] synthesized Bi$_2$WO$_6$/C@Cu$_2$O Z-scheme photocatalysts for TC degradation. The wrapped carbon layer on Cu$_2$O avoided the photo-corrosion of Cu$_2$O. Furthermore, the oxygen-containing groups in the carbon layer decreased interfacial resistance and promoted electron transfer. The degradation rate constant of the ternary composite was 2.8 times higher than that of pure Bi$_2$WO$_6$.

4.3. The Composite of Bi$_2$WO$_6$/Au or Ag-Based Materials/Other Materials

Among the Bi$_2$WO$_6$/Ag-based materials, Wang et al. [104] prepared a Ag$_2$CO$_3$/AgBr/Bi$_2$WO$_6$ ternary photocatalyst via a precipitation method. When used for the degradation of Rh B, the degradation rate of the ternary composite was 95.1% within 60 min under solar illumination, and the degradation efficiency was much higher than that of each component. Gang et al. [105] synthesized a Ag/AgBr/Bi$_2$WO$_6$ composite via the oil/water self-assembly method. In this ternary composite, Ag/AgBr was uniformly dispersed on the Bi$_2$WO$_6$ surface, which extended the visible-light absorption range for the surface plasmonic resonance (SPR) effect of Ag. Moreover, the composite accelerated the separation of photogenerated charges. When utilized for the degradation of Rh B and phenol, the ternary composite presented optimum photocatalytic activity under visible light, much better than Ag/AgBr and pure Bi$_2$WO$_6$. Jin et al. [106] prepared a Au@TiO$_2$/Bi$_2$WO$_6$ composite via a sol-gel method followed by hydrothermal reaction. In this ternary system, core-shell structured Au@TiO$_2$ nanoparticles were dispersed on flower-like Bi$_2$WO$_6$ nanosheets. The formation of a Z-scheme heterojunction and SPR effect of Au promoted the generation, separation, and interfacial transfer of photogenerated charge carriers. When served for the degradation of sulfamethoxazole (SMX) and TCH under visible light, the degradation rate was 96.9% and 95.0% within 75 min, respectively. Moreover, the degradation rate constant was 7.2 times and 1.9 times higher than that of pure Bi$_2$WO$_6$, respectively.

4.4. The Composite of Bi$_2$WO$_6$/Bi-Series Semiconductors/Other Materials

In the field of Bi$_2$WO$_6$/Bi-series semiconductors, Zhu et al. [107] prepared a magnetic Bi$_2$WO$_6$/BiOI@Fe$_3$O$_4$ ternary composite for the photodegradation of TC. The optimized Bi$_2$WO$_6$/BiOI@5%Fe$_3$O$_4$ sample showed the highest TC degradation rate of 97%, much higher than that of pure Bi$_2$WO$_6$ (63%). Moreover, the spent powder can be magnetically recycled, and the recycled sample also exhibited good photocatalytic activity. Combining the electrostatic spinning technique, Chen et al. [108] fabricated 1D magnetic flower-like CoFe$_2$O$_4$@Bi$_2$WO$_6$@BiOBr photocatalysts for the degradation of Rh B. The resulting flower-like heterojunction enhanced the specific surface area and accelerated the separation of photogenerated charge carriers. Consequently, the ternary composite degraded 92.08% Rh B within 3 h.

4.5. Other Composites

In view of the high surface area and tight interfacial contact of 2D nanomaterials [109–111], Sharma et al. [112] prepared a 2D-2D-2D ZnO/Bi$_2$WO$_6$/Ti$_3$C$_2$ ternary composite photocatalyst via two-step electrostatic assembly. The optimized ZBT05 containing 5 wt% Ti$_3$C$_2$ delivered the highest degradation rate (~77%) for the removal of ciprofloxacin (CFX) within 160 min due to the enhanced photogenerated charge carrier separation caused by the generated ternary interface. Besides ternary composites, the composite photocatalysts composed of four or five components were reported for enhancing photocatalytic activity. In this respect, Ma et al. [113] prepared a GO-modified Ag/Ag$_2$O/BiPO$_4$/Bi$_2$WO$_6$ multi-component composite photocatalyst and investigated the photocatalytic activity for the degradation of Rh B and amoxicillin (AMX). The composite exhibited a small size, fast charge transfer efficiency, and extended light absorption range, which presented enhanced photocatalytic activity for the degradation of AMX, Rh B, and E. coli under visible light irradiation.

5. Conclusions and Prospects

To sum up, the advances of Bi$_2$WO$_6$-based photocatalysts are summarized in this review, including morphology control, the surface modification and heteroatom doping of Bi$_2$WO$_6$, Bi$_2$WO$_6$-based binary composites, and Bi$_2$WO$_6$-based ternary composites. The most popular synthesis method of Bi$_2$WO$_6$ is the hydrothermal or solvothermal method, and the reaction temperature and time heavily affect the microstructure and photocatalytic performance of Bi$_2$WO$_6$. The surfactants of CTAB and PVP were used to adjust the microstructure of Bi$_2$WO$_6$. Furthermore, Au-decorated Bi$_2$WO$_6$ hollow microspheres were synthesized to utilize the SPR effect of Au nanoparticles. Heteroatom doping can be used to enhance the photoactivity of Bi$_2$WO$_6$. Among various dopants, N, F, Cl, and I serve as non-metal dopants for doping Bi$_2$WO$_6$. In addition, Fe, Ti, Sr, Er, La, Au, Ag, and Mo are used to fabricate metal-doped Bi$_2$WO$_6$. Besides single atom doping, (La, Mo) co-doped Bi$_2$WO$_6$ was reported to enhance the photoactivity of Bi$_2$WO$_6$ by adjusting the particle size and lattice spacing. In view of the limited photocatalytic activity of single Bi$_2$WO$_6$, the development of Bi_2WO_6-based binary and ternary composites has become a major topic for constructing high-performance photocatalysts. Bi_2WO_6-based binary composites show a wide research range for the diversity of alternative materials. The existing Bi_2WO_6-based binary composites can be classified into six types: Bi_2WO_6/carbon or MOF composite, Bi_2WO_6/g-C_3N_4 composite, Bi_2WO_6/metal oxides composite, Bi_2WO_6/metal sulfides composite, Bi_2WO_6/Bi-series composite, and Bi_2WO_6/metal tungstates composite. Due to the diversity of target organic pollutants, and the difference of pollutant concentration, light source or powder, and catalyst dosage, it is very difficult to compare the photocatalytic activity of different Bi_2WO_6-based binary composites. Compared to other semiconductors, g-C_3N_4 and metal oxides are widely used to hybridize with Bi_2WO_6, and the resulting Bi_2WO_6/g-C_3N_4 and Bi_2WO_6/metal oxides composites deliver enhanced photodegradation efficiency, which is much better than each component. Besides Bi_2WO_6-based binary composites, lots of Bi_2WO_6-based ternary composites were developed as high-performance photocatalysts. The commonly used components include carbon materials, g-C_3N_4, BiOX, AgBr, Ag_2CO_3, Ag_2O, Cu_2O, TiO_2, ZnO, Ti_3C_2, Bi_2MoO_6, $BiPO_4$, and Au/Ag nanoparticles. According to the material type, binary Bi_2WO_6/g-C_3N_4, Bi_2WO_6/carbon materials, Bi_2WO_6/Au or Ag-based materials, and Bi_2WO_6/Bi-series semiconductors were fabricated for further hybridizing with the third component, and they present outstanding photocatalytic activity for the formation of double heterostructures and the synergistic effect of three components. In addition, a GO modified Ag/Ag_2O/$BiPO_4$/Bi_2WO_6 multi-component composite was synthesized to further improve photocatalytic activity.

Based on the summary above, abundant progress has been achieved in Bi_2WO_6-based photocatalysts. However, some urgent problems still exist, such as the controllable microstructure, the suitable component and ratio optimization, and the photocatalytic mechanism of different Bi_2WO_6-based composites. Aiming to solving the three problems mentioned above, we put forward the following promising research trends:

(1) The controllable synthesis and microstructure optimization of Bi_2WO_6 and Bi_2WO_6-based composite. The ideal microstructures of photocatalysts include hierarchical hollow structures, flowers, or spheres with a high specific surface area. Moreover, binary or ternary composites should have a strong interfacial binding strength, and the ratio optimization of different components is a major task.

(2) The selection of suitable candidate semiconductor photocatalysts. The selection of semiconductors should consider the band gap feature of Bi_2WO_6, and the resulting Bi_2WO_6-based composite should form a Z-scheme, S-scheme heterojunction, or double heterojunctions. In addition, the heteroatom doping and introduction of noble metal nanoparticles can be adopted as an effective strategy for enhancing photocatalytic activity.

(3) The combination of theoretical calculation and experimental results clarify the photocatalytic mechanism. The photocatalytic mechanism of the Bi_2WO_6-based composite is the difficulty for designing high-performance hybrid photocatalysts. Besides the traditional characterization techniques, theory computations should be paid more attention for clarifying the photocatalytic mechanism.

Author Contributions: Conceptualization, H.J. and X.H.; validation, J.H., C.D. and B.L.; writing—original draft preparation, H.J.; writing—review and editing, X.H. and B.L. All authors have read and agreed to the published version of the manuscript.

Funding: This work was funded by Guangdong Provincial Key Laboratory of Battery Recycling and Reuse (2021B1212050002) and the Innovation Team of Universities of Guangdong Province (2022KCXTD030).

Institutional Review Board Statement: Not applicable.

Informed Consent Statement: Not applicable.

Conflicts of Interest: The authors declare no conflict of interest.

References

1. Chen, T.; Liu, L.; Hu, C.; Huang, H. Recent advances on Bi_2WO_6-based photocatalysts for environmental and energy applications. *Chin. J. Catal.* **2021**, *42*, 1413–1438. [CrossRef]
2. Orimolade, B.O.; Idris, A.O.; Feleni, U.; Mamba, B. Recent advances in degradation of pharmaceuticals using Bi_2WO_6 mediated photocatalysis—A comprehensive review. *Environ. Pollut.* **2021**, *289*, 117891. [CrossRef] [PubMed]
3. Duan, Z.; Zhu, Y.; Hu, Z.; Zhang, J.; Liu, D.; Luo, X.; Gao, M.; Lei, L.; Wang, X.; Zhao, G. Micro-patterned $NiFe_2O_4$/Fe-TiO_2 composite films: Fabrication, hydrophilicity and application in visible-light-driven photocatalysis. *Ceram. Int.* **2020**, *46*, 27080–27091. [CrossRef]
4. Chen, J.; Chen, X.; Li, N.; Liang, Y.; Yu, C.; Yao, L.; Lai, Y.; Huang, Y.; Chen, H.; Chen, Y.; et al. Enhanced photocatalytic activity of $La_{1-x}Sr_xCoO_3$/Ag_3PO_4 induced by the synergistic effect of doping and heterojunction. *Ceram. Int.* **2021**, *47*, 19923–19933. [CrossRef]
5. Xiong, J.; Li, W.; Zhao, K.; Li, W.; Cheng, G. Engineered zinc oxide nanoaggregates for photocatalytic removal of ciprofloxacin with structure dependence. *J. Nanopart. Res.* **2020**, *22*, 155. [CrossRef]
6. Wang, H.; Sun, T.; Xu, N.; Zhou, Q.; Chang, L. 2D sodium titanate nanosheet encapsulated Ag_2O-TiO_2 p-n heterojunction photocatalyst: Improving photocatalytic activity by the enhanced adsorption capacity. *Ceram. Int.* **2021**, *47*, 4905–4913. [CrossRef]
7. Hong, X.; Li, Y.; Wang, X.; Long, J.; Liang, B. Carbon nanosheet/MnO_2/BiOCl ternary composite for degradation of organic pollutants. *J. Alloy. Compd.* **2022**, *891*, 162090. [CrossRef]
8. Khedr, T.M.; Wang, K.; Kowalski, D.; El-Sheikh, S.M.; Abdeldayem, H.M.; Ohtani, B.; Kowalska, E. Bi_2WO_6-based Z-scheme photocatalysts: Principles, mechanisms and photocatalytic applications. *J. Environ. Chem. Eng.* **2022**, *10*, 107838. [CrossRef]
9. Lai, M.T.L.; Lai, C.W.; Lee, K.M.; Chook, S.W.; Yang, T.C.K.; Chong, S.H.; Juan, J.C. Facile one-pot solvothermal method to synthesize solar active Bi_2WO_6 for photocatalytic degradation of organic dye. *J. Alloy. Compd.* **2019**, *801*, 502–510. [CrossRef]
10. Helen Selvi, M.; Reddy Vanga, P.; Ashok, A. Photocatalytic application of Bi_2WO_6 nanoplates structure for effective degradation of methylene blue. *Optik* **2018**, *173*, 227–234. [CrossRef]
11. Guo, X.; Wu, D.; Long, X.; Zhang, Z.; Wang, F.; Ai, G.; Liu, X. Nanosheets-assembled Bi_2WO_6 microspheres with efficient visible-light-driven photocatalytic activities. *Mater. Charact.* **2020**, *163*, 110297. [CrossRef]
12. Bai, J.; Zhang, B.; Xiong, T.; Jiang, D.; Ren, X.; Lu, P.; Fu, M. Enhanced visible light driven photocatalytic performance of Bi_2WO_6 nano-catalysts by introducing oxygen vacancy. *J. Alloy. Compd.* **2021**, *887*, 161297. [CrossRef]
13. Chankhanittha, T.; Somaudon, V.; Photiwat, T.; Hemavibool, K.; Nanan, S. Preparation, characterization, and photocatalytic study of solvothermally grown CTAB-capped Bi_2WO_6 photocatalyst toward photodegradation of Rhodamine B dye. *Opt. Mater.* **2021**, *117*, 111183. [CrossRef]
14. Zhou, Y.; Lv, P.; Zhang, W.; Meng, X.; He, H.; Zeng, X.; Shen, X. Pristine Bi_2WO_6 and hybrid Au-Bi_2WO_6 hollow microspheres with excellent photocatalytic activities. *Appl. Surf. Sci.* **2018**, *457*, 925–932. [CrossRef]
15. Hoang, L.H.; Phu, N.D.; Peng, H.; Chen, X.B. High photocatalytic activity N-doped Bi_2WO_6 nanoparticles using a two-step microwave-assisted and hydrothermal synthesis. *J. Alloy. Compd.* **2018**, *744*, 228–233. [CrossRef]
16. Chen, Y.; Zhang, F.; Guan, S.; Shi, W.; Wang, X.; Huang, C.; Chen, Q. Visible light degradation of tetracycline by hierarchical nanoflower structured fluorine-doped Bi_2WO_6. *Mater. Sci. Semicon. Proc.* **2022**, *140*, 106385. [CrossRef]
17. Phuruangrat, A.; Dumrongrojthanath, P.; Thongtem, S.; Thongtem, T. Hydrothermal synthesis of I-doped Bi_2WO_6 for using as a visible-light-driven photocatalyst. *Mater. Lett.* **2018**, *224*, 67–70. [CrossRef]
18. Hu, T.; Li, H.; Du, N.; Hou, W. Iron-Doped Bismuth Tungstate with an Excellent Photocatalytic Performance. *ChemCatChem* **2018**, *10*, 3040–3048. [CrossRef]
19. Arif, M.; Zhang, M.; Yao, J.; Yin, H.; Li, P.; Hussain, I.; Liu, X. Layer-assembled 3D Bi_2WO_6 hierarchical architectures by Ti-doping for enhanced visible-light driven photocatalytic and photoelectrochemical performance. *J. Alloy. Compd.* **2019**, *792*, 878–893. [CrossRef]
20. Maniyazagan, M.; Hussain, M.; Kang, W.S.; Kim, S.J. Hierarchical Sr-Bi_2WO_6 photocatalyst for the degradation of 4-nitrophenol and methylene blue. *J. Ind. Eng. Chem.* **2022**, *110*, 168–177. [CrossRef]
21. Qiu, Y.; Lu, J.; Yan, Y.; Niu, J. Enhanced visible-light-driven photocatalytic degradation of tetracycline by 16% Er^{3+}-Bi_2WO_6 photocatalyst. *J. Hazard. Mater.* **2022**, *422*, 126920. [CrossRef] [PubMed]
22. Ning, J.; Zhang, J.; Dai, R.; Wu, Q.; Zhang, L.; Zhang, W.; Yan, J.; Zhang, F. Experiment and DFT study on the photocatalytic properties of La-doped Bi_2WO_6 nanoplate-like materials. *Appl. Surf. Sci.* **2022**, *579*, 152219. [CrossRef]
23. Phuruangrat, A.; Buapoon, S.; Bunluesak, T.; Suebsom, P.; Wannapop, S.; Thongtem, T.; Thongtem, S. Hydrothermal preparation of Au-doped Bi_2WO_6 nanoplates for enhanced visible-light-driven photocatalytic degradation of rhodamine B. *Solid State Sci.* **2022**, *128*, 106881. [CrossRef]
24. Phu, N.D.; Hoang, L.H.; Van Hai, P.; Huy, T.Q.; Chen, X.B.; Chou, W.C. Photocatalytic activity enhancement of Bi_2WO_6 nanoparticles by Ag doping and Ag nanoparticles modification. *J. Alloy. Compd.* **2020**, *824*, 153914. [CrossRef]
25. Longchin, P.; Sakulsermsuk, S.; Wetchakun, K.; Wetchakun, N. Synergistic effect of La and Mo co-doping on the enhanced photocatalytic activity of Bi_2WO_6. *Mater. Lett.* **2021**, *305*, 130779. [CrossRef]
26. Zhao, Y.; Liang, X.; Hu, X.; Fan, J. rGO/Bi_2WO_6 composite as a highly efficient and stable visible-light photocatalyst for norfloxacin degradation in aqueous environment. *J. Colloid Interface Sci.* **2021**, *589*, 336–346. [CrossRef] [PubMed]

27. Arya, M.; Kaur, M.; Kaur, A.; Singh, S.; Devi, P.; Kansal, S.K. Hydrothermal synthesis of rGO-Bi$_2$WO$_6$ heterostructure for the photocatalytic degradation of levofloxacin. *Opt. Mater.* **2020**, *107*, 110126. [CrossRef]
28. Yue, L.; Wang, S.; Shan, G.; Wu, W.; Qiang, L.; Zhu, L. Novel MWNTs–Bi$_2$WO$_6$ composites with enhanced simulated solar photoactivity toward adsorbed and free tetracycline in water. *Appl. Catal. B-Environ.* **2015**, *176–177*, 11–19. [CrossRef]
29. Yan, F.; Wang, Y.; Yi, C.; Xu, J.; Wang, B.; Ma, R.; Xu, M. Construction of carbon dots modified Cl-doped Bi$_2$WO$_6$ hollow microspheres for boosting photocatalytic degradation of tetracycline under visible light irradiation. *Ceram. Int.* **2022**, *in press*. [CrossRef]
30. Liang, W.; Pan, J.; Duan, X.; Tang, H.; Xu, J.; Tang, G. Biomass carbon modified flower-like Bi$_2$WO$_6$ hierarchical architecture with improved photocatalytic performance. *Ceram. Int.* **2020**, *46*, 3623–3630. [CrossRef]
31. Wang, T.; Liu, S.; Mao, W.; Bai, Y.; Chiang, K.; Shah, K.; Paz-Ferreiro, J. Novel Bi$_2$WO$_6$ loaded N-biochar composites with enhanced photocatalytic degradation of rhodamine B and Cr(VI). *J. Hazard. Mater.* **2020**, *389*, 121827. [CrossRef] [PubMed]
32. Mao, W.; Zhang, L.; Liu, Y.; Wang, T.; Bai, Y.; Guan, Y. Facile assembled N, S-codoped corn straw biochar loaded Bi$_2$WO$_6$ with the enhanced electron-rich feature for the efficient photocatalytic removal of ciprofloxacin and Cr(VI). *Chemosphere* **2021**, *263*, 127988. [CrossRef] [PubMed]
33. Kaur, M.; Mehta, S.K.; Devi, P.; Kansal, S.K. Bi$_2$WO$_6$/NH$_2$-MIL-88B(Fe) heterostructure: An efficient sunlight driven photocatalyst for the degradation of antibiotic tetracycline in aqueous medium. *Adv. Powder Technol.* **2021**, *32*, 4788–4804. [CrossRef]
34. He, Y.; Wang, D.; Li, X.; Fu, Q.; Yin, L.; Yang, Q.; Chen, H. Photocatalytic degradation of tetracycline by metal-organic frameworks modified with Bi$_2$WO$_6$ nanosheet under direct sunlight. *Chemosphere* **2021**, *284*, 131386. [CrossRef]
35. Li, Q.; Li, L.; Long, X.; Tu, Y.; Ling, L.; Gu, J.; Hou, L.; Xu, Y.; Liu, N.; Li, Z. Rational design of MIL-88A(Fe)/Bi$_2$WO$_6$ heterojunctions as an efficient photocatalyst for organic pollutant degradation under visible light irradiation. *Opt. Mater.* **2021**, *118*, 111260. [CrossRef]
36. Tu, Y.; Ling, L.; Li, Q.; Long, X.; Liu, N.; Li, Z. Greatly enhanced photocatalytic activity over Bi$_2$WO$_6$ by MIL-53(Fe) modification. *Opt. Mater.* **2020**, *110*, 110500. [CrossRef]
37. Zhang, X.; Yuan, N.; Chen, T.; Li, B.; Wang, Q. Fabrication of hydrangea-shaped Bi$_2$WO$_6$/ZIF-8 visible-light responsive photocatalysts for degradation of methylene blue. *Chemosphere* **2022**, *307*, 135949. [CrossRef]
38. Dai, X.; Feng, S.; Wu, W.; Zhou, Y.; Ye, Z.; Wang, Y.; Cao, X. Photocatalytic Degradation of Tetracycline by Z-Scheme Bi$_2$WO$_6$/ZIF-8. *J. Inorg. Organomet. Polym. Mater.* **2022**, *32*, 2371–2383. [CrossRef]
39. Qi, S.; Zhang, R.; Zhang, Y.; Liu, X.; Xu, H. Preparation and photocatalytic properties of Bi$_2$WO$_6$/g-C$_3$N$_4$. *Inorg. Chem. Commun.* **2021**, *132*, 108761. [CrossRef]
40. Zhao, Y.; Liang, X.; Wang, Y.; Shi, H.; Liu, E.; Fan, J.; Hu, X. Degradation and removal of Ceftriaxone sodium in aquatic environment with Bi$_2$WO$_6$/g-C$_3$N$_4$ photocatalyst. *J. Colloid Interface Sci.* **2018**, *523*, 7–17. [CrossRef]
41. Chen, J.; Yang, Q.; Zhong, J.; Li, J.; Hu, C.; Deng, Z.; Duan, R. In-situ construction of direct Z-scheme Bi$_2$WO$_6$/ g-C$_3$N$_4$ composites with remarkably promoted solar-driven photocatalytic activity. *Mater. Chem. Phys.* **2018**, *217*, 207–215. [CrossRef]
42. Zhu, D.; Zhou, Q. Novel Bi$_2$WO$_6$ modified by N-doped graphitic carbon nitride photocatalyst for efficient photocatalytic degradation of phenol under visible light. *Appl. Catal. B-Environ.* **2020**, *268*, 118426. [CrossRef]
43. Wang, Y.; Jiang, W.; Luo, W.; Chen, X.; Zhu, Y. Ultrathin nanosheets g-C$_3$N$_4$@Bi$_2$WO$_6$ core-shell structure via low temperature reassembled strategy to promote photocatalytic activity. *Appl. Catal. B-Environ.* **2018**, *237*, 633–640. [CrossRef]
44. Zhang, M.; Zhang, Y.; Tang, L.; Zeng, G.; Wang, J.; Zhu, Y.; Feng, C.; Deng, Y.; He, W. Ultrathin Bi$_2$WO$_6$ nanosheets loaded g-C$_3$N$_4$ quantum dots: A direct Z-scheme photocatalyst with enhanced photocatalytic activity towards degradation of organic pollutants under wide spectrum light irradiation. *J. Colloid Interface Sci.* **2019**, *539*, 654–664. [CrossRef] [PubMed]
45. Gordanshekan, A.; Arabian, S.; Solaimany Nazar, A.R.; Farhadian, M.; Tangestaninejad, S. A comprehensive comparison of green Bi$_2$WO$_6$/g-C$_3$N$_4$ and Bi$_2$WO$_6$/TiO$_2$ S-scheme heterojunctions for photocatalytic adsorption/ degradation of Cefixime: Artificial neural network, degradation pathway, and toxicity estimation. *Chem. Eng. J.* **2023**, *451*, 139067. [CrossRef]
46. Li, W.; Ding, X.; Wu, H.; Yang, H. In-situ hydrothermal synthesis of TiO$_2$/Bi$_2$WO$_6$ heterojunction with enhanced photocatalytic activity. *Mater. Lett.* **2018**, *227*, 272–275. [CrossRef]
47. Chen, G.; Wang, Y.; Shen, Q.; Xiong, X.; Ren, S.; Dai, G.; Wu, C. Fabrication of TiO$_2$ nanofibers assembled by Bi$_2$WO$_6$ nanosheets with enhanced visible light photocatalytic activity. *Ceram. Int.* **2020**, *46*, 21304–21310. [CrossRef]
48. Lu, Q.; Dong, C.; Wei, F.; Li, J.; Wang, Z.; Mu, W.; Han, X. Rational fabrication of Bi$_2$WO$_6$ decorated TiO$_2$ nanotube arrays for photocatalytic degradation of organic pollutants. *Mater. Res. Bull.* **2022**, *145*, 111563. [CrossRef]
49. Wang, Q.; Li, H.; Yu, X.; Jia, Y.; Chang, Y.; Gao, S. Morphology regulated Bi$_2$WO$_6$ nanoparticles on TiO$_2$ nanotubes by solvothermal Sb^{3+} doping as effective photocatalysts for wastewater treatment. *Electrochim. Acta* **2020**, *330*, 135167. [CrossRef]
50. Sun, M.; Yao, Y.; Ding, W.; Anandan, S. N/Ti^{3+} co-doping biphasic TiO$_2$/Bi$_2$WO$_6$ heterojunctions: Hydrothermal fabrication and sonophotocatalytic degradation of organic pollutants. *J. Alloy. Compd.* **2020**, *820*, 153172. [CrossRef]
51. Liu, J.; Luo, Z.; Han, W.; Zhao, Y.; Li, P. Preparation of ZnO/Bi$_2$WO$_6$ heterostructures with improved photocatalytic performance. *Mater. Sci. Semicon. Proc.* **2020**, *106*, 104761. [CrossRef]
52. Koutavarapu, R.; Babu, B.; Reddy, C.V.; Reddy, I.N.; Reddy, K.R.; Rao, M.C.; Aminbhavi, T.M.; Cho, M.; Kim, D.; Shim, J. ZnO nanosheets-decorated Bi$_2$WO$_6$ nanolayers as efficient photocatalysts for the removal of toxic environmental pollutants and photoelectrochemical solar water oxidation. *J. Environ. Manag.* **2020**, *265*, 110504. [CrossRef] [PubMed]

53. Chen, X.; Li, J.; Chen, F. Photocatalytic degradation of MB by novel and environmental ZnO/Bi$_2$WO$_6$-CC hierarchical heterostructures. *Mater. Charact.* **2022**, *189*, 111961. [CrossRef]
54. Zhao, F.; Gao, D.; Zhu, X.; Dong, Y.; Liu, X.; Li, H. Rational design of multifunctional C/N-doped ZnO/Bi$_2$WO$_6$ Z-scheme heterojunction for efficient photocatalytic degradation of antibiotics. *Appl. Surf. Sci.* **2022**, *587*, 152780. [CrossRef]
55. Salari, H.; Yaghmaei, H. Z-scheme 3D Bi$_2$WO$_6$/MnO$_2$ heterojunction for increased photoinduced charge separation and enhanced photocatalytic activity. *Appl. Surf. Sci.* **2020**, *532*, 147413. [CrossRef]
56. Mallikarjuna, K.; Kim, H. Bandgap-tuned ultra-small SnO$_2$-nanoparticle-decorated 2D-Bi$_2$WO$_6$ nanoplates for visible-light-driven photocatalytic applications. *Chemosphere* **2021**, *263*, 128185. [CrossRef]
57. Zhu, B.; Song, D.; Jia, T.; Sun, W.; Wang, D.; Wang, L.; Guo, J.; Jin, L.; Zhang, L.; Tao, H. Effective Visible Light-Driven Photocatalytic Degradation of Ciprofloxacin over Flower-like Fe$_3$O$_4$/Bi$_2$WO$_6$ Composites. *ACS Omega* **2021**, *6*, 1647–1656. [CrossRef]
58. Zhang, X.; Zhang, H.; Yu, J.; Wu, Z.; Zhou, Q. Preparation of flower-like Co$_3$O$_4$ QDs/Bi$_2$WO$_6$ p-n heterojunction photocatalyst and its degradation mechanism of efficient visible-light-driven photocatalytic tetracycline antibiotics. *Appl. Surf. Sci.* **2022**, *585*, 152547. [CrossRef]
59. Koutavarapu, R.; Syed, K.; Pagidi, S.; Jeon, M.Y.; Rao, M.C.; Lee, D.Y.; Shim, J. An effective CuO/Bi$_2$WO$_6$ heterostructured photocatalyst: Analyzing a charge-transfer mechanism for the enhanced visible-light-driven photocatalytic degradation of tetracycline and organic pollutants. *Chemosphere* **2022**, *287*, 132015. [CrossRef]
60. Chen, X.; Li, Y.; Li, L. Facet-engineered surface and interface design of WO$_3$/Bi$_2$WO$_6$ photocatalyst with direct Z-scheme heterojunction for efficient salicylic acid removal. *Appl. Surf. Sci.* **2020**, *508*, 144796. [CrossRef]
61. Liu, G.; Cui, P.; Liu, X.; Wang, X.; Liu, G.; Zhang, C.; Liu, M.; Chen, Y.; Xu, S. A facile preparation strategy for Bi$_2$O$_4$/Bi$_2$WO$_6$ heterojunction with excellent visible light photocatalytic activity. *J. Solid State Chem.* **2020**, *290*, 121542. [CrossRef]
62. Chao, P.Y.; Chang, C.J.; Lin, K.S.; Wang, C.F. Synergistic effects of morphology control and calcination on the activity of flower-like Bi$_2$WO$_6$-Bi$_2$O$_3$ photocatalysts prepared by an ionic liquid-assisted solvothermal method. *J. Alloy. Compd.* **2021**, *883*, 160920. [CrossRef]
63. Qin, Y.; Li, H.; Lu, J.; Ding, Y.; Ma, C.; Liu, X.; Meng, M.; Yan, Y. Fabrication of Bi$_2$WO$_6$/In$_2$O$_3$ photocatalysts with efficient photocatalytic performance for the degradation of organic pollutants: Insight into the role of oxygen vacancy and heterojunction. *Adv. Powder Technol.* **2020**, *31*, 2890–2900. [CrossRef]
64. Tang, Q.Y.; Chen, W.F.; Lv, Y.R.; Yang, S.Y.; Xu, Y.H. Z-scheme hierarchical Cu$_2$S/Bi$_2$WO$_6$ composites for improved photocatalytic activity of glyphosate degradation under visible light irradiation. *Sep. Purif. Technol.* **2020**, *236*, 116243. [CrossRef]
65. Zhang, Y.; Ju, P.; Hao, L.; Zhai, X.; Jiang, F.; Sun, C. Novel Z-scheme MoS$_2$/Bi$_2$WO$_6$ heterojunction with highly enhanced photocatalytic activity under visible light irradiation. *J. Alloy. Compd.* **2021**, *854*, 157224. [CrossRef]
66. Xie, T.; Liu, Y.; Wang, H.; Wu, Z. Layered MoSe$_2$/Bi$_2$WO$_6$ composite with P-N heterojunctions as a promising visible-light induced photocatalyst. *Appl. Surf. Sci.* **2018**, *444*, 320–329. [CrossRef]
67. Kumar, G.; Dutta, R.K. Fabrication of plate-on-plate SnS$_2$/Bi$_2$WO$_6$ nanocomposite as photocatalyst for sunlight mediated degradation of antibiotics in aqueous medium. *J. Phys. Chem. Solids* **2022**, *164*, 110639. [CrossRef]
68. Su, M.; Sun, H.; Tian, Z.; Zhao, Z.; Li, P. Z-scheme 2D/2D WS$_2$/Bi$_2$WO$_6$ heterostructures with enhanced photocatalytic performance. *Appl. Catal. A-Gen.* **2022**, *631*, 118485. [CrossRef]
69. Su, Y.; Xu, X.; Li, R.; Luo, X.; Yao, H.; Fang, S.; Peter Homewood, K.; Huang, Z.; Gao, Y.; Chen, X. Design and fabrication of a CdS QDs/Bi$_2$WO$_6$ monolayer S-scheme heterojunction configuration for highly efficient photocatalytic degradation of trace ethylene in air. *Chem. Eng. J.* **2022**, *429*, 132241. [CrossRef]
70. Zhang, Y.; Lin, Y.; Liu, F. Preparation and photocatalytic performance of CdS@Bi$_2$WO$_6$ hybrid nanocrystals. *J. Alloy. Compd.* **2021**, *889*, 161668. [CrossRef]
71. Xu, F.; Xu, C.; Chen, H.; Wu, D.; Gao, Z.; Ma, X.; Zhang, Q.; Jiang, K. The synthesis of Bi$_2$S$_3$/2D-Bi$_2$WO$_6$ composite materials with enhanced photocatalytic activities. *J. Alloy. Compd.* **2019**, *780*, 634–642. [CrossRef]
72. He, Z.; Siddique, M.S.; Yang, H.; Xia, Y.; Su, J.; Tang, B.; Wang, L.; Kang, L.; Huang, Z. Novel Z-scheme In$_2$S$_3$/Bi$_2$WO$_6$ core-shell heterojunctions with synergistic enhanced photocatalytic degradation of tetracycline hydrochloride. *J. Clean. Prod.* **2022**, *339*, 130634. [CrossRef]
73. Lu, X.; Che, W.; Hu, X.; Wang, Y.; Zhang, A.; Deng, F.; Luo, S.; Dionysiou, D.D. The facile fabrication of novel visible-light-driven Z-scheme CuInS$_2$/Bi$_2$WO$_6$ heterojunction with intimate interface contact by in situ hydrothermal growth strategy for extraordinary photocatalytic performance. *Chem. Eng. J.* **2019**, *356*, 819–829. [CrossRef]
74. Shangguan, X.Y.; Fang, B.L.; Xu, C.X.; Tan, Y.; Chen, Y.G.; Xia, Z.J.; Chen, W. Fabrication of direct Z-scheme FeIn$_2$S$_4$/Bi$_2$WO$_6$ hierarchical heterostructures with enhanced photocatalytic activity for tetracycline hydrochloride photodagradation. *Ceram. Int.* **2021**, *47*, 6318–6328. [CrossRef]
75. Guo, M.; Zhou, Z.; Yan, S.; Zhou, P.; Miao, F.; Liang, S.; Wang, J.; Cui, X. Bi$_2$WO$_6$-BiOCl heterostructure with enhanced photocatalytic activity for efficient degradation of oxytetracycline. *Sci. Rep.* **2020**, *10*, 18401. [CrossRef]
76. Liang, Z.; Zhou, C.; Yang, J.; Mo, Q.; Zhang, Y.; Tang, Y. Visible light responsive Bi$_2$WO$_6$/BiOCl heterojunction with enhanced photocatalytic activity for degradation of tetracycline and rohdamine B. *Inorg. Chem. Commun.* **2018**, *93*, 136–139. [CrossRef]

77. Liu, K.; Zhang, H.; Muhammad, Y.; Fu, T.; Tang, R.; Tong, Z.; Wang, Y. Fabrication of n-n isotype BiOBr- Bi_2WO_6 heterojunctions by inserting Bi_2WO_6 nanosheets onto BiOBr microsphere for the superior photocatalytic degradation of Ciprofloxacin and tetracycline. *Sep. Purif. Technol.* **2021**, *274*, 118992. [CrossRef]
78. Ren, X.; Wu, K.; Qin, Z.; Zhao, X.; Yang, H. The construction of type II heterojunction of Bi_2WO_6/BiOBr photocatalyst with improved photocatalytic performance. *J. Alloy. Compd.* **2019**, *788*, 102–109. [CrossRef]
79. Chen, X.; Zhao, B.; Ma, J.; Liu, L.; Luo, H.; Wang, W. The BiOBr/Bi/Bi_2WO_6 photocatalyst with SPR effect and Z-scheme heterojunction synergistically degraded RhB under visible light. *Opt. Mater.* **2021**, *122*, 111641. [CrossRef]
80. He, J.; Liu, Y.; Wang, M.; Wang, Y.; Long, F. Ionic liquid-hydrothermal synthesis of Z-scheme BiOBr/Bi_2WO_6 heterojunction with enhanced photocatalytic activity. *J. Alloy. Compd.* **2021**, *865*, 158760. [CrossRef]
81. Huang, X.; Guo, Q.; Yan, B.; Liu, H.; Chen, K.; Wei, S.; Wu, Y.; Wang, L. Study on photocatalytic degradation of phenol by BiOI/Bi_2WO_6 layered heterojunction synthesized by hydrothermal method. *J. Mol. Liq.* **2021**, *322*, 114965. [CrossRef]
82. Zheng, X.; Chu, Y.; Miao, B.; Fan, J. Ag-doped Bi_2WO_6/BiOI heterojunction used as photocatalyst for the enhanced degradation of tetracycline under visible-light and biodegradability improvement. *J. Alloy. Compd.* **2022**, *893*, 162382. [CrossRef]
83. Qiang, Z.; Liu, X.; Li, F.; Li, T.; Zhang, M.; Singh, H.; Huttula, M.; Cao, W. Iodine doped Z-scheme $Bi_2O_2CO_3$/Bi_2WO_6 photocatalysts: Facile synthesis, efficient visible light photocatalysis, and photocatalytic mechanism. *Chem. Eng. J.* **2021**, *403*, 126327. [CrossRef]
84. Wu, R.; Song, H.; Luo, N.; Ji, G. Hydrothermal preparation of 3D flower-like $BiPO_4$/Bi_2WO_6 microsphere with enhanced visible-light photocatalytic activity. *J. Colloid Interface Sci.* **2018**, *524*, 350–359. [CrossRef] [PubMed]
85. Zhang, Y.; Xu, C.; Wan, F.; Zhou, D.; Yang, L.; Gu, H.; Xiong, J. Synthesis of flower-like $Bi_2Sn_2O_7$/Bi_2WO_6 hierarchical composites with enhanced visible light photocatalytic performance. *J. Alloy. Compd.* **2019**, *788*, 1154–1161. [CrossRef]
86. Tao, R.; Li, X.; Li, X.; Liu, S.; Shao, C.; Liu, Y. Discrete heterojunction nanofibers of $BiFeO_3$/Bi_2WO_6: Novel architecture for effective charge separation and enhanced photocatalytic performance. *J. Colloid Interface Sci.* **2020**, *572*, 257–268. [CrossRef]
87. Teng, P.; Li, Z.; Gao, S.; Li, K.; Bowkett, M.; Copner, N.; Liu, Z.; Yang, X. Fabrication of one-dimensional Bi_2WO_6/$CuBi_2O_4$ heterojunction nanofiber and its photocatalytic degradation property. *Opt. Mater.* **2021**, *121*, 111508. [CrossRef]
88. Wang, L.; Yang, G.; Wang, D.; Lu, C.; Guan, W.; Li, Y.; Deng, J.; Crittenden, J. Fabrication of the flower-flake- like $CuBi_2O_4$/Bi_2WO_6 heterostructure as efficient visible-light driven photocatalysts: Performance, kinetics and mechanism insight. *Appl. Surf. Sci.* **2019**, *495*, 143521. [CrossRef]
89. Kumar, P.; Verma, S.; Korošin, N.Č.; Žener, B.; Štangar, U.L. Increasing the photocatalytic efficiency of $ZnWO_4$ by synthesizing a Bi_2WO_6/$ZnWO_4$ composite photocatalyst. *Catal. Today* **2022**, *397–399*, 278–285. [CrossRef]
90. Miao, B.; Chu, Y.; Zheng, X.; Su, H. Sb_2WO_6/Bi_2WO_6 composite photocatalyst prepared by one-step hydrothermal method: Simple synthesis and excellent visible-light photocatalytic performance. *Mater. Sci. Semicon. Proc.* **2021**, *125*, 105636. [CrossRef]
91. Ni, Z.; Shen, Y.; Xu, L.; Xiang, G.; Chen, M.; Shen, N.; Li, K.; Ni, K. Facile construction of 3D hierarchical flower- like Ag_2WO_4/Bi_2WO_6 Z-scheme heterojunction photocatalyst with enhanced visible light photocatalytic activity. *Appl. Surf. Sci.* **2022**, *576*, 151868. [CrossRef]
92. Zhang, R.; Zeng, K. A novel flower-like dual Z-scheme BiSI/Bi_2WO_6/g-C_3N_4 photocatalyst has excellent photocatalytic activity for the degradation of organic pollutants under visible light. *Diam. Relat. Mater.* **2021**, *115*, 108343. [CrossRef]
93. Sun, H.; Zou, C.; Tang, W. Designing double Z-scheme heterojunction of g-C_3N_4/Bi_2MoO_6/Bi_2WO_6 for efficient visible-light photocatalysis of organic pollutants. *Colloids Surf. A* **2022**, *654*, 130105. [CrossRef]
94. Zhou, K.; Liu, Y.; Hao, J. One-pot hydrothermal synthesis of dual Z-scheme BiOBr/g-C_3N_4/Bi_2WO_6 and photocatalytic degradation of tetracycline under visible light. *Mater. Lett.* **2020**, *281*, 128463. [CrossRef]
95. Chu, Y.; Fan, J.; Wang, R.; Liu, C.; Zheng, X. Preparation and immobilization of Bi_2WO_6/BiOI/g-C_3N_4 nanoparticles for the photocatalytic degradation of tetracycline and municipal waste transfer station leachate. *Sep. Purif. Technol.* **2022**, *300*, 121867. [CrossRef]
96. Hu, H.; Kong, W.; Wang, J.; Liu, C.; Cai, Q.; Kong, Y.; Zhou, S.; Yang, Z. Engineering 2D compressed layered g-C_3N_4 nanosheets by the intercalation of $BiVO_4$-Bi_2WO_6 composites for boosting photocatalytic activities. *Appl. Surf. Sci.* **2021**, *557*, 149796. [CrossRef]
97. Wu, K.; Song, S.; Wu, H.; Guo, J.; Zhang, L. Facile synthesis of Bi_2WO_6/C_3N_4/Ti_3C_2 composite as Z-scheme photocatalyst for efficient ciprofloxacin degradation and H_2 production. *Appl.Catal. A-Gen.* **2020**, *608*, 117869. [CrossRef]
98. Li, Q.; Lu, M.; Wang, W.; Zhao, W.; Chen, G.; Shi, H. Fabrication of 2D/2D g-C_3N_4/Au/Bi_2WO_6 Z-scheme photocatalyst with enhanced visible-light-driven photocatalytic activity. *Appl. Surf. Sci.* **2020**, *508*, 144182. [CrossRef]
99. Jia, J.; Zhang, X.; Jiang, C.; Huang, W.; Wang, Y. Visible-light-driven nitrogen-doped carbon quantum dots decorated g-C_3N_4/Bi_2WO_6 Z-scheme composite with enhanced photocatalytic activity and mechanism insight. *J. Alloy. Compd.* **2020**, *835*, 155180. [CrossRef]
100. Guan, Z.; Li, X.; Wu, Y.; Chen, Z.; Huang, X.; Wang, D.; Yang, Q.; Liu, J.; Tian, S.; Chen, X.; et al. AgBr nanoparticles decorated 2D/2D GO/Bi_2WO_6 photocatalyst with enhanced photocatalytic performance for the removal of tetracycline hydrochloride. *Chem. Eng. J.* **2021**, *410*, 128283. [CrossRef]
101. Mengting, Z.; Kurniawan, T.A.; Yanping, Y.; Avtar, R.; Othman, M.H.D. 2D Graphene oxide (GO) doped p-n type BiOI/Bi_2WO_6 as a novel composite for photodegradation of bisphenol A (BPA) in aqueous solutions under UV-vis irradiation. *Mater. Sci. Eng. C-Mater.* **2020**, *108*, 110420. [CrossRef] [PubMed]

102. Tian, J.; Zhu, Z.; Liu, B. Novel Bi_2MoO_6/Bi_2WO_6/MWCNTs photocatalyst with enhanced photocatalytic activity towards degradation of RB-19 under visible light irradiation. *Colloid. Surf. A* **2019**, *581*, 123798. [CrossRef]
103. Niu, J.; Song, Z.; Gao, X.; Ji, Y.; Zhang, Y. Construction of Bi_2WO_6 composites with carbon-coated Cu_2O for effective degradation of tetracycline. *J. Alloy. Compd.* **2021**, *884*, 161292. [CrossRef]
104. Wang, X.; Liu, X.; Li, H.; Yang, Y.; Ren, Y. $Ag_2CO_3/AgBr/Bi_2WO_6$ nanocomposite: Synthesis and solar photocatalytic activity. *Inorg. Chem. Commun.* **2021**, *132*, 108826. [CrossRef]
105. Gan, W.; Zhang, J.; Niu, H.; Bao, L.; Hao, H.; Yan, Y.; Wu, K.; Fu, X. Fabrication of $Ag/AgBr/Bi_2WO_6$ hierarchical composites with high visible light photocatalytic activity. *Chem. Phys. Lett.* **2019**, *737*, 136830. [CrossRef]
106. Jin, K.; Qin, M.; Li, X.; Wang, R.; Zhao, Y.; Wang, H. Z-scheme $Au@TiO_2/Bi_2WO_6$ heterojunction as efficient visible-light photocatalyst for degradation of antibiotics. *J. Mol. Liq.* **2022**, *364*, 120017. [CrossRef]
107. Mengting, Z.; Kurniawan, T.A.; Yanping, Y.; Dzarfan Othman, M.H.; Avtar, R.; Fu, D.; Hwang, G.H. Fabrication, characterization, and application of ternary magnetic recyclable $Bi_2WO_6/BiOI@Fe_3O_4$ composite for photodegradation of tetracycline in aqueous solutions. *J. Environ. Manag.* **2020**, *270*, 110839. [CrossRef]
108. Chen, Y.; Su, X.; Ma, M.; Hou, Y.; Lu, C.; Wan, F.; Ma, Y.; Xu, Z.; Liu, Q.; Hao, M.; et al. One- dimensional magnetic flower-like $CoFe_2O_4@Bi_2WO_6@BiOBr$ composites for visible-light catalytic degradation of Rhodamine B. *J. Alloy. Compd.* **2022**, *929*, 167297. [CrossRef]
109. Rao, C.; Zhou, L.; Pan, Y.; Lu, C.; Qin, X.; Sakiyama, H.; Muddassir, M.; Liu, J. The extra-large calixarene-based MOFs-derived hierarchical composites for photocatalysis of dye: Facile syntheses and contribution of carbon species. *J. Alloy. Compd.* **2022**, *897*, 163178. [CrossRef]
110. Jin, J.C.; Wang, J.; Guo, J.; Yan, M.H.; Wang, J.; Srivastava, D.; Kumar, A.; Sakiyama, H.; Muddassir, M.; Pan, Y. A 3D rare cubane-like tetramer Cu(II)-based MOF with 4-fold dia topology as an efficient photocatalyst for dye degradation. *Colloid. Surf. A* **2023**, *656*, 130475. [CrossRef]
111. Li, L.; Zou, J.; Han, Y.; Liao, Z.; Lu, P.; Nezamzadeh-Ejhieh, A.; Liu, J.; Peng, Y. Recent advances in Al(iii)/In(iii)-based MOFs for the detection of pollutants. *New J. Chem.* **2022**, *46*, 19577–19592. [CrossRef]
112. Sharma, V.; Kumar, A.; Kumar, A.; Krishnan, V. Enhanced photocatalytic activity of two dimensional ternary nanocomposites of $ZnO-Bi_2WO_6-Ti_3C_2$ MXene under natural sunlight irradiation. *Chemosphere* **2022**, *287*, 132119. [CrossRef] [PubMed]
113. Ma, Q.; Ming, J.; Sun, X.; Liu, N.; Chen, G.; Yang, Y. Visible light active graphene oxide modified $Ag/Ag_2O/BiPO_4/Bi_2WO_6$ for photocatalytic removal of organic pollutants and bacteria in wastewater. *Chemosphere* **2022**, *306*, 135512. [CrossRef] [PubMed]

Article

In Situ Fabrication of N-Doped ZnS/ZnO Composition for Enhanced Visible-Light Photocatalytic H_2 Evolution Activity

Jinhua Xiong [1,2,*], Xuxu Wang [2], Jinling Wu [2], Jiaming Han [2], Zhiyang Lan [2] and Jianming Fan [2,*]

[1] State Key Laboratory of Photocatalysis on Energy and Environment, Fuzhou University, Fuzhou 350002, China
[2] Fujian Provincial Key Laboratory of Clean Energy Materials, Longyan University, Longyan 364000, China
* Correspondence: xjh970996937@sina.com (J.X.); jmfan1989@163.com (J.F.); Tel.: +86-0597-2790525 (J.X.)

Abstract: For achieving the goal of peaking carbon dioxide emissions and achieving carbon neutrality, developing hydrogen energy, the green and clean energy, shows a promising perspective for solving the energy and ecological issues. Herein, firstly, we used the hydrothermal method to synthesize the ZnS(en)$_{0.5}$ as the precursor. Then, ZnS/ZnO composite was obtained by the in situ transformation of ZnS(en)$_{0.5}$ with heat treatment under air atmosphere. The composition, optical property, morphology, and structural properties of the composite were characterized by X-ray photoemission spectroscopy (XPS), Ultraviolet-visible absorption spectra (Uv-vis Abs), Scanning electron microscopy (SEM) and Transmission electron microscopy image (TEM). Moreover, the content of ZnO in ZnS/ZnO was controlled via adjustment of the calcination times. The visible-light response of ZnS/ZnO originated from the in situ doping of N during the transformation of ZnS(en)$_{0.5}$ to ZnS/ZnO under heat treatment, which was verified well by XPS. Photocatalytic hydrogen evolution experiments demonstrated that the sample of ZnS/ZnO-0.5 h with 6.9 wt% of ZnO had the best H_2 evolution activity (1790 μmol/h/g) under visible light irradiation (λ > 400 nm), about 7.0 and 12.3 times that of the pure ZnS and ZnO, respectively. The enhanced activities of the ZnS/ZnO composites were ascribed to the intimated hetero-interface between components and efficient transfer of photo-generated electrons from ZnS to ZnO.

Keywords: ZnS/ZnO; N-doped; photocatalytic H_2 evolution; heterojunction

1. Introduction

Hydrogen, as a clear and renewable energy, has been considered as a candidate for future energy use. Photocatalysis is a powerful technology for transforming solar energy into chemical energy [1–3]. The search for and design of photocatalysts are the key ways for achieving the highly efficient photocatalytic hydrogen evolution [4,5].

ZnS is one of the well-known photocatalysts for H_2 production, but suffers from a ultraviolet response, a low surface area and a repressed carriers migration [6]. As reported by our previous work [7], designing ZnS with a porous plate-like structure and N-doping in crystal lattice is a feasible approach to broaden the optical absorption and increase the surface areas of ZnS. Therefore, to further enhance the photocatalytic hydrogen evolution activity of N-doped porous ZnS nanoplates, one's attentions should be focused on the issues of carriers separation. Over the past decades, engineering heterojunctions in photocatalysts have been proved to be a promising route for promoting carriers separation [8,9]. One of the most vital factors for the effective separation of electron–hole pairs involves the ohmic contact interface and energy band structures of different components in photocatalyst [10,11]. Fortunately, ZnO, as another important II-VI group semiconductor, has similar physical and chemical properties to ZnS [12]. Their energy band structures are generally staggered, forming a so-called type II heterojunction, which facilitates the separation of carriers [13,14]. More importantly, hexagonal ZnS and ZnO are isomorphous compounds and a mutual transformation between ZnS and ZnO can be executed via anion exchange [15]. Hence, the preparation of the composition of ZnS/ZnO by partial sulfuration of ZnO or oxidation of ZnS could form a firm heterojunction with a superior ohmic contact interface [16–18] This has been further confirmed by recent reports. As Wang et al. reported [19], an all-solid-state Z-Scheme ZnS-ZnO heterostructure photocatalyst was prepared via in situ sulfurization of ZnO sheets. The obtained ZnS-ZnO showed a remarkable enhancement of photocatalytic H_2 evolution activity, but pursuing visible-light activity was still in demand. Additionally, Cheng et al. reported a ZnO/ZnS heteronanostructure for photocatalytic H_2 evolution under visible light (λ >420 nm) [20]. The method for preparing ZnO/ZnS suffered from a tedious process, including synthesis of Zn-MOF, partial sulfuration of Zn-MOF, high temperature treatment for in-situ transformation of ZnS@Zn-MOF into ZnS@C, and further oxidation of ZnS@C in air into ZnS/ZnO. The visible-light photocatalytic H_2 evolution activity of the ZnS/ZnO was not more desirable. Therefore, engineering a ZnS/ZnO heterojunction photocatalyst with visible-light response for photocatalytic H_2 evolution is still imperative and challenging. For addressing these terms, this work provides a feasible and simple way to fabricate a ZnS/ZnO photocatalyst for H_2 evolution with visible-light response and a heterojunction structure simultaneously.

Herein, N-doped ZnS/ZnO composition was prepared by the in situ transformation of $ZnS(en)_{0.5}$ with heat treatment under air atmosphere. The content of ZnO in ZnS/ZnO was able to be controlled via adjustment of the calcination times. Photocatalytic hydrogen evolution experiments demonstrated that the sample of ZnS/ZnO-0.5 h with 6.9 wt% of ZnO had the best H_2 evolution activity (1790 μmol/h/g) under visible light irradiation (λ > 400 nm), about 7.0 and 12.3 times that of the pure ZnS and ZnO, respectively. The enhanced activities of the ZnS/ZnO composites were ascribed to the formation of heterojunction between ZnS and ZnO.

2. Results and Discussion

2.1. Structure and Morphology

The X-ray diffraction (XRD) diffraction pattern of the as-prepared precursor is shown in Figure 1A. As shown, the pattern perfectly matched the referenced date of $ZnS(en)_{0.5}$ (CCDC No. 200433), indicating that $ZnS(en)_{0.5}$ was synthesized successfully [21]. The strongest peak, at about 10°, was assigned to the stacking direction (a axis) of a S-Zn-S layer slab. The scanning electron microscopy (SEM) image in Figure 1B and transmission electron microscopy image (TEM) in Figure 1C show that $ZnS(en)_{0.5}$ had a slab morphology with various lateral sizes and thicknesses, consistent with the crystal structure of $ZnS(en)_{0.5}$. Meanwhile, the selected area electron diffraction (SAED) image (inset in Figure 1C) revealed that $ZnS(en)_{0.5}$ had a polycrystalline structure. The first diffraction ring in the SAED image had a radius of around 3.22 nm, which was ascribed to the diffraction of (002) plane of $ZnS(en)_{0.5}$. Moreover, the high-resolution TEM (HRTEM) image in Figure 1D further confirmed that the bulk $ZnS(en)_{0.5}$ consisted of a nanocrystal of $ZnS(en)_{0.5}$ [7].

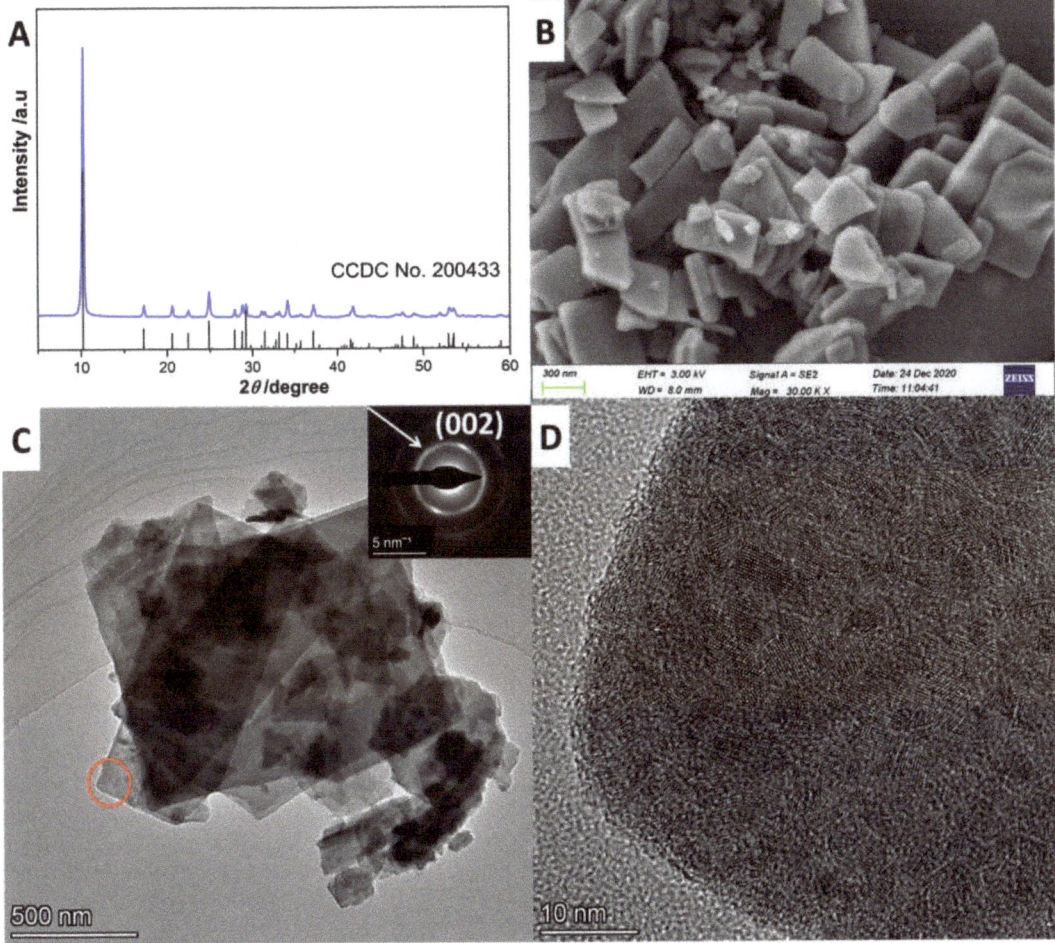

Figure 1. (**A**) XRD diffraction patterns, (**B**) SEM image, (**C**) TEM image and (**D**) HRTEM image of the as-prepared ZnS(en)$_{0.5}$. Inset in C is the SAED image.

Figure 2 shows the XRD patterns of the samples obtained via calcination of ZnS(en)$_{0.5}$ at 500 °C for different times. When time was 5 min, the obtained sample, defined as ZnS, was assigned to pure hexagonal ZnS. The diffraction peaks at 27.0°, 28.6°, 30.6°, 39.5°, 47.6°, 51.9°, 56.4° were attributed to the (100), (002), (101), (102), (110), (103) and (112) planes of hexagonal ZnS (PDF#02-1310), respectively. Alongside increasing to 10 min, the XRD diffraction patterns of the sample (ZnS/ZnO-10 min) exhibited three extra weak diffraction peaks at 31.7°, 34.3° and 36.2°, which corresponded to the (100), (002) and (101) planes of hexagonal ZnO (PDF#05-0664), respectively, indicating the beginning of the transformation from ZnS to ZnO. With a further increase in the heat treatment time, the intensity of diffraction peaks assigned to ZnS and ZnO tended to decay and enhance, respectively. Until the time extended to 5 h, the obtained sample was pure hexagonal ZnO.

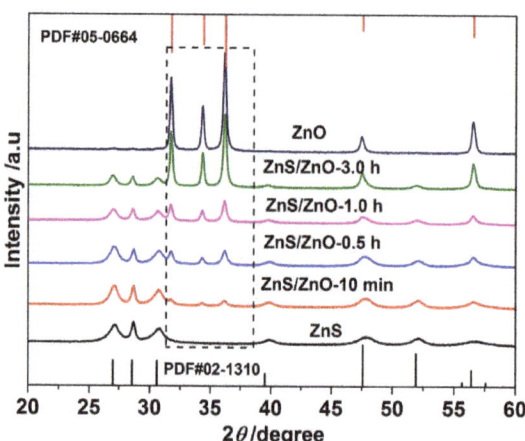

Figure 2. XRD diffractions patterns of the samples obtained via calcination of ZnS(en)$_{0.5}$ at 500 °C for different times including 5 min, 10 min, 0.5 h, 1.0 h, 3.0 h and 5.0 h.

The morphologies of the samples were obtained via SEM observation (Figure 3). As shown in Figure 3A, ZnS had a plate-like morphology. Meanwhile, some tiny nanoparticles and pores in nanoscale were observed on ZnS nanoplates for ZnS/ZnO-10 min (Figure 3B). As confirmed by XRD analysis, these nanoparticles should be ZnO. Furthermore, along with the increasing time of heat treatment for the sample of ZnS/ZnO-30 min (Figure 3C), ZnS/ZnO-1 h (Figure 3D) and ZnS/ZnO-3 h (Figure 3E), the amount and the size of the ZnO nanoparticles on the ZnS surface increased and grew up. As for the sample of ZnO (Figure 3F), only nanoparticles with several tens of nanometer were seen. Furthermore, based on the XRD diffraction dates at 2θ = 36.20 (Figure 2), the ZnO particle sizes in ZnS/ZnO compositions were calculated via the Debye–Scherrer Formula of $D = K\gamma/B\cos\theta$. The obtained ZnO particle sizes in the samples of ZnS/ZnO-10 min, ZnS/ZnO-0.5 h, ZnS/ZnO-1 h, ZnS/ZnO-3h and ZnO were 13.9, 16.4, 18.1, 23.2 and 24.3 nm, respectively. The morphology and structure of ZnS/ZnO composition, and ZnS/ZnO-0.5 h as a representative sample, was further confirmed by TEM, HAADF-STEM mapping and HRTEM, (shown in Figure 4. TEM image in Figure 4A demonstrated that ZnS/ZnO-0.5 h had a nanoporous plate-like morphology with nanoparticles anchored on. The inset in Figure 4A shows that the surface nanoparticle had a size mainly ranging from 10–70 nm. The HAADF-STEM and mapping images in Figure 4B verified that Zn, S, O, C and N were uniformly dispersed on ZnS/ZnO-0.5 h. The elements of C and N should originate from the residual carbon and the N-doping in composition, because of the decomposition of ethanediamine (en) in the precursor of ZnS(en)$_{0.5}$ under calcination [7]. The existence of O might indicate that partial ZnS was transformed into ZnO. Figure 4C is the HRTEM image of the surface nanoparticle in site 1 in Figure 3A. It shows the nanoparticle-owned clear lattice fringes with a distance of 0.28 nm, which was assigned to the (100) plane of ZnO [22]. It further confirmed ZnS that was transformed into ZnO. Figure 4D shows the HRTEM image of the nanoplate's counterpart in site 2 in Figure 4A. As shown, it also exhibited distinct lattice fringes with a distance of 0.33 nm, corresponding to the (100) plane of hexagonal ZnS [23]. Additionally, some defects were found, possibly arising from the N-doping, thus leading to the local disorder of ZnS.

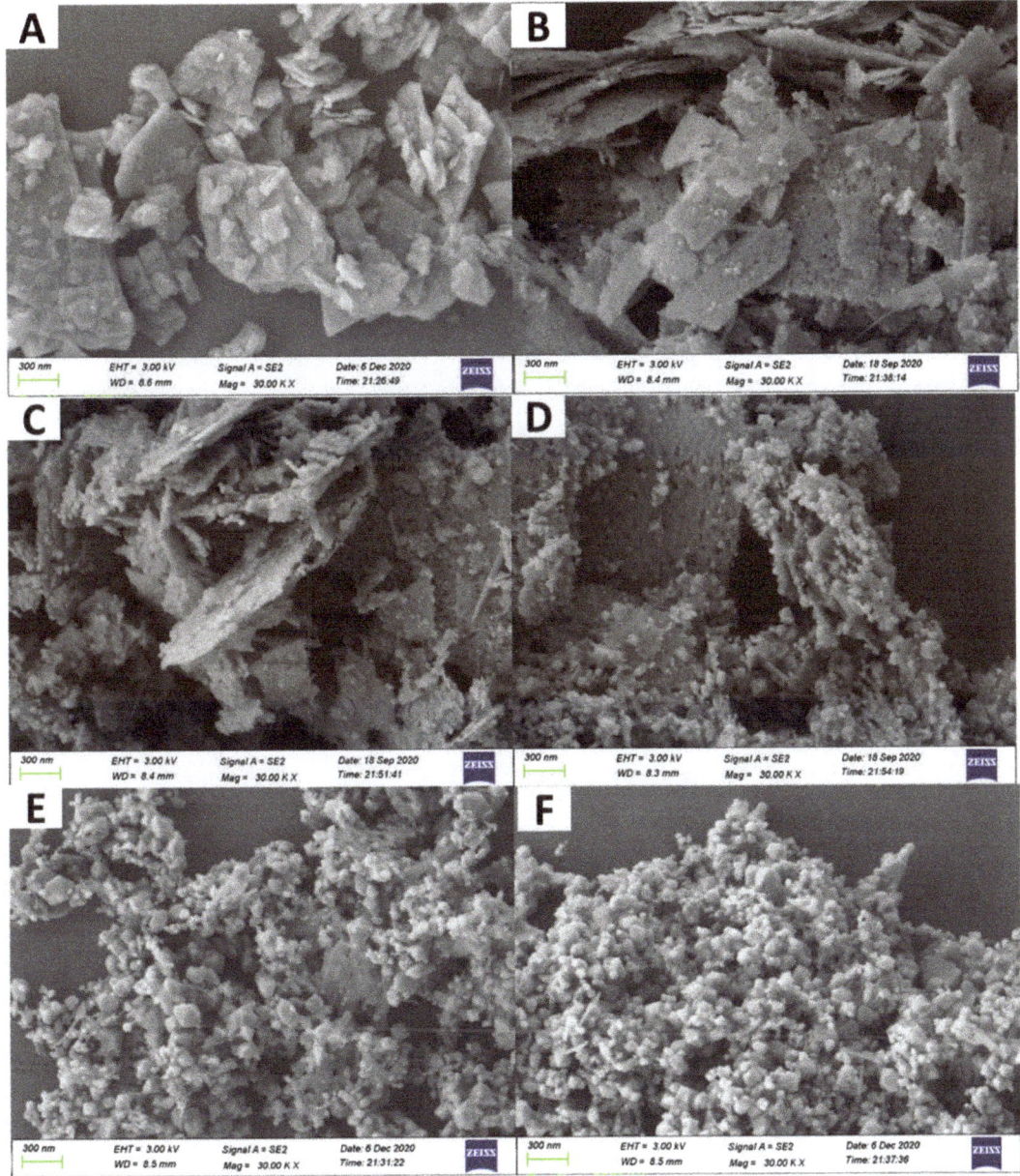

Figure 3. SEM images of the as-prepared samples, (**A**) ZnS, (**B**) ZnS/ZnO-10 min, (**C**) ZnS/ZnO-0.5 h, (**D**) ZnS/ZnO-1 h, (**E**) ZnS/ZnO-3 h, (**F**) ZnO.

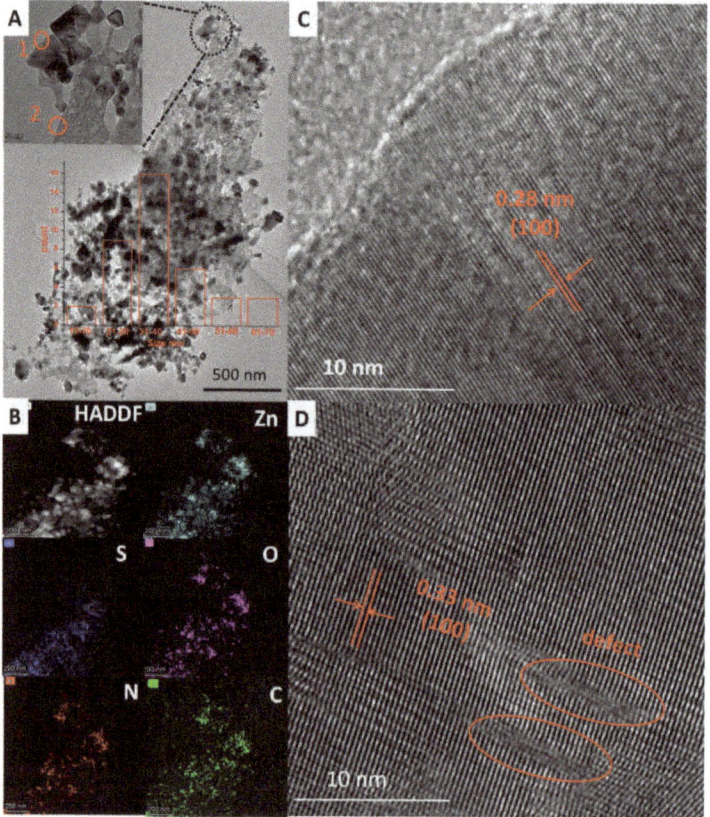

Figure 4. (**A**) TEM, (**B**) HAADF-STEM and Mapping images of ZnS/ZnO-0.5 h, (**C,D**) HRTEM images in site 1 and site 2 in A, respectively. Inset in A is the amount of the nanoparticles in different sizes.

2.2. Analysis of Components, BET Surface Area and Energy Band Structure

To further confirm the surface elemental components and valence states of samples, X-ray photoemission spectroscopy (XPS) was carried out. As shown in Figure 5A, the Zn $2p_{3/2}$ and Zn $2p_{1/2}$ of ZnO, ZnS/ZnO-0.5 h, and ZnS located at 1021.48 eV and 1044.58 eV, 1021.78 eV, 1044.88 eV, 1021.88 eV, and 1044.98 eV, respectively, corresponding to the binding energy of Zn^{2+} [24]. The XPS spectra of S^{2-} 2p (Figure 5B) showed that the binding energy of S $2p_{3/2}$ and $2p_{1/2}$ for ZnS and ZnS/ZnO-0.5 h were around 161.78 eV and 162.98 eV. No signal of S 2p was detected in ZnO, demonstrating that ZnS was completely transformed into ZnO, thus matching the analysis results of the XRD diffraction. Figure 5C shows O 1s spectra. The binding energies of O 1s, located at about 530.18 eV and 531.78 eV, were assigned to O^{2-} of Zn-O and surface O-H [16], respectively. Moreover, the binding energy of O 1s of Zn-O (529.98 eV) in ZnS/ZnO reduced by 0.2 eV, compared with that of Zn-O in pure ZnO. The negative shift of binding energy of O 1s in ZnS/ZnO was attributed to the formation of a heterojunction between ZnS and ZnO [19], which resulted in a transfer of electron density from ZnS to ZnO, thus leading to an enhanced electron density of the ZnO surface [25]. Furthermore, as shown in Figure 5B and C, during the transformation of ZnS→ZnO, the content of O and S was increased and decreased, respectively. Based on the calculation of the peak area of S 2p of Zn-S and O 1s of Zn-O, the ratio of S/O was 11.3, meaning that the weight percentage of ZnO in the sample of ZnS/ZnO-0.5 h was about 6.9 wt%. Furthermore, Figure 5D shows that ZnS, ZnS/ZnO, and ZnO all exhibited the N

1s spectrum located at 399.58 eV, 399.78 eV and 400.38 eV, respectively, due to N-doping, which was the origination of the visible-light photocatalytic activity of these photocatalysts. Noticeably, the binding energy of N 1s gradually had a positive shift. In terms of ZnS, the N 1s arose from the Zn-N bond. During ZnS→ZnO, the oxidation of Zn-N resulted in the increasing binding energy of N, benefiting the generation of ZnO derived from ZnS directly.

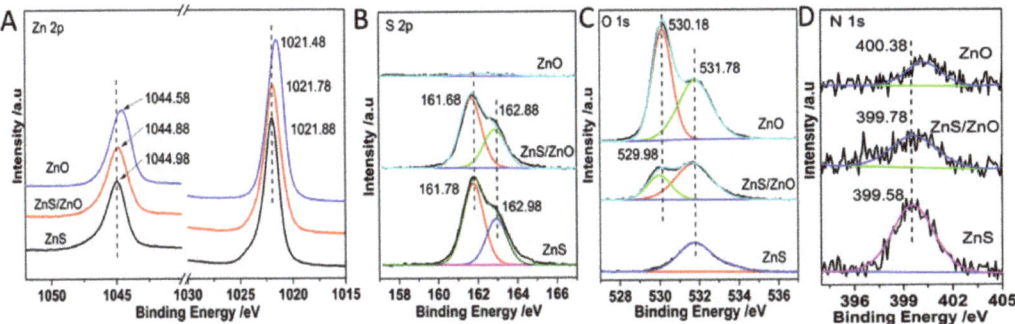

Figure 5. The XPS spectra for the samples. (**A**) Zn 2p, (**B**) S 2p, (**C**) O 1s and (**D**) N 1s. ZnS/ZnO represented the sample of ZnS/ZnO-0.5 h.

Figure 6 shows the specific surface areas and pore size distributions of ZnS, ZnS/ZnO-0.5 h and ZnO, respectively, which were obtained via Brunauer–Emmett–Tell (BET) N_2 adsorption–desorption isothermal measurements. As shown in Figure 6A, the three samples all had a typical IV adsorption isotherm, proving the existence of mesopores. The BET surface areas were 88.2, 33.9 and 14.2 m^2/g for ZnS, ZnS/ZnO-0.5 h and ZnO, respectively. Along with the transformation from ZnS to ZnO, a decrease in surface areas was noticed, which should arise from the destruction of the nanoporous plate-like structure of ZnS and the block effect of newborn ZnO nanoparticles. The pore size distribution curves in Figure 6B show that the average pore sizes and the pore volumes were enlarged and decreased, respectively, which further confirmed the collapse of the mesoporous structure, consistent with the SEM and TEM observations discussed above.

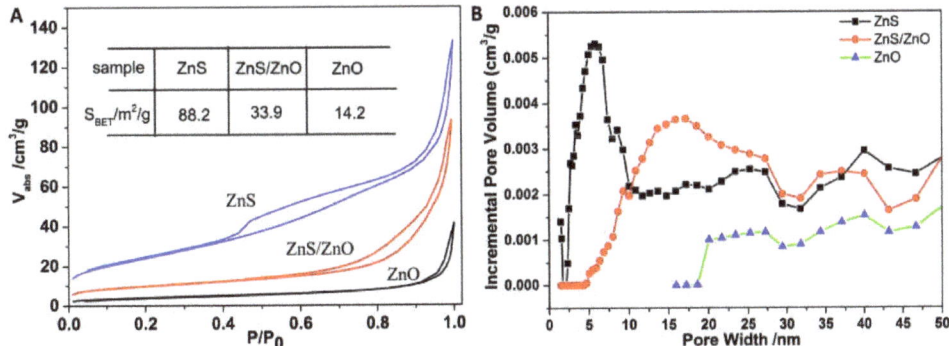

Figure 6. (**A**) BET absorption at different P/P_0 and (**B**) pore size distribution curves of ZnS, ZnS/ZnO and ZnO. ZnS/ZnO represented the sample of ZnS/ZnO-0.5 h.

Figure 7A shows the optical absorption property of the samples. As shown, ZnS, ZnS/ZnO-0.5 h and ZnO displayed a band-edge absorption (λ_{abs}) around 398, 412 and 427 nm, corresponding to the band gap (E_g) of 3.16, 3.01 and 2.90 eV estimated via the empirical equation of $1240/\lambda_{abs}$ [26], respectively. Noticeably, the three samples had an obvious tailing absorption in the visible region (400–700 nm), indicating that the samples were visible-light response photocatalysts. However, with the transformation from ZnS to ZnO, the samples displayed a decay of spectral absorption in the visible region, which was attributed to the decreased content of doping N atoms. Furthermore, the CB edge position of ZnS and ZnO was evaluated by a Mott–Schottky plots test (Figure 7B). As shown, the conduction band edge (E_{CB}) of ZnS and ZnO located at -1.01 V and -0.94 V (vs. SCE, pH = 7), respectively, were higher than the H_2 evolution potential (-0.66 V, vs. SCE, pH = 7), meaning that the photocatalysts were powerful enough for the photocatalytic reduction of H_2O for H_2 evolution. Additionally, based on $E_g = E_{VB} - E_{CB}$ [27], the valence band edge (E_{VB}) of ZnS and ZnO is located at the position of 2.15 V and 1.96 V. Based on the analysis of energy band edge, the composition of ZnS/ZnO-0.5 h might form the type I heterojunction [8], which could promote the separation of a photogenerated carrier.

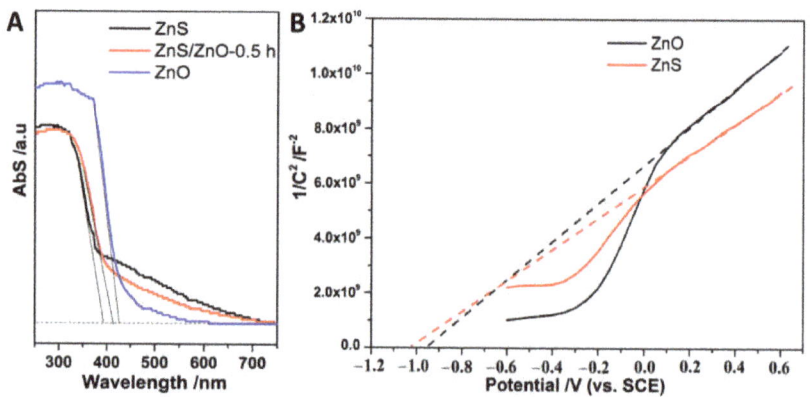

Figure 7. (**A**) The Uv-vis Abs spectra of ZnS, ZnS/ZnO-0.5 h and ZnO; (**B**) Mott–Schottky plots of ZnO and ZnS.

2.3. Photocatalytic Activity and Stability

Figure 8A,B shows the time course of H_2 evolution activity and the corresponding H_2 evolution rates of the as-prepared samples. As shown, the hydrogen evolution activity exhibited a volcano-like variation with the increasing content of ZnO in photocatalysts. The average rates of H_2 evolution for pure ZnS and ZnO were only 255 and 145 µmol/h/g, far lower than that of the ZnS/ZnO composites, 1600, 1790, 1605 and 670 µmol/h/g for ZnS/ZnO-10 min, ZnS/ZnO-0.5 h, ZnS/ZnO-1 h and ZnS/ZnO-3 h, respectively. These results demonstrated that ZnS/ZnO composites had advantages in photocatalytic H_2 evolution, compared with single ZnS or ZnO, and should be ascribed to an efficient separation of photogenerated carriers because of the formation of a heterojunction and type I energy band edge alignment between a ZnS and ZnO component. Figure 8C shows the stability of photocatalytic hydrogen evolution over ZnS/ZnO-0.5 h. As shown, the hydrogen evolution activity decreased by about 15% after 4 recycle tests. The decay of activity was ascribed to the destruction of the interface structure of the photocatalyst, which was confirmed by the XRD diffraction patterns of ZnS/ZnO-0.5 h after recycle photocatalytic tests. As shown in Figure 8D, compared with the XRD diffraction patterns of ZnS/ZnO-0.5 h before the reaction, the diffraction peaks at 31.7°, 34.3° and 36.2° assigned to the diffraction peaks of ZnO became weaker, because ZnO was transformed into ZnS under the condition of SO_3^{2-}/S^{2-} as the sacrificial reagent. With the transformation of ZnO to ZnS, the heterostructure of ZnS/ZnO suffered from a damage, leading to the decay of photocatalytic hydrogen evolution activity, which further demonstrated the importance of the heterojunction for separation of carrier and enhanced photocatalytic activity. Moreover, as shown in Table 1, the N-doped ZnS/ZnO heterojunction photocatalyst in this work displayed a competitive activity for H_2 evolution under visible light, as compared with the reported ZnS/ZnO compositions.

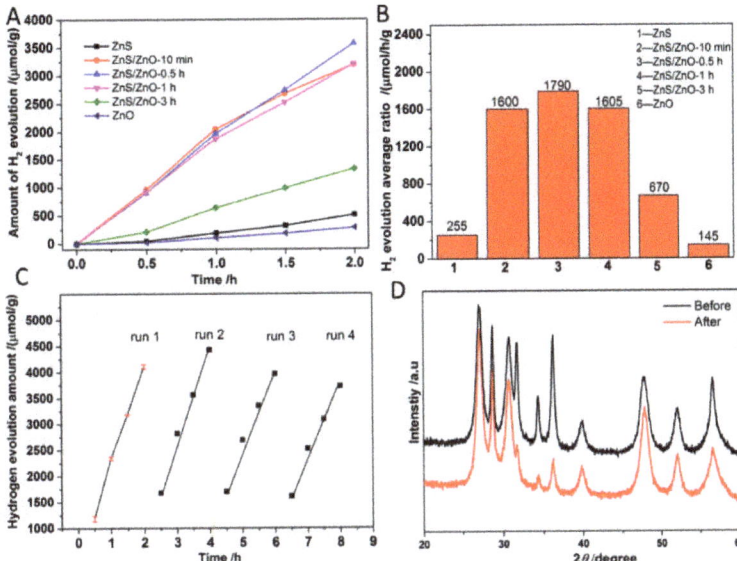

Figure 8. (**A**) and (**B**) Time course of H_2 evolution activity and the corresponding average rates of H_2 evolution over the as-prepared samples, (**C**) recycle tests for photocatalytic reaction over ZnS/ZnO-0.5 h, (**D**) the XRD diffraction patterns before and after photocatalytic reaction.

Table 1. Comparison of the photocatalytic H_2 evolution activities of ZnS/ZnO.

Photocatalyst	Sacrificial Agent in Aqueous Solution	H_2 Evolution Activity /$\mu mol \cdot g^{-1} \cdot h^{-1}$	Light/nm	Reference
N-doped ZnS/ZnO-Pt%	0.1 M S^{2-}/0.1 M SO_3^{2-}	1790	$\lambda > 400$	This work
ZnS/ZnO	CH_3OH 50 % v/v	1242	254	[15]
ZnS/ZnO@CT	5% lactic acid	37.1	400–780	[28]
ZnS@ZnO	0.1 M S^{2-}/0.1 M SO_3^{2-}	≈4600	Xenon lamp	[18]
ZnS-ZnO	0.25 M S^{2-}/0.35 M SO_3^{2-}	22	$\lambda > 420$	[29]
Pt/ZnS-ZnO	0.1 M S^{2-}/0.1 M SO_3^{2-}	10,700	Xenon lamp	[19]
ZnS/ZnO	0.45 M S^{2-}/0.45 M SO_3^{2-}	374	$\lambda > 400$	[30]
ZnS/ZnO	0.1 M S^{2-}/0.1 M SO_3^{2-}	≈250	$\lambda > 420$	[31]
ZnS-ZnO	0.4 M S^{2-}	494.8	Xenon lamp	[32]
Pt/ZnS@ZnO	Water	87.6	Xenon lamp	[33]
ZnO/ZnS	0.1 M S^{2-}/0.1 M SO_3^{2-}	415.3	$\lambda > 420$	[20]

3. Materials and Methods

3.1. Materials

The $H_2PtCl_6 \cdot 6H_2O$, (A. R., Sinopharm Chemical Reagent Co. (SCRC, Shanghai, China), ethanol (EtOH, A. R., SCRC), deionized water (home-made), ethanediamine (en, A. R., SCRC), thiourea (A. R., SCRC), $ZnCl_2$ (A. R., SCRC.). N,N-dimethylformamide (DMF, A. R., SCRC), $Na_2S \cdot H_2O$ and Na_2SO_3 (A. R., SCRC.).

3.2. Preparation of Photocatalysts

Synthesis of ZnS(en)$_{0.5}$: 0.191 g of $ZnCl_2$, 0.64 g of thiourea and 50 mL of ethanediamine were added into a 100 mL teflon-lined autoclave, and the mixture was stirred for 1 h.

Subsequently, the autoclave was sealed and maintained at 160 °C for 12 h and naturally air cooled. The resulting white solid products were centrifuged, washed with absolute ethanol and distilled water several times, and then dried at 40 °C overnight.

Synthesis of ZnS/ZnO composition: The as-prepared 40 mg of ZnS(en)$_{0.5}$ was spread into a combustion boat with a capacity of 5 mL. Then, the sample was rapidly put into a muffle furnace with a temperature of 500 °C for some time (t) and taken out immediately. The samples, calcinated at different times (t = 5 min, 10 min, 0.5 h, 1 h, 3 h and 5 h), were respectively defined as ZnS, ZnS/ZnO-5 min, ZnS/ZnO-10 min, ZnS/ZnO-0.5 h, ZnS/ZnO-1 h, ZnS/ZnO-3h and ZnO.

3.3. Characterization

Structure and Morphology: XRD patterns were recorded on a X'Pert3 Powder (PANalytical, Almelo, Netherlands) X-ray diffractometer with Cu Kα radiation operated at 40 kV and 40 mA. To obtain the transmission electron microscopy (TEM) images, high-resolution (HR) TEM images and STEM-EDX mapping, the samples were dropped on a Mo grid and operated on a Talos F200S (Thermo, Waltham, USA). X-ray photoelectron spectroscopy (XPS) measurements were performed on a ThermoFischer system (Thermo, Waltham, USA) with a monochromatic Al Kα source. XPS dates were calibrated by C1s = 284.8 eV. Ultraviolet-visible absorption spectra was obtained using UV-2600 (Shimadzu, Tyoto, Japan), BaSO4 as the reference. Field-emission scanning electron microscopy (FESEM, Carl Zeiss Sigma 300, Oberkochen, Germany) was used to determine the morphology of the samples. The Brunauer–Emmett–Teller (BET) surface area was measured with an TriStar II Plus apparatus (Micromeritics Instrument Corp, Atlanta, USA).

Electrochemical measurements: The working electrode was prepared on fluorinedoped tin oxide (FTO) glass, which was cleaned by sonication in acetone, ethanol and deionized water for 30 min each. Next, 5 mg of photocatalyst powder was dispersed in 0.5 mL of dimethylformamide (DMF) under sonication for 2 h to produce slurry. Then, 10 μL of the as-prepared slurry was spread onto the conductive surface of the FTO glass to form a photocatalyst film with an area of 0.25 cm^2. After air drying naturally, the uncoated parts of the electrode were isolated with an epoxy resin. Subsequently, the electrodes were put into an oven at 100 °C for 2 h. For the Mott–Schottky experiment, the potential ranged from −0.6 to 0.6 V with an increase in voltage of 50 mV, and the amplitude was 5 mV under the frequency of 500 Hz. The measurement was also performed in a conventional three electrode cell, using a Pt wire and a SCE electrode as the counter electrode and reference electrode, respectively. The electrolyte was 0.2 M of Na_2SO_4 aqueous solution without additive (pH = 6.8).

Photocatalytic tests: The photocatalytic reactions were carried out in a photocatalytic hydrogen evolution system (MC-H20II, Merry Change Co., Beijing, China). For this, 20 mg photocatalyst was suspended in 50 mL of (0.1 M) Na_2SO_3/(0.1 M) Na_2S aqueous solution. Next, 1%Pt was introduced into the reaction system via in situ photodeposition of H_2PtCl_6. The suspension was then thoroughly degassed and irradiated with visible light (λ > 400 nm) by using a 300 W Xenon lamp (PLS-SXE300D, Perfectlight Co., Beijing, China). H_2 was detected at set intervals, automatically, by an online gas chromatograph. For photocatalytic recycle tests over ZnS/ZnO-0.5 h, three parallel photocatalytic H_2 evolution tests were firstly performed. Then, the photocatalyst was recovered after every parallel test. The recovered photocatalyst was used for next run test.

4. Conclusions

N-doped ZnS/ZnO composite with heterostructure was prepared successfully by the in situ transformation of ZnS(en)$_{0.5}$ with heat treatment under air atmosphere. The content of ZnO in ZnS/ZnO was able to be controlled via adjustment of the calcination time. SEM demonstrated ZnO nanoparticles were dispersed on the ZnS surface. TEM further verified that the surface ZnO nanoparticles had sizes ranging from 10 to70 nm and anchored on porous ZnS nanoplate firmly. XPS verified that the N was doped into ZnS/ZnO during the in situ transformation of ZnS(en)$_{0.5}$ to ZnS/ZnO, and a heterojunction was formed between ZnS and ZnO. Photocatalytic hydrogen evolution experiments demonstrated that the sample of ZnS/ZnO-0.5 h with 6.9 wt% of ZnO had the best H$_2$ evolution activity (1790 μmol/h/g) under visible light irradiation ($\lambda > 400$ nm), about 7.0 and 12.3 times of that of the pure ZnS and ZnO, respectively. The enhanced activities of the ZnS/ZnO composites were ascribed to the intimated hetero-interface between components and efficient transfer of photo-generated electrons from ZnS to ZnO. However, although the N-doped ZnS/ZnO obtained via the in situ transformation method achieved visible-light photocatalytic H$_2$ evolution activity, the catalytic stability and absolute H$_2$ evolution activity of ZnS/ZnO should be improved further for meeting the potential demand of hydrogen energy. Furthermore, the as-prepared N-doped ZnS/ZnO could be used for the photoanode material of dye sensitized solar cells because of the existence of the heterojunction with staggered conduction band edges and nanoporous structure, which availed the transmission of photoelectrons and absorption of dye molecules, thus improving the photoelectric conversion efficiency.

Author Contributions: Conceptualization, validation, project administration, writing-review, funding acquisition, J.X.; investigation, data curation and methodology, X.W. and J.W.; formal analysis, visualization, J.H. and Z.L., funding acquisition, editing and supervision, J.F. All authors have read and agreed to the published version of the manuscript.

Funding: This research was funded by National Natural Science Foundation of China (21802063) and Science and Technology Planning Project of Longyan (2019LYF13005).

Institutional Review Board Statement: Not applicable.

Informed Consent Statement: Not applicable.

Data Availability Statement: The data presented in this study are available on request from the corresponding author.

Conflicts of Interest: The authors declare no conflict of interest.

Sample Availability: Samples of the compounds are not available from the authors.

References

1. Liu, Y.; Mao, J.; Huang, Y.; Qian, Q.; Luo, Y.; Xue, H.; Yang, S. Pt-chitosan-TiO$_2$ for efficient photocatalytic hydrogen evolution via ligand-to-metal charge transfer mechanism under visible light. *Molecules* **2022**, *27*, 4673. [CrossRef]
2. Sultanov, F.; Daulbayev, C.; Azat, S.; Kuterbekov, K.; Bekmyrza, K.; Bakbolat, B.; Bigaj, M.; Mansurov, Z. Influence of metal oxide particles on bandgap of 1D photocatalysts based on SrTiO$_3$/PAN fibers. *Nanomaterials* **2020**, *10*, 1734. [CrossRef] [PubMed]
3. Daulbayev, C.; Sultanov, F.; Korobeinyk, A.V.; Yeleuov, M.; Azat, S.; Bakbolat, B.; Umirzakov, A.; Mansurov, Z. Bio-waste-derived few-layered graphene/ SrTiO$_3$/PAN as efficient photocatalytic system for water splitting. *Appl. Surf. Sci.* **2021**, *549*, 149176. [CrossRef]
4. Wang, Y.; Wang, H.; Li, Y.; Zhang, M.; Zheng, Y. Designing a 0D/1D s-scheme heterojunction of cadmium selenide and polymeric carbon nitride for photocatalytic water splitting and carbon dioxide reduction. *Molecules* **2022**, *27*, 6286. [CrossRef] [PubMed]
5. Li, X.; Yu, J.; Low, J.; Fang, Y.; Xiao, J.; Chen, X. Engineering heterogeneous semiconductors for solar water splitting. *J. Mater. Chem. A* **2015**, *3*, 2485–2534.
6. Lee, G.-J.; Wu, J.J. Recent developments in ZnS photocatalysts from synthesis to photocatalytic applications—A review. *Powder Technol.* **2017**, *318*, 8–22. [CrossRef]
7. Xiong, J.; Li, Y.; Lu, S.; Guo, W.; Zou, J.; Fang, Z. Controllable sulphur vacancies confined in nanoporous zns nanoplates for visible-light photocatalytic hydrogen evolution. *Chem. Commun.* **2021**, *57*, 8186–8189. [CrossRef] [PubMed]
8. Low, J.; Yu, J.; Jaroniec, M.; Wageh, S.; Al-Ghamdi, A.A. Heterojunction photocatalysts. *Adv. Mater.* **2017**, *29*, 1601694. [CrossRef] [PubMed]
9. Di Liberto, G.; Cipriano, L.A.; Tosoni, S.; Pacchioni, G. Rational design of semiconductor heterojunctions for photocatalysis. *Chem. Eur. J.* **2021**, *27*, 13306–13317. [CrossRef]
10. Yang, H. A short review on heterojunction photocatalysts: Carrier transfer behavior and photocatalytic mechanisms. *Mater. Res. Bull.* **2021**, *142*, 111406. [CrossRef]
11. Bai, S.; Xiong, Y. Some recent developments in surface and interface design for photocatalytic and electrocatalytic hybrid structures. *Chem. Commun.* **2015**, *51*, 10261–10271. [CrossRef] [PubMed]

12. Ong, C.B.; Ng, L.Y.; Mohammad, A.W. A review of zno nanoparticles as solar photocatalysts: Synthesis, mechanisms and applications. *Renew. Sustain. Energy Rev.* **2018**, *81*, 536–551. [CrossRef]
13. Wang, Y.; Wang, Q.; Zhan, X.; Wang, F.; Safdar, M.; He, J. Visible light driven type ii heterostructures and their enhanced photocatalysis properties: A review. *Nanoscale* **2013**, *5*, 8326–8339. [CrossRef] [PubMed]
14. Marschall, R. Semiconductor composites: Strategies for enhancing charge carrier separation to improve photocatalytic activity. *Adv. Funct. Mater.* **2014**, *24*, 2421–2440. [CrossRef]
15. Piña-Pérez, Y.; Aguilar-Martínez, O.; Acevedo-Peña, P.; Santolalla-Vargas, C.E.; Oros-Ruíz, S.; Galindo-Hernández, F.; Gómez, R.; Tzompantzi, F. Novel zns-zno composite synthesized by the solvothermal method through the partial sulfidation of zno for h2 production without sacrificial agent. *Appl. Catal. B Environ.* **2018**, *230*, 125–134. [CrossRef]
16. Li, X.; Li, X.; Zhu, B.; Wang, J.; Lan, H.; Chen, X. Synthesis of porous zns, zno and zns/zno nanosheets and their photocatalytic properties. *RSC Adv.* **2017**, *7*, 30956–30962. [CrossRef]
17. Guo, P.; Jiang, J.; Shen, S.; Guo, L. Zns/zno heterojunction as photoelectrode: Type ii band alignment towards enhanced photoelectrochemical performance. *Int. J. Hydrog. Energy* **2013**, *38*, 13097–13103. [CrossRef]
18. Luan, Q.; Chen, Q.; Zheng, J.; Guan, R.; Fang, Y.; Hu, X. Construction of 2d-zns@zno z-scheme heterostructured nanosheets with a highly ordered zno core and disordered zns shell for enhancing photocatalytic hydrogen evolution. *ChemNanoMat* **2020**, *6*, 470–479. [CrossRef]
19. Wang, X.; Cao, Z.; Zhang, Y.; Xu, H.; Cao, S.; Zhang, R. All-solid-state z-scheme pt/zns-zno heterostructure sheets for photocatalytic simultaneous evolution of H_2 and O_2. *Chem. Eng. J.* **2020**, *385*, 123782. [CrossRef]
20. Zhao, X.; Feng, J.; Liu, J.; Lu, J.; Shi, W.; Yang, G.; Wang, G.; Feng, P.; Cheng, P. Metal–organic framework-derived zno/zns heteronanostructures for efficient visible-light-driven photocatalytic hydrogen production. *Adv. Sci.* **2018**, *5*, 1700590. [CrossRef] [PubMed]
21. Ouyang, X.; Tsai, T.-Y.; Chen, D.-H.; Huang, Q.-J.; Cheng, W.-H.; Clearfield, A. Ab initio structure study from in-house powder diffraction of a novel zns(en)$_{0.5}$ structure with layered wurtzite zns fragment. *Chem. Commun.* **2003**, 886–887. [CrossRef] [PubMed]
22. Zeng, W.; Ren, Y.; Zheng, Y.; Pan, A.; Zhu, T. In-situ copper doping with zno/zns heterostructures to promote interfacial photocatalysis of microsized particles. *ChemCatChem* **2021**, *13*, 564–573. [CrossRef]
23. Fang, Z.; Weng, S.; Ye, X.; Feng, W.; Zheng, Z.; Lu, M.; Lin, S.; Fu, X.; Liu, P. Defect engineering and phase junction architecture of wide-bandgap zns for conflicting visible light activity in photocatalytic h2 evolution. *ACS Appl. Mater. Inter.* **2015**, *7*, 13915–13924. [CrossRef]
24. Lonkar, S.P.; Pillai, V.V.; Alhassan, S.M. Facile and scalable production of heterostructured zns-zno/graphene nano-photocatalysts for environmental remediation. *Sci. Rep.* **2018**, *8*, 13401. [CrossRef] [PubMed]
25. Fan, J.; Xiong, J.; Liu, D.; Tang, Y.; He, S.; Hu, Z. A cathodic photocorrosion-assisted strategy to construct a cds/pt heterojunction photocatalyst for enhanced photocatalytic hydrogen evolution. *New J. Chem.* **2021**, *45*, 10315–10324. [CrossRef]
26. Xiong, J.; Wen, L.; Jiang, F.; Liu, Y.; Liang, S.; Wu, L. Ultrathin hnb 3 o 8 nanosheet: An efficient photocatalyst for the hydrogen production. *J. Mater. Chem. A* **2015**, *3*, 20627–20632. [CrossRef]
27. Yu, K.; Huang, H.-B.; Zeng, X.-Y.; Xu, J.-Y.; Yu, X.-T.; Liu, H.-X.; Cao, H.-L.; Lü, J.; Cao, R. Cdzns nanorods with rich sulphur vacancies for highly efficient photocatalytic hydrogen production. *Chem. Commun.* **2020**, *56*, 7765–7768. [CrossRef]
28. Huang, H.-B.; Yu, K.; Wang, J.-T.; Zhou, J.-R.; Li, H.-F.; Lü, J.; Cao, R. Controlled growth of zns/zno heterojunctions on porous biomass carbons via one-step carbothermal reduction enables visible-light-driven photocatalytic H_2 production. *Inorg. Chem. Front.* **2019**, *6*, 2035–2042. [CrossRef]
29. Wang, Z.; Cao, S.-W.; Loo, S.C.J.; Xue, C. Nanoparticle heterojunctions in zns–zno hybrid nanowires for visible-light-driven photocatalytic hydrogen generation. *CrystEngComm* **2013**, *15*, 5688–5693. [CrossRef]
30. Zhao, H.; Dong, Y.; Jiang, P.; Wu, X.; Wu, R.; Chen, Y. Facile preparation of a zns/zno nanocomposite for robust sunlight photocatalytic H_2 evolution from water. *RSC Adv.* **2015**, *5*, 6494–6500. [CrossRef]
31. Zhou, Q.; Li, L.; Xin, Z.; Yu, Y.; Wang, L.; Zhang, W. Visible light response and heterostructure of composite cds@ zns–zno to enhance its photocatalytic activity. *J. Alloy. Compd.* **2020**, *813*, 152190. [CrossRef]
32. Hong, E.; Kim, J.H. Oxide content optimized ZnS–ZnO heterostructures via facile thermal treatment process for enhanced photocatalytic hydrogen production. *Int. J. Hydrog. Energy* **2014**, *39*, 9985–9993. [CrossRef]
33. Ji, X.; Xu, H.; Liang, S.; Gan, L.; Zhang, R.; Wang, X. 3D ordered macroporous Pt/ZnS@ZnO core-shell heterostructure for highly effective photocatalytic hydrogen evolution. *Int. J. Hydrog. Energy* **2022**, *47*, 17640–17649. [CrossRef]

M-Carboxylic Acid Induced Formation of New Coordination Polymers for Efficient Photocatalytic Degradation of Ciprofloxacin

Jian Li [1,2], Xiaojia Wang [1] and Yunyin Niu [1,*]

1 Green Catalysis Center, and College of Chemistry, Zhengzhou University, Zhengzhou 450001, China
2 College of Ecology and Environment, Zhengzhou University, Zhengzhou 450001, China
* Correspondence: niuyy@zzu.edu.cn

Abstract: Four new 2–3D materials were designed and synthesized by hydrothermal methods, namely, {[(L1·Cu·2H$_2$O) (4,4-bipy)$_{0.5}$] (β-Mo$_8$O$_{26}$)$_{0.5}$·H$_2$O} (**1**), {[(L1·Cu)$_2$·(4,4-bipy)] (Mo$_5$O$_{16}$)} (**2**), {Co(L1)$_2$}$_n$ (**3**), and {[(L1)$_2$][β-Mo$_8$O$_{26}$]$_{0.5}$·5H$_2$O} (**4**). [L1 = 5-(4-aminopyridine) isophthalic acid]. The degradation of ciprofloxacin (CIP) in water by compounds **1–4** was studied under visible light. The experimental results show that compounds **1–4** have obvious photocatalytic degradation effect on CIP. In addition, for compound **1**, the effects of temperature, pH, and adsorbent dosage on photocatalytic performance were also investigated. The stability of compound **1** was observed by a cycle experiment, indicating that there was no significant change after three cycles of CIP degradation.

Keywords: ciprofloxacin; coordination polymers; hydrothermal method; photocatalytic degradation

1. Introduction

With the development of medical level, antibiotics have made great contributions to the prevention and treatment of diseases. Antibiotics can enter aquatic environments through various ways. The increase of their use and the water solubility, stability, and volatility made them present a "lasting" state in aquatic environments, resulting in a series of environments problems. Antibiotic wastewater is not facile to degrade, and its high biological toxicity makes it a ticklish problem in the field of sewage treatment. In recent years, high concentrations of antibiotics have been detected in many countries and regions.

Ciprofloxacin (CIP) antibiotics (Figure 1) are widely used in human and veterinary medicine because of their strong bactericidal ability, and lower toxicity and side effects [1,2]. However, the extensive use of CIP has also caused environmental pollution and posed a threat to the ecosystem and to human health. The removal of CIP from environment has become a mandatory issue already [3,4]. Unfortunately, the conventional chemical and physical methods lack enough efficiency for its removal and the new biochemical treatment method often produces byproducts that are more harmful.

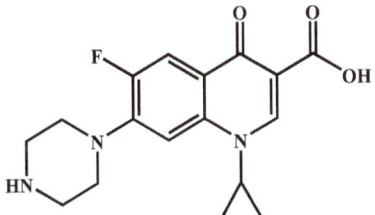

Figure 1. The atomic structure of CIP.

Nowadays, the migration behavior and degradation mode of antibiotics in aquatic environments have become a research hotspot. Photochemical degradation is an important way for the migration and transformation of antibiotics in aquatic environments [5], which has a significant impact on the environmental toxicological effects of these substances. In recent years, photocatalytic technology has been widely used in water pollution control, which shows the important application prospect and potential of photocatalytic technology [6,7]. The photocatalytic degradation processes provided an ideal technique for the transportation and degradation of CIP [8–10]. Compared with the methods mentioned above, using photocatalytic technology can degrade CIP efficiently without secondary pollution.

As a new type of functional molecular materials, coordination polymers are formed by the self-assembly process of organic ligands and metal ions. In the relevant research literature [11–16], they are also called porous coordination polymers (PCP) or organic-inorganic hybrid materials. Coordination polymers materials developed rapidly and became a research hotspot in the fields of chemistry [5], environment [17], materials [18], etc.. The diversity of metal ions and organic ligands, as well as the different coordination modes between them [13,19], determine the structural diversity of these materials. The synthetic fibers of coordination polymers, synthetic methods, the specific configuration formed by the inherent ligands between organic ligands and metal ions lead to the diversity of the functions. The unique structure and function also show broad application prospects in the selective adsorption and catalytic degradation of toxic and harmful substances [20]. Due to the advantages of high specific surface area, self-assembled structures, and evenly distributed active sites, coordination polymer materials have great potential in the fields of catalyst preparation, adsorption, gas storage, and photocatalytic degradation [11,12,14,16,21]. In the published articles on the photocatalytic degradation of ciprofloxacin, most of the catalysts are doped or loaded modified materials [8–10]. Therefore, we imagine whether materials with excellent degradation performance can be directly obtained through a simple one-step reaction.

We have been working on the exploration of organic-inorganic hybrid materials for a long time [5,13,22,23]. Carboxylic acid ligands usually have good metal ion binding ability. Carboxylate is a hard base, which can form strong coordination bonds with various common metal ions. Moreover, carboxylate has negative charges, which can neutralize the positive charges of metal ions and metal clusters, so that the pores of porous complexes do not need to contain counter anions, which is conducive to improving the stability of the structure. Based on this, we chose the unreported ligand 5-(4-aminopyridine) isophthalic acid (Scheme S1) to react with metal salts by hydrothermal method and obtained four new novel organic-inorganic hybrid materials **1–4**. It is worth mentioning that compounds **1–4** are directly obtained by hydrothermal reaction, without further doping and modification. Compounds **1–4** were systematically characterized by infrared spectroscopy (IR), elemental analysis, and powder X-ray diffraction (PXRD). Under simulated light, compounds **1–4** have a good degradation effect on CIP and show a good photocatalytic performance. In addition, the effects of temperature, pH, and catalyst dosage on the degradation of CIP by compound **1** were investigated. Cyclic experiments show the workability of the photocatalyst.

2. Results

2.1. Crystal Structure

X-ray single-crystal structural analysis indicated that compound **1** crystallized in triclinic system (space group P-1). As shown in Figure 2a, its asymmetric structural unit is composed of an L1 ligand, a Cu (II) ion, half of a 4,4-bipyridine molecule, a terminal coordination water molecule and a μ- coordination water molecule, half a $[\beta\text{-}Mo_8O_{26}]^{4-}$. Cu coordinate to a O atom in the one carboxylate of L1 ligand, N atom in the 4,4-bipyridine molecule, a terminal O and two μ-O atoms in the coordination water molecule, and an O in $[\beta\text{-}Mo_8O_{26}]^{4-}$, forming an octahedral hexacoordinate mode. The bond length around the Cu ion ranges from 1.912 (6)–2.649 (6) Å (Cu1-O) and 1.972 (7) Å (Cu1-N), the bond angel around Cu ion ranges from N-Cu-O = 45.947 (12)–92.47 (12)°, and O-Cu-O = 44.585 (84)–60.224 (88)°.

L1 ligand acts as a terminal ligand to coordinate with Cu atom. Via the coordination and connected mode with organic and polymolybdate ligands, these $Cu_2(H_2O)_4$ dimer can be infinitely extended in the space to form a 3D network structure. Figure 2b is the stacking diagram of **1**.

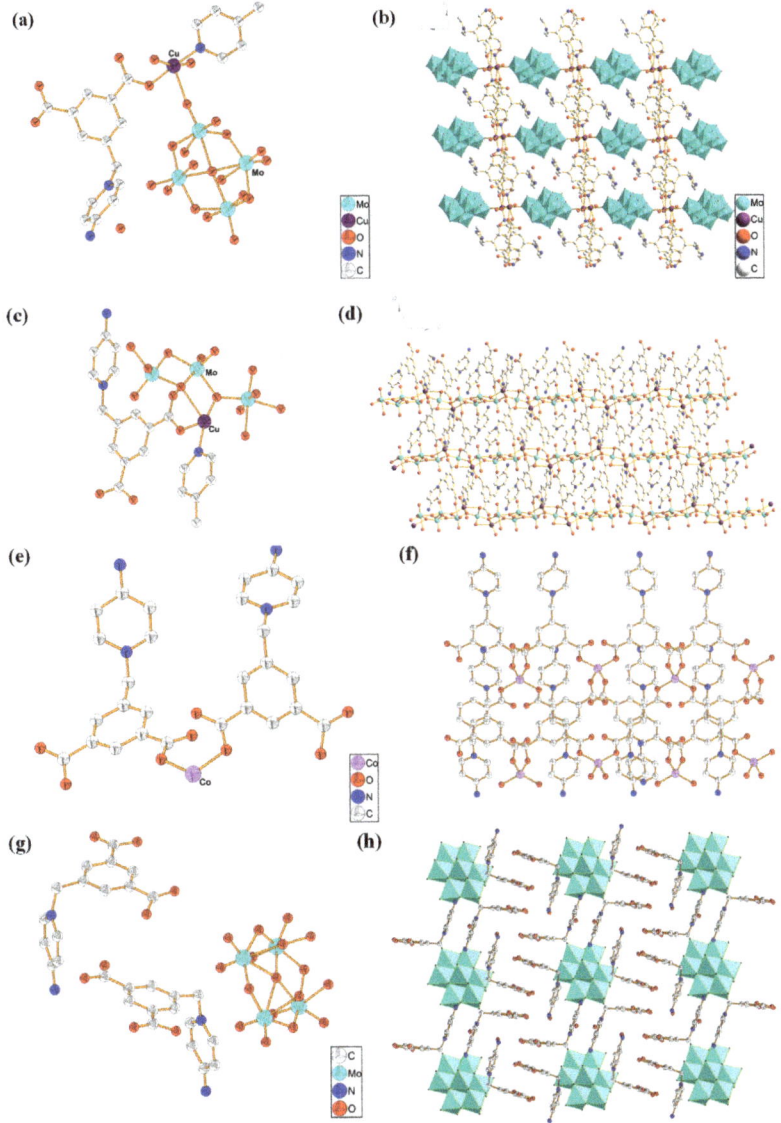

Figure 2. (a) Monomer structure diagram of compound **1** (with the H atom omitted), (b) Stacking diagram of compound **1**, (c) Structural unit diagram of compound **2**, (d) Stacking diagram of compound **2**, (e) Asymmetric structural unit diagram of compound **3**, (f) Stacking diagram of compound **3**, (g) Asymmetric structural unit diagram of compound **4**, (h) Stacking diagram of compound **4**.

X-ray single-crystal structural analysis indicated that compound **2** crystallized in monoclinic system with space group I2/C. As shown in Figure 2c, the smallest structural unit of compound **2** is composed of two Cu (II) ions, two L1 ligands, a 4,4-bipyridine molecule, and a $[Mo_5O_{16}]^{2-}$ anion cluster. Both Cu (II) ions present a hexacoordinated octahedral configuration formed by the coordination with two O atoms from two L1 ligands, three O atoms from three $[Mo_5O_{16}]^{2-}$ anion clusters, and a N atom from a 4,4-bipyridine molecule. Interestingly the two carboxylate groups in L1 ligand act as different bridging modes; one links a Cu and a Mo, while another links two Cu atoms and a Mo atom. Via the coordination and connected mode with organic and polymolybdate ligands, these $Cu_2(H_2O)_4$ dimers can be infinitely extended in the space to form a 3D network structure. Figure 2d is the stacking diagram of **2**.

Compound **3** belongs to the orthorhombic system and pbcn space group. As shown in Figure 2e, the asymmetric structural unit of compound **3** is composed of one Co (II) ion and two L1 ligands. In compound **3**, each Co is tetra-coordinated by coordinating with four O atoms from four different L1 ligands. The main bond length and bond angle around the Co (II) atom are Co1-O2 = 1.954 (5) Å, Co1-O3 = 1.946 (5) Å, O2-Co1-O3 = 40.323 (141)°, O3-Co1-O3 = 107.5 (3)°, respectively. L1 ligand acts as a bridging ligand to link two different Co centers. Via the coordination and connected mode, compound **3** forms a 1D chain structure. Figure 2f is the stacking diagram of **3**.

X-ray single-crystal structural analysis indicated that compound **4** crystallized in triclinic system, P-1 space group. The asymmetric structural unit of compound **4** consists of half $[β-Mo_8O_{26}]^{4-}$, two L1 ligands, and five free H_2O molecules (Figure 2g). Figure 2h is the stacking diagram of compound **4**, where it can be seen that $[β-Mo_8O_{26}]^{4-}$ anion clusters are filled among the organic cationic ligands. There is not coordination bond between the organic cationic ligand L1 and the $[β-Mo_8O_{26}]^{4-}$ anion cluster; they form a 2D supramolecular structure through an electrostatic interaction, intermolecular force, and hydrogen bond (C-H ··· O).

Compounds **1–4** are all obtained by hydrothermal synthesis, but different coordination rules are shown between ligand L1 and metals. In compound **1**, one carboxylate on ligand L1 is coordinated with Cu, while in compound **2**, one carboxylate is coordinated with Cu, and the other carboxylate is coordinated with Mo. In compound **3**, Co is coordinated by two carboxylate groups. In compound **4**, there is no coordination contact between ligand L1 and Mo, but a charge balanced supramolecular compound is formed.

2.2. XRD Analysis

The PXRD pattern of compounds **1–4** were recorded and compared with the simulated single-crystal diffraction data in order to affirm the purity of the compounds. For compounds **1–4**, the position of the peaks are basically consistent with the simulated patterns generated from the results of the single crystal diffraction data, indicating the purity of products (Figure 3a–d). The difference in reflection intensities between the simulated and experimental patterns was due to the variation in the preferred orientation of the powder sample during the collection of the experimental PXRD data. Crystal data for compounds **1–4** were summarized in detail in Table S1. Selected bond lengths and bond angles were put in Table S2.

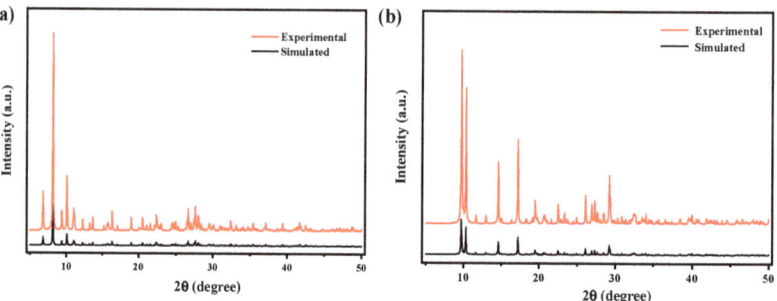

Figure 3. *Cont.*

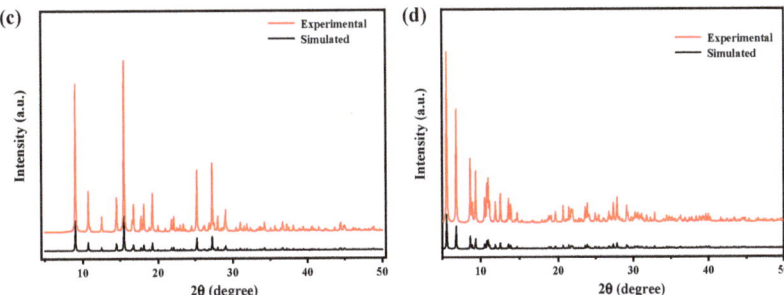

Figure 3. (a) PXRD pattern of compound **1**, (b) PXRD pattern of compound **2**, (c) PXRD pattern of compound **3**, (d) PXRD pattern of compound **4**.

2.3. TG Analysis

In order to investigate the thermal stability of compounds **1–4**, TG analysis was performed. As shown in Figure 4, compounds **1–4** remained substantially unchanged from room temperature to 300 °C. The pyrolysis process of compounds **1**, **2**, and **4** are very similar, probably because of the similar ligand constitution containing L1 and molybdate. Compared with other compounds, the pyrolysis process of compound **3** is quite different, which may be because there is no molybdic acid in its structural composition.

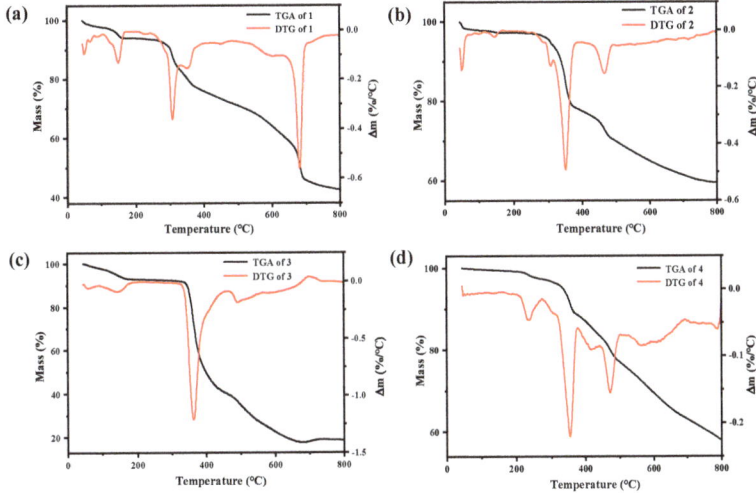

Figure 4. The TG and DTG curve of compounds **1(a)**, **2(b)**, **3(c)**, and **4(d)**.

2.4. Band Gap Analysis

As shown in Figure 5, the band gap values of compounds **1–4** are 2.12 eV, 1.89 eV, 2.01 eV, and 2.27 eV, respectively. This shows that compounds **1–4** are expected to be semiconductors when exposed to visible light and have potential photocatalytic activity. Bandgap is an important characteristic parameter of semiconductors. Its size is related to the crystal structure and the bonding properties of atoms. The diffuse reflectance UV-Vis spectra of Compounds **1-4** can be seen in Figure S1. The different bandgap values of compounds **1–4** may be caused by their different crystal structure and the bonding properties. Based on the size of band gap and the reported literature [13], we speculated that the compounds **1–4** might have potential photocatalytic activity, so we carried out subsequent experimental exploration.

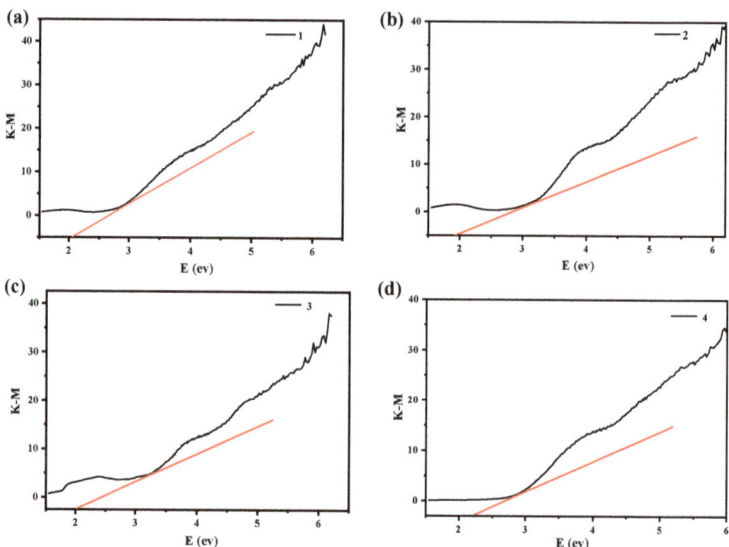

Figure 5. Optical band gap of compounds **1(a)**, **2(b)**, **3(c)**, and **4(d)**.

2.5. Photocatalytic Activity

Ciprofloxacin (CIP) antibiotics are widely used in human and veterinary medicine because of their strong bactericidal ability, lower toxicity, and side effects [24,25]. However, this also led to an increase in the concentration of CIP antibiotics in aquatic environments, resulting in a series of aquatic environments pollution. Therefore, CIP was used as the target degradation product for photocatalytic degradation experiment to explore the photocatalytic performance of compounds **1–4**. Take compound **1** as an example. First, the initial absorbance of CIP was surveyed in the range of 300–450 nm. Then, 0.94 mol% of compound **1** was added to an aqueous solution of CIP. Before lighting, the mixture was magnetically stirred in the dark for 30 min to achieve the adsorption equilibrium between compound **1** and the CIP solution. A 500 W mercury lamp was chosen as a visible light source, and during the photocatalytic reaction process, 3 mL of the suspension was taken out of the mixed solution at regular intervals, and the supernatant was analyzed on a UV-Vis spectrophotometer after centrifugation. The photocatalytic activity of compounds **1–4** were measured by degrading the aqueous solution of CIP under visible light. In the experiment without a catalyst, no significant degradation of CIP was observed, but CIP began to degrade after the addition of compounds, indicating that the effect of light on CIP degradation is negligible, and compounds **1–4** can be used as photocatalysts for CIP. For the reusability test, the supernatant was poured out after the degradation reaction is complete, and fresh CIP solution (20 mL, 25 mg/L) was added to the mixture. Subsequently, the photocatalytic reaction was continuously, magnetically stirred under the irradiation of a 500 W high-pressure xenon lamp. This operation was repeated three times [13].

The calculation formulas of degradation efficiency [26] and removal rate are as follows: C_0 is the CIP concentration when the illumination time is 0, and C_t is the CIP concentration when the illumination time is t. A_0 and A_t are the absorbance of CIP when the illumination time is 0 and t, respectively. It can be seen from the Figure 6a that CIP is hardly degraded under visible light without catalyst. After adding compounds **1–4**, respectively, the degradation rates reached 86.95%, 67.18%, 62.02%, and 59.34%, respectively. This showed that compounds **1–4** have a degrading effect on CIP solution under visible light. It can be seen from Figure 6b that the reaction rate constants of ciprofloxacin degradation by compounds **1–4** are K = 0.00444 min^{-1}, K = 0.00373 min^{-1}, K = 0.00391 min^{-1} and K = 0.00382 min^{-1} respectively, indicating that compound **1** has a faster degradation rate of CIP than compounds **2**, **3**, and **4**. We found that the degradation effect of compound **1** obtained by our one-step reaction is comparable with the reported modified $Bi_2Ti_2O_7/TiO_2/RGO$ composite [8]. Based on this, in the follow-up research work, we used compound **1** as a catalyst to study the effects of pH, temperature, and catalyst amount on the catalytic degradation of CIP.

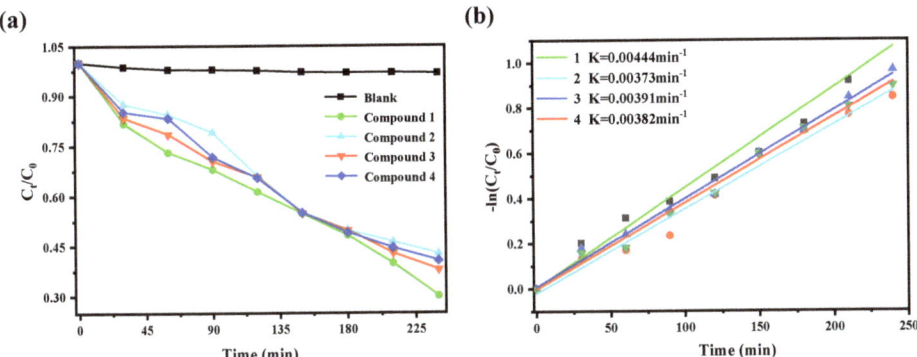

Figure 6. (**a**) The catalytic efficiency of different catalysts on CIP, (**b**) Quasi−first order kinetics of ciprofloxacin degradation by different catalysts.

The pH value of the solution is an important factor affecting the photocatalytic performance [27–29]. The photocatalytic degradation of CIP solution by compound **1** at different pH was studied. Before the formal experiment, adjust the pH with nitric acid and sodium hydroxide to prepare CIP solutions with pH values of 3, 5, 7, and 9, respectively. It can be seen from the Figure 7a that when the pH value is lower than 7, the degradation rate of CIP by compound **1** increased with the increase of pH value. When the pH values were 5 and 7, the final degradation efficiency reached 64.47% and 69.88%, respectively. It can also be seen from Figure 7b that when the pH value is 7, the degradation rate constant (k = 0.00444 min^{-1}) of CIP by compound **1** is the largest. The results showed that the neutral condition was more suitable for the degradation of CIP by compound **1**.

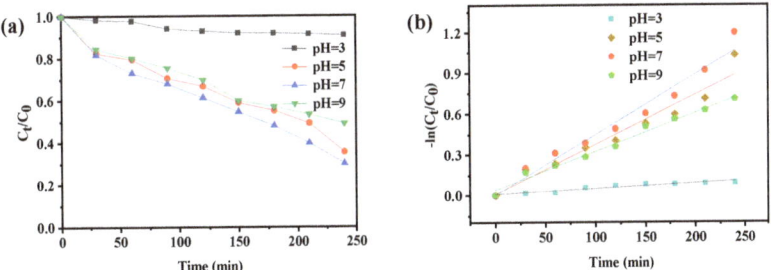

Figure 7. (**a**) Degradation rate of compound **1** to ciprofloxacin under different pH conditions, (**b**) Quasi-first order kinetic reaction rate at different pH values.

Temperature also has a certain influence on the catalytic degradation effect of the catalyst [30–32]. Based on this, in this study, we explored the degradation effect of compound **1** on CIP at different temperatures. It can be seen from the Figure 8a that the degradation effect increased first and then decreased with the increase of temperature. At different temperatures, the final degradation rates of CIP reached 69.88%, 78.9%, and 71.25%, respectively. The results showed that the degradation effect of compound **1** on CIP increased and decreased slightly with the increase of temperature, so the effect of temperature on the photocatalytic degradation of CIP by compound **1** was not significant. The rate constant of k = 0.00444 min^{-1} under 30 °C, k = 0.00574 min^{-1} under 40 °C and k = 0.00446 min^{-1} under 50 °C, the degradation rate of CIP by compound **1** was slightly higher at 40 °C.

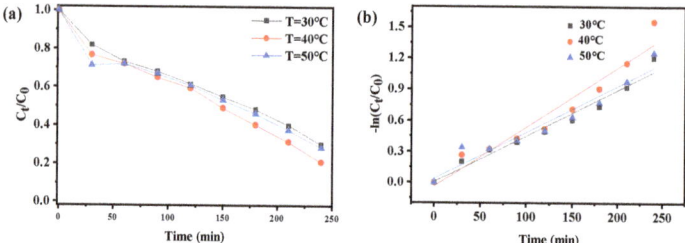

Figure 8. (**a**) Degradation rate of compound **1** to ciprofloxacin under different temperatures, (**b**) Quasi−first order kinetic reaction rate at different temperatures.

In the degradation experiment, we also explored the influence of the amount of catalyst on the catalytic effect. CIP solution was placed under constant stirring conditions, and the degradation of CIP was studied by adding different doses of compound **1**. It can be seen from the Figure 9a that in the CIP/compound **1** system, when the amount of catalyst is from 5 mg to 20 mg, the degradation efficiency of CIP first increases and then decreases. The reason for this state is that the aggregation of excess compound **1** particles hinder and inhibit the scattering and transmission of light in the solution, while the organics adsorbed on the photocatalyst will reduce the utilization of light [33–35]. After adding different doses of compound **1**, the final degradation rates reached 81.13%, 86.95%, and 69.97%, respectively. It can also be seen from Figure 9b that among the three doses, when the amount of compound **1** is 10 mg, the degradation rate of CIP is the highest.

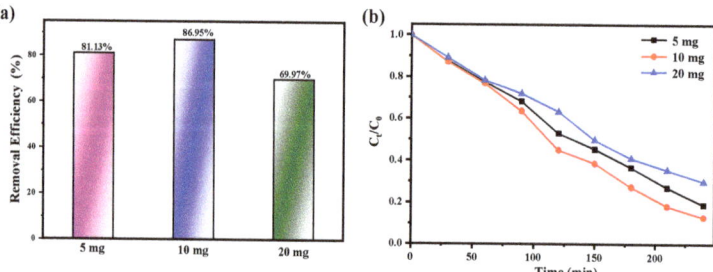

Figure 9. (**a**) Effect of catalyst **1** dosage, (**b**) Degradation rate of compound **1** to ciprofloxacin under different doses.

To investigate the practical value of compound **1** in photocatalytic degradation of CIP, we explored the regeneration and stability of compound **1**. The photocatalytic stability of compound **1** is shown in Figure 10. After the photocatalytic degradation of CIP solution finished, compound **1** was collected. The surface was cleaned with deionized water to remove the residual CIP in the previous experiment, and then put the collected compound **1** into the fresh CIP solution to start a new cycle. The stability and reusability of compound **1** for degradation of CIP was investigated by three consecutive cycles. In the cycle experiment, all experimental conditions were exactly the same as the first experiment. As shown in Figure 10, the degradation efficiency of CIP has not decreased significantly. After three cycles, the degradation efficiency can still reach 80.47%. The results showed that compound **1** can be used as a stable photocatalyst for the photocatalytic degradation of CIP. The stability of compound **1** before and after the photocatalytic reaction was further verified by scanning electron microscope analysis, and the results are shown in Figure 11.

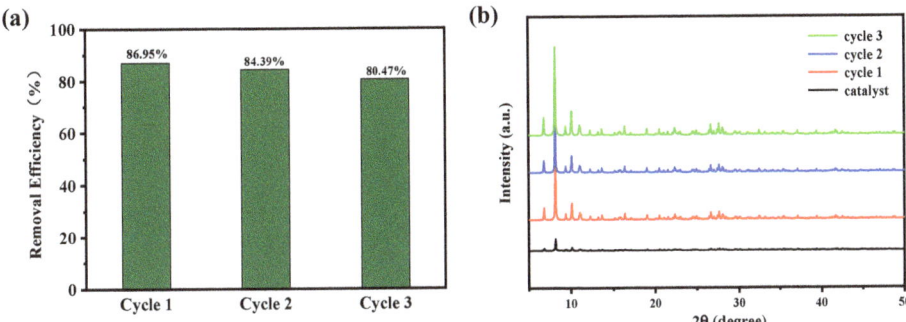

Figure 10. (a) Cycle test of compound **1** to degrade CIP at 30 °C, (b) PXRD patterns of compound **1** before and after the catalytic reaction.

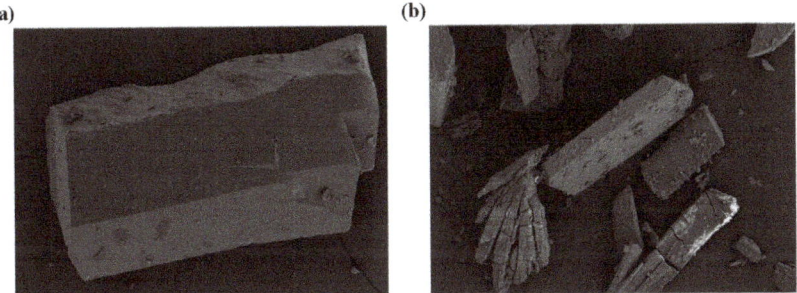

Figure 11. The SEM crystal diagram of compound **1** before (**a**) and after (**b**) the catalytic reaction.

To further investigate the internal mechanism behind compound **1** for CIP degradation, active substance capture experiments were performed. In order to clarify the active substances produced by the catalyst in the catalytic degradation of CIP, different free radical scavengers were added to the photocatalytic reaction system under the same light conditions. Specifically, three active substances of EDTA-2Na (capture h^+), BQ (capture $·O_2^-$) and IPA (capture $·OH$) are mainly used in the photocatalytic process [36,37]. Figure 12 shows that after adding BQ and IPA to the above solution, the degradation rates of CIP by BQ and IPA are 45.38% and 31.45%, respectively. When EDTA-2Na is added to the reaction system, it can greatly inhibit the photocatalytic degradation of CIP, which indicated that h+ plays an important role. Although the addition of BQ interfered with photocatalytic activity, the photocatalytic degradation rate only decreased to 45.38%, indicating that ·OH is not a critical reactant.

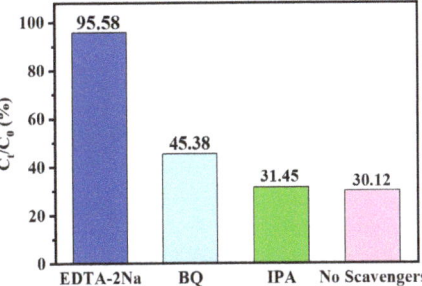

Figure 12. Trapping experiment of active species during the photocatalytic degradation of CIP over catalyst **1** under visible light.

3. Conclusions

In summary, four new compounds were synthesized by hydrothermal method. They were characterized by single-crystal X-ray diffraction, elemental analysis, IR, powder X-ray diffraction and TG analysis, they have good catalytic potential for photodegradating CIP. Among them, compound **1** has a faster degradation rate of CIP. The stability of compound **1** was observed by a cycle experiment, indicating that there was no significant change after three cycles of CIP degradation. In subsequent studies, in order to make our research more comprehensive, it is sensible to perform a hot test to check heterogeneity of the reaction. Moreover, it is expected to explore the size and shape effects on the photocatalytic property.

4. Materials and Methods

4.1. Materials

The ligand L1 was synthesized (Scheme S1) similarly to the reference method [33]. All other reagents for the synthesis and analysis were commercially available and used without further treatment.

4.2. Methods

4.2.1. Synthesis Methods

Synthesis of Compound 1

A solution of L1 (0.0086 g, 0.025 mmol), $(NH_4)_6Mo_7O_{24} \cdot 4H_2O$ (0.0309 g, 0.025 mmol), $CuCl_2 \cdot 2H_2O$ (0.0043 g, 0.025 mmol), 4,4-bipyridine (0.0038 g, 0.025 mmol), and H_2O (10 mL) was stirred under ambient conditions, and the pH was adjusted to 5 with HCl (2 mol/L), then sealed in a Teflonlined steel autoclave, heated at 120 °C for 3 days, and cooled to room temperature. The resulting product was recovered by filtration, washed with distilled water, and dried in air. Yield: 45%. IR (KBr, cm^{-1}): 3361.96 (m), 3224.37 (m), 1875.07 (w), 1698.07 (m), 1651.87 (s), 1111.22 (w), 1035.15 (m), 945.63 (m), 919.62 (m), 668.59 (m), 506.36 (m), 471.46 (m). Elemental Anal. Calc. for $C_{19}H_{23}CuMo_4N_3O_{20}$ (772.54): C, 29.54; H, 2.98; N, 5.44. Found: C, 29.57; H, 3.01; N, 5.47.

A similar procedure was followed to prepare compounds **2–4** [see the Supporting Information (SI)].

4.2.2. Characterization Methods

The infrared spectrum was measured on Shimazu IR 435 spectrometer, in the form of a KBr disk (4000–400 cm^{-1}). Elemental analyses (C, H, and N) were performed on a FLASH EA 1112 elemental analyzer. Amodel NETZSCHTG209 thermal analyzer was used to record simultaneous TG curves in flowing air atmosphere of 20 mL·min^{-1} at a heating rate of 5 °C·min^{-1} in the temperature range 25–800 °C using platinum crucibles. Suitable single crystals of **1–4** were carefully selected under an optical microscope and glued to thin glass fibers. The crystallographic data of compounds are acquired on a Bruker APEX-II area detector diffractometer equipped with graphite monochromatic Cu-Kα radiation (λ = 1.54184 Å) at 293(2) K. The structure was refined with full-matrix least-squares techniques on F^2 using the OLEX2 program package. The CCDC reference numbers are 2,085,864, 2,124,165, 2,085,863, and 2,085,866 for compounds **1**, **2**, **3**, and **4** respectively.

Supplementary Materials: The following supporting information can be downloaded at: https://www.mdpi.com/article/10.3390/molecules27227731/s1, Scheme S1, Synthesis route of the cationic template L1; Figure S1, The diffuse reflectance UV-Vis spectra of Compounds **1–4**; Table S1, Crystal data and structure refinement details for **1–4**; Table S2, Bond length (Å) and bond angle data of compounds **1–4** (°).

Author Contributions: Conceptualization, J.L.; software, X.W.; validation, J.L. and X.W.; formal analysis, J.L. and Y.N.; investigation, Y.N.; supervision, Y.N.; writing-original draft preparation, J.L. and Y.N. All authors have read and agreed to the published version of the manuscript.

Funding: This research received no external funding.

Institutional Review Board Statement: Not applicable.

Informed Consent Statement: Not applicable.

Data Availability Statement: All data generated or analyzed during this study are included in this article.

Conflicts of Interest: The authors declare no conflict of interest.

References

1. Babić, S.; Periša, M.; Škorić, I. Photolytic degradation of norfloxacin, enrofloxacin and ciprofloxacin in various aqueous media. *Chemosphere* **2013**, *91*, 1635–1642. [CrossRef] [PubMed]
2. Antonin, V.S.; Santos, M.C.; Garcia-Segura, S.; Brillas, E. Electrochemical incineration of the antibiotic ciprofloxacin in sulfate medium and synthetic urine matrix. *Water Res.* **2015**, *83*, 31–41. [CrossRef] [PubMed]
3. Porras, J.; Bedoya, C.; Silva-Agredo, J.; Santamaría, A.; Fernández, J.J.; Torres-Palma, R.A. Role of humic substances in the degradation pathways and residual antibacterial activity during the photodecomposition of the antibiotic ciprofloxacin in water. *Water Res.* **2016**, *94*, 1–9. [CrossRef] [PubMed]
4. An, T.; Yang, H.; Li, G.; Song, W.; Cooper, W.J.; Nie, X. Kinetics and mechanism of advanced oxidation processes (AOPs) in degradation of ciprofloxacin in water. *Appl. Catal. B: Environ.* **2010**, *94*, 288–294. [CrossRef]
5. Wang, X.; Zhu, G.; Wang, C.; Niu, Y. Effective degradation of tetracycline by organic-inorganic hybrid materials induced by triethylenediamine. *Environ. Res.* **2021**, *198*, 111253. [CrossRef]
6. Ali, I.; Han, G.; Kim, J. Reusability and photocatalytic activity of bismuth-TiO2 nanocomposites for industrial wastewater treatment. *Environ. Res.* **2019**, *170*, 222–229. [CrossRef]
7. Qiu, H.; Fang, S.; Huang, G.; Bi, J. A novel application of In2S3 for visible-light-driven photocatalytic inactivation of bacteria: Kinetics, stability, toxicity and mechanism. *Environ. Res.* **2020**, *190*, 110018. [CrossRef]
8. Li, W.; Zuo, Y.; Jiang, L.; Yao, D.; Chen, Z.; He, G.; Chen, H. Bi2Ti2O7/TiO2/RGO composite for the simulated sunlight-driven photocatalytic degradation of ciprofloxacin. *Mater. Chem. Phys.* **2020**, *256*, 123650. [CrossRef]
9. Núñez-Salas, R.E.; Hernández-Ramírez, A.; Santos-Lozano, V.; Hinojosa-Reyes, L.; Guzmán-Mar, J.L.; Gracia-Pinilla, M.Á.; Maya-Treviño, M.d.L. Synthesis, characterization, and photocatalytic performance of FeTiO3/ZnO on ciprofloxacin degradation. *J. Photochem. Photobiol. A Chem.* **2021**, *411*, 113186. [CrossRef]
10. Alhokbany, N.S.; Mousa, R.; Naushad, M.; Alshehri, S.M.; Ahamad, T. Fabrication of Z-scheme photocatalysts g-C3N4/Ag3PO4/chitosan for the photocatalytic degradation of ciprofloxacin. *Int. J. Biol. Macromol.* **2020**, *164*, 3864–3872. [CrossRef]
11. Jiang, W.; Li, Z.; Liu, C.; Wang, D.; Yan, G.; Liu, B.; Che, G. Enhanced visible-light-induced photocatalytic degradation of tetracycline using BiOI/MIL-125(Ti) composite photocatalyst. *J. Alloy. Compd.* **2021**, *854*, 157166. [CrossRef]
12. Qiao, G.-Y.; Li, S.-M.; Niu, Y.-Y. One new copper iodide coordination polymer directed by 4-pyridyl dithioether ligand: Syntheses, structures, and photocatalysis. *Main Group Chem.* **2020**, *19*, 217–225. [CrossRef]
13. Qiao, X.; Wang, C.; Niu, Y. N-Benzyl HMTA induced self-assembly of organic-inorganic hybrid materials for efficient photocatalytic degradation of tetracycline. *J. Hazard. Mater.* **2020**, *391*, 122121. [CrossRef] [PubMed]
14. Wang, C.; Kim, J.; Tang, J.; Kim, M.; Lim, H.; Malgras, V.; You, J.; Xu, Q.; Li, J.; Yamauchi, Y. New Strategies for Novel MOF-Derived Carbon Materials Based on Nanoarchitectures. *Chem* **2020**, *6*, 19–40. [CrossRef]
15. Xue, Y.; Zhao, G.; Yang, R.; Chu, F.; Chen, J.; Wang, L.; Huang, X. 2D metal-organic framework-based materials for electrocatalytic, photocatalytic and thermocatalytic applications. *Nanoscale* **2021**, *13*, 3911–3936. [CrossRef]
16. Yang, Z.; Guo, Z.; Zhang, J.; Hu, Y. The development and application of metal-organic frameworks in the field of photocatalysis. *Res. Chem. Intermed.* **2021**, *47*, 325–343. [CrossRef]
17. Dao, X.-Y.; Sun, W.-Y. Single- and mixed-metal–organic framework photocatalysts for carbon dioxide reduction. *Inorg. Chem. Front.* **2021**, *8*, 3178–3204. [CrossRef]
18. Wang, X.; Qiao, G.; Zhu, G.; Li, J.; Guo, X.; Liang, Y.; Niu, Y. Preparation of 2D supramolecular material doping with TiO2 for degradation of tetracycline. *Environ. Res.* **2021**, *202*, 111689. [CrossRef]
19. Yu, D.; Shao, Q.; Song, Q.; Cui, J.; Zhang, Y.; Wu, B.; Ge, L.; Wang, Y.; Zhang, Y.; Qin, Y.; et al. A solvent-assisted ligand exchange approach enables metal-organic frameworks with diverse and complex architectures. *Nat. Commun.* **2020**, *11*, 927. [CrossRef]
20. Chaemchuen, S.; Alam Kabir, N.; Zhou, K.; Verpoort, F. Metal–organic frameworks for upgrading biogas via CO2 adsorption to biogas green energy. *Chem. Soc. Rev.* **2013**, *42*, 9304–9332. [CrossRef]
21. Wang, X.-J.; Qiao, X.-Y.; Niu, Y.-Y. Synthesis, characterization, adsorption and catalytic activity of a polyoxometalate supramolecule templated by arylmethylamine. *Main Group Chem.* **2020**, *19*, 187–198. [CrossRef]
22. Niu, Y.; Song, Y.; Hou, H.; Zhu, Y. Synthesis, Structure, and Large Optical Limiting Effect of the First Coordination Polymeric Cluster Based on an {I@[AgI(inh)]6} Hexagram Block. *Inorg. Chem.* **2005**, *44*, 2553–2559. [CrossRef] [PubMed]
23. Li, J.; Liu, Z.; Liu, Y.; Liu, J.; Li, Y.; Qiao, X.; Huang, W.; Niu, Y. POM-based metal-organic compounds: Assembly, structures and properties. *Main Group Chem.* **2021**, *20*, 575–592. [CrossRef]

24. Fang, H.; Oberoi, A.S.; He, Z.; Khanal, S.K.; Lu, H. Ciprofloxacin-degrading *Paraclostridium* sp. isolated from sulfate-reducing bacteria-enriched sludge: Optimization and mechanism. *Water Res.* **2021**, *191*, 116808. [CrossRef] [PubMed]
25. Karimi-Maleh, H.; Ayati, A.; Davoodi, R.; Tanhaei, B.; Karimi, F.; Malekmohammadi, S.; Orooji, Y.; Fu, L.; Sillanpää, M. Recent advances in using of chitosan-based adsorbents for removal of pharmaceutical contaminants: A review. *J. Clean Prod.* **2021**, *291*, 125880. [CrossRef]
26. Chen, F.; Yang, Q.; Sun, J.; Yao, F.; Wang, S.; Wang, Y.; Wang, X.; Li, X.; Niu, C.; Wang, D.; et al. Enhanced Photocatalytic Degradation of Tetracycline by AgI/BiVO$_4$ Heterojunction under Visible-Light Irradiation: Mineralization Efficiency and Mechanism. *ACS Appl. Mater. Interfaces* **2016**, *8*, 32887–32900. [CrossRef]
27. Lei, X.; Wang, J.; Shi, Y.; Yao, W.; Wu, Q.; Wu, Q.; Zou, R. Constructing novel red phosphorus decorated iron-based metal organic framework composite with efficient photocatalytic performance. *Appl. Surf. Sci.* **2020**, *528*, 146963. [CrossRef]
28. Li, C.; Che, H.; Huo, P.; Yan, Y.; Liu, C.; Dong, H. Confinement of ultrasmall CoFe$_2$O$_4$ nanoparticles in hierarchical ZnIn$_2$S$_4$ microspheres with enhanced interfacial charge separation for photocatalytic H$_2$ evolution. *J. Colloid Interface Sci.* **2021**, *581*, 764–773. [CrossRef]
29. Wu, J.; Wang, W.; Tian, Y.; Song, C.; Qiu, H.; Xue, H. Piezotronic effect boosted photocatalytic performance of heterostructured BaTiO$_3$/TiO$_2$ nanofibers for degradation of organic pollutants. *Nano Energy* **2020**, *77*, 105122. [CrossRef]
30. Qian, Q.; Asinger, P.A.; Lee, M.J.; Han, G.; Rodriguez, K.M.; Lin, S.; Benedetti, F.M.; Wu, A.X.; Chi, W.S.; Smith, Z.P. MOF-Based Membranes for Gas Separations. *Chem. Rev.* **2020**, *120*, 8161–8266. [CrossRef]
31. Chen, X.; Peng, X.; Jiang, L.; Yuan, X.; Fei, J.; Zhang, W. Photocatalytic removal of antibiotics by MOF-derived Ti^{3+}- and oxygen vacancy-doped anatase/rutile TiO$_2$ distributed in a carbon matrix. *Chem. Eng. J.* **2022**, *427*, 130945. [CrossRef]
32. Ji, Z.; Di, Z.; Li, H.; Zou, S.; Wu, M.; Hong, M. A flexible Zr-MOF with dual stimulus responses to temperature and guest molecules. *Inorg. Chem. Commun.* **2021**, *128*, 108597. [CrossRef]
33. Chaturvedi, G.; Kaur, A.; Kansal, S.K. CdS-Decorated MIL-53(Fe) Microrods with Enhanced Visible Light Photocatalytic Performance for the Degradation of Ketorolac Tromethamine and Mechanism Insight. *J. Phys. Chem. C* **2019**, *123*, 16857–16867. [CrossRef]
34. Zhang, X.; Wang, X.; Chai, J.; Xue, S.; Wang, R.; Jiang, L.; Wang, J.; Zhang, Z.; Dionysiou, D.D. Construction of novel symmetric double Z-scheme BiFeO$_3$/CuBi$_2$O$_4$/BaTiO$_3$ photocatalyst with enhanced solar-light-driven photocatalytic performance for degradation of norfloxacin. *Appl. Catal. B.* **2020**, *272*, 119017. [CrossRef]
35. Huang, D.; Li, J.; Zeng, G.; Xue, W.; Chen, S.; Li, Z.; Deng, R.; Yang, Y.; Cheng, M. Facile construction of hierarchical flower-like Z-scheme AgBr/Bi$_2$WO$_6$ photocatalysts for effective removal of tetracycline: Degradation pathways and mechanism. *Chem. Eng. J.* **2019**, *375*, 121991. [CrossRef]
36. Zhu, Q.; Sun, Y.; Na, F.; Wei, J.; Xu, S.; Li, Y.; Guo, F. Fabrication of CdS/titanium-oxo-cluster nanocomposites based on a Ti32 framework with enhanced photocatalytic activity for tetracycline hydrochloride degradation under visible light. *Appl. Catal. B: Environ.* **2019**, *254*, 541–550. [CrossRef]
37. Gan, H.; Yi, F.; Zhang, H.; Qian, Y.; Jin, H.; Zhang, K. Facile ultrasonic-assisted synthesis of micro-nanosheet structure Bi$_4$Ti$_3$O$_{12}$/g-C$_3$N$_4$ composites with enhanced photocatalytic activity on organic pollutants. *Chin. J. Chem. Eng.* **2018**, *26*, 2628–2635. [CrossRef]

Article

Modification of Polymeric Carbon Nitride with Au–CeO$_2$ Hybrids to Improve Photocatalytic Activity for Hydrogen Evolution

Linzhu Zhang [1,2], Lu Chen [1,2], Yuzhou Xia [1,2], Zhiyu Liang [1,2], Renkun Huang [1,2], Ruowen Liang [1,2,*] and Guiyang Yan [1,2,*]

[1] Province University Key Laboratory of Green Energy and Environment Catalysis, Ningde Normal University, Ningde 352100, China
[2] Fujian Provincial Key Laboratory of Featured Materials in Biochemical Industry, Ningde Normal University, Ningde 352100, China
* Correspondence: t1629@ndnu.edu.cn (R.L.); ygyfjnu@163.com (G.Y.); Tel.: +86-593-296427 (R.L.); +86-593-2565503 (G.Y.)

Abstract: The construction of a multi-component heterostructure for promoting the exciton splitting and charge separation of conjugated polymer semiconductors has attracted increasing attention in view of improving their photocatalytic activity. Here, we integrated Au nanoparticles (NPs) decorated CeO$_2$ (Au–CeO$_2$) with polymeric carbon nitride (PCN) via a modified thermal polymerization method. The combination of the interfacial interaction between PCN and CeO$_2$ via N-O or C-O bonds, with the interior electronic transmission channel built by the decoration of Au NPs at the interface between CeO$_2$ and PCN, endows CeAu–CN with excellent efficiency in the transfer and separation of photo-induced carriers, leading to the enhancement of photochemical activity. The amount-optimized CeAu–CN nanocomposites are capable of producing ca. 80 µmol· H$_2$ per hour under visible light irradiation, which is higher than that of pristine CN, Ce–CN and physical mixed CeAu and PCN systems. In addition, the photocatalytic activity of CeAu–CN remains unchanged for four runs in 4 h. The present work not only provides a sample and feasible strategy to synthesize highly efficient organic polymer composites containing metal-assisted heterojunction photocatalysts, but also opens up a new avenue for the rational design and synthesis of potentially efficient PCN-based materials for efficient hydrogen evolution.

Keywords: CeAu–CN heterostructure; interfacial interaction; interior electronic transmission channel; photocatalytic hydrogen production

1. Introduction

With the future acceleration of environmental pollution and the energy crisis, photocatalytic water splitting has been the subject of intense research as it can replace fossil fuels by providing clean and renewable energy [1,2]. Following the pioneering work of Honda and Fujishima on a photoelectrochemical cell equipped with Pt-TiO$_2$, considerable progress has been made in this field [3,4]. In particular, this is because photocatalysts play a vital role in water splitting systems; a large number of photocatalysts with appropriate band structure, visible light response and good photochemical stability have been designed and developed [5].

Polymeric carbon nitride (PCN), a new generation metal-free conjugated polymeric photocatalysts, has become a research hotspot for water splitting and CO$_2$ reduction due to its unique advantages and significant physical and optical characteristics [6–10]. Given the potential application of PCN, improving the corresponding properties of materials is still an enormous challenge, since the methods of controlling the separation of photogenerated charge carriers are not well realized. To this end, developing a PCN-based photocatalytic system with higher solar-to-hydrogen energy (STH) conversion efficiency is one of the most promising ways to improve the separation and transfer of photogenerated charge carriers [11]. To date, many studies have proved that constructing reasonable and novel heterostructures can efficiently decrease exciton binding energy and enhance charge separation for PCN-based materials [12,13]. For example, recently, researchers have successfully constructed g-C_3N_4-based heterostructure composites, such as metal oxide–g-C_3N_4 [14,15], polymer–g-C_3N_4 [16] and sulfide–g-C_3N_4 [17,18] to enhance the photocatalytic activity through facilitating the separation of photogenerated charges and holes. In a typical example, Zheng et al. constructed a CdSe–CN S-scheme heterojunction via a linker-assisted hybridization approach, and showed that CdSe–CN exhibited superior photocatalytic reactivity to CdSe and CN in water splitting and CO_2 conversion [19]. However, the efficiency of photogenerated carrier transportation has been limited owing to the interfacial effects between two different materials. Therefore, the rational design of PCN- based composites which would enable them to effectively adjust the interface charge transfer toward heterojunctions still is a huge challenge.

As a popular rare oxide, CeO_2 nanomaterials possess good electrical conductivity, abundant redox chemistry, and plentiful surface oxygen vacancies due to the valence change between Ce^{3+} and Ce^{4+} oxidation states [20–22]. These excellent properties of CeO_2 make it more likely to generate strong electron interaction with other materials, consequently, this will potentially improve the catalytic performance. For example, the $Fe(OH)_3$–CeO_2 composite, due to its tight interface effect, promoted the separation efficiency of photogenerated charges, thus showing excellent photocatalytic activity of water oxidation [23]. Additionally, Dong Lin's group have undertaken a lot of related studies on CeO_2–g-C_3N_4 complexes, which showed that the construction of CeO_2–g-C_3N_4 heterojunctions provided an internal charge transport channel, leading to faster separation of electron–hole pairs and better photocatalytic activity than CeO_2 and g-C_3N_4 [24–26]. Moreover, the properties of cerium can also be regulated by doping or acting as a carrier of various metals such as Ag, Pt, or Au, although the activity and selectivity of photocatalysts depend to a large extent on the type and concentration of dopants [27–29]. In a typical case, Primo et al. reported that the deposition of Au nanoparticles at low loading increases the photocatalytic activity of ceria more than the WO_3 owing to its unique electronic structure [30]. Encouraged by the advantages of CeO_2 and the excellent electrical conductivity of Au, the photocatalytic activity of CeO_2-Au-PCN can be expected to be further improved. However, such three-component systems based on PCN have seldom been reported so far due to the great challenge in their synthesis.

Based on the above considerations, herein, novel Au NPs decorated CeO_2 coupled with PCN ternary composites were designed and successfully synthesized via a modified thermal polymerization method, aiming to promote charge separation and improve the hydrogen evolution of PCN. FTIR and XPS results demonstrated a strong interfacial effect between PCN and CeAu through N-O or C-O bonds. The Au NPs can serve as a suitable electron acceptor and transfer channel, and CeO_2 as a proper-level electron-accepting platform. Therefore, Au NPs provide a direct interior pathway to introduce photogenerated electrons from PCN into CeO_2, which assists with the efficient transfer and separation of photogenerated electrons and holes. Not surprisingly, the ternary composites of CeO_2–Au–PCN show strikingly ameliorated photocatalytic H_2 evolution compared to the pristine and two-component CeO_2–PCN system.

2. Results and Discussion

X-ray diffraction (XRD) was applied to phase analysis and crystal structure determination of samples. As presented in Figure 1, the diffraction peaks of pure CN located at 13.1° and 27.2° are attributed to the (100) and (002) planes, respectively. The diffraction peaks of CeAu at 2θ of 28.6°, 33.1°, 47.6°, 56.4°, 59.1°, 69.4°, 76.7°, 79.1° marked with "*" can be retrieved as well-crystallized face centered cubic structure of CeO_2 JCPDS (no.34-0394) [31], whereas the other peaks at 38.2°, 44.4°, 64.8° marked with "●" can be assigned to the face-centered cubic structure of Au corresponding to (JCPDS no. 89-3697), indicating the formation of Au NPs on the surface of CeO$_2$ [32]. For the CeAu–CN samples, diffraction peaks corresponding to CN, CeO$_2$ and Au can be observed with the increasing CeAu content in composite, which confirms the successful formation of the three-component CeAu–CN system. The low intensity of the CeO$_2$ and Au diffraction peaks can be observed owing to the low content of CeAu randomly distributed on the surface of CN. Nevertheless, compared with the pure CN, the (100) peak of x% CeAu–CN becomes much weaker, and almost disappears with the increasing CeAu content in composite. The phenomenon is similar to the previous report of disorders in the arrangement of in-plane structural motifs caused by doping oxygen atoms in CN [33]. Moreover, the gradual weakening and widening of the 27.2° peak is probably because of the structure fluctuation owing to the addition of CeAu.

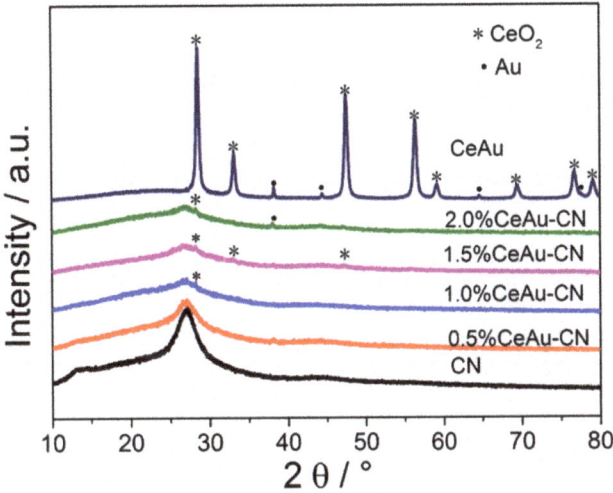

Figure 1. The XRD patterns of CN, x% CeAu–CN (x = 0.5, 1.0, 1.5, 2.0), and CeAu.

The chemical structure information of the as-prepared materials can be well confirmed by FTIR spectra. In Figure 2, it can be seen that both pure CN and x% CeAu–CN composite materials exhibit similar vibrational modes of triazine heterocyclic ring molecular in polymeric carbon nitride. The characteristic peaks at 1200–1600 cm^{-1} belong to the stretching vibrations of CN heterocycles, while another significant characteristic peak at 810 cm^{-1} is caused by the breathing vibration of the triazine units. In addition, a weak broad peak at 2900–3300 cm^{-1} is attributed to the presence of the free amino group (e.g., NH$_2$ or NH) and absorbed H$_2$O molecules on the surface of the CN-based polymer. Surprisingly, from the partial magnification of FTIR spectra we can clearly see that a new weak band at ~985 cm^{-1} became more and more obvious with the increasing of CeAu content in x% CeAu–CN composites, probably due to the existence of the stretching vibration of the N-O group [34]. Because of this discovery, it is speculated that the O atom in CeO$_2$ is hybridized with the N atom of CN, indicating the formation of the solid interfacial interaction between the CeAu and CN. The tight interfacial effect contributes to charge transfer, thereby improving the photocatalytic activity to some extent.

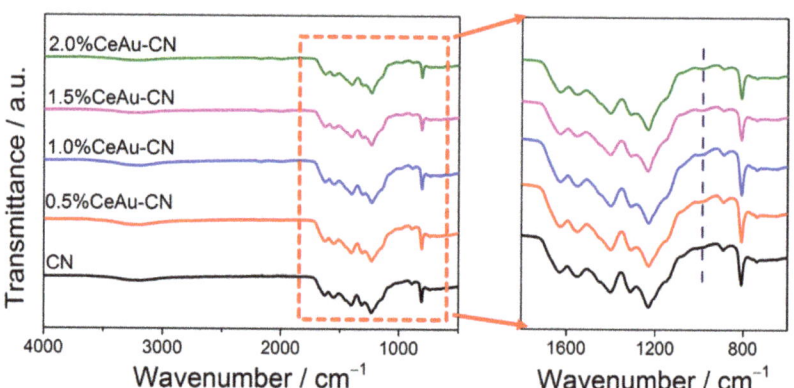

Figure 2. FTIR spectra and the partial magnified graph of CN and x% CeAu–CN (x = 0.5, 1.0, 1.5, 2.0).

For the optical absorption properties of the samples to be investigated, the UV–Vis absorption measurement was carried out. As displayed in Figure 3a, CN shows an obvious absorption in the visible light region, while CeO_2 exhibits optical absorption only in the ultraviolet light region ($\lambda \leq 400$ nm), owing to the intrinsic nature of the samples. Notably, the visible light absorption of CeAu is significantly improved with marked absorption peaks at ca. 550 nm, which is known by common sense to be caused by the SPR effect of the loaded Au nanoparticles. Additionally, in Figure 3b, it can be seen that the optical edges gradually shift to the red wavelength region with the increase in the CeAu content in x% CeAu–CN compositions, but no significant light absorption caused by the SPR effect can be observed, which is mainly because the content of CeAu is too low compared with the urea precursor. Commonly, the band-gap energies of CN and CeO_2 are calculated according to the following formula [35]:

$$Ahv = (\alpha hv - E_g)^n \quad (1)$$

Here, α is the absorbance, h is the Planck's constant, v is the photon frequency, and E_g is the photonic energy band gap. Wherein, the value of n is related to the type of electronic transitions to which the samples belong. For CN belonging to the indirect bandgap, the value of n is 2, while for CeO_2 belonging to the direct bandgap, the value of n is 0.5. Therefore, the band gap of CN and CeO_2 are estimated from the Tauc plot (Figure 3a, inset) to be 2.65 eV and 3.20 eV, respectively.

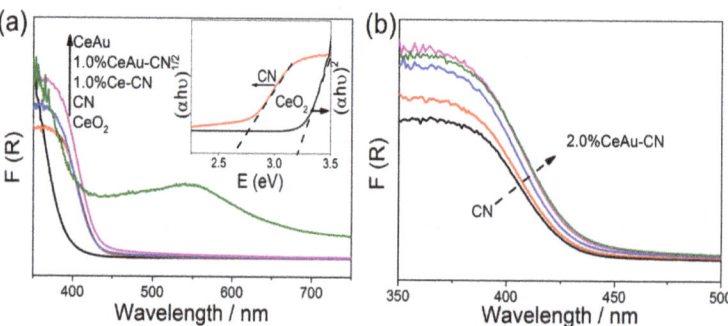

Figure 3. (a,b) UV–Vis DRS spectra of CN, CeO_2, CeAu, Ce–CN and x% CeAu–CN samples (x = 0.5, 1.0, 1.5, 2.0).

The more detailed morphology and texture information of the samples were investigated by using scanning electron microscopy (SEM) and transmission electron microscopy

(TEM). As shown in Figure S1, the SEM image 1.0% CeAu–CN sample exhibits a nanosheet with more porous holes and wrinkles relative to CN, which may benefit the increase in the BET surface areas of the sample. Meanwhile, Figure 4a distinctly shows some irregular CeAu nanoparticles randomly distributed on the surface of the CN photocatalyst. Together with Figure 4b, we can see the obvious lattice fringes with a d spacing of 0.31 nm are consistent with the value for the CeO_2 (111) plane. With further enlargement of the selected area, the lattice-fringe spacing of 0.24 nm is observed, which is in agreement with the (111) plane of Au. Furthermore, the EDX element mapping images (Figure 4c) clearly display well-defined spatial distribution of C, N, Ce, O and Au for the 1.0% CeAu–CN photocatalyst. This results, together with XRD analysis, reveal the successful construction of the ternary CeO_2-Au–CN hybrids. Notably, there is close contact between CeO_2, Au, and CN, which is conducive to the charge transfer across their interface.

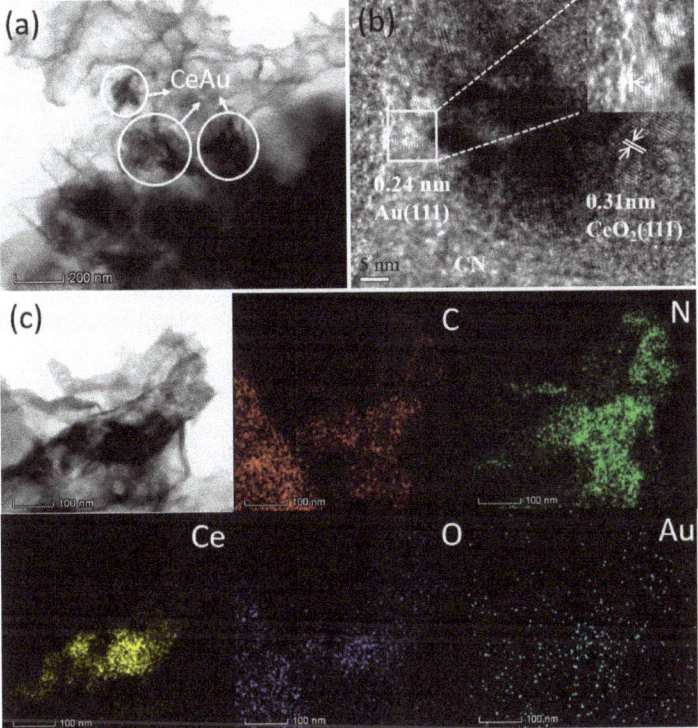

Figure 4. (a) TEM and (b) HRTEM of 1.0% CeAu–CN; (c) STEM image and element mapping images of C, N, Ce, O and Au for the 1.0% CeAu–CN photocatalyst.

Having confirmed the morphology and structure information, we further examined the chemical compositions and element valence states of pristine CN, 1.0% CeAu–CN and CeAu materials by X-ray photoelectron spectroscopy (XPS). Not surprisingly, CN and 1.0% CeAu–CN show similar high-resolution C1s and N1s core-level XPS spectra. The C 1s spectra in Figure 5a can be resolved into three peaks located at ~284.6, 285.9, and 288.2 eV corresponding to the typical impurity carbon, carbon atoms in C-O, and sp^2-hybridized carbon in N-containing aromatic ring (N-C=N), respectively. Moreover, the formation of C-O bonds may be related to oxygen-containing intermediates produced during the pyrolysis of urea or the lattice oxygen in the cerium oxide [36]. Figure 5b displays the XPS spectrum of N1s, which is mainly divided into four peaks, among which the peaks at 398.7 and 399.8 eV ascribed to sp^2-hybridized nitrogen in the form of C=N-C and tertiary nitrogen N-(C)3 groups, respectively, and both of them are the main substructure units forming the tri-s-triazine heterocyclic ring. The other two peaks, located at 401.1 and 404.2 eV, are attributed either to the surface uncondensed C-N-H functional groups and charging effects or the positive charge localization in the heterocycles, respectively. Figure 5c shows the high-resolution XPS spectrum of Ce3d, by using a Gaussian fitting method, the Ce3d core level XPS fitted plot at ~885.4 and 903.6 eV can be assigned to 3d5/2 and 3d3/2 spin-orbit states, respectively, which ascertains the presence of both Ce^{4+} and Ce^{3+} in the CeAu, in agreement with previous studies [37,38]. Additionally, the weak peak of Ce3d can also be detected in the 1.0% CeAu–CN composites, though its content is shallow, which indicates the existence of Ce^{4+} and Ce^{3+} in the samples. Usually, the presence of Ce^{3+} is accompanied by the generation of oxygen vacancies (Ov), which play a vital role in enhancing visible light absorption as well as the interfacial interaction of ceria-based materials. Furthermore, the high resolution XPS spectra of O 1 s were displayed in Figure S3, for CeAu, the peak observed at 529.5 eV represents the lattice oxygen of CeO_2. However, for 1.0% CeAu–CN composites, the O 1s spectra can be devolved into two peaks, one at 529.5 eV corresponds to the lattice oxygen of CeO_2, another main peak at 531.7 eV is assigned to the C3-N+-O- species formed by the hybridization of CN and CeO_2 or the O-H groups attach on the surface of the samples, which is consistent with the FTIR analysis results [39]. High-resolution XPS spectra of Au 4f orbital fitted to two peaks located at 84 and 87.7 eV corresponded Au4f7/2 and 4f5/2, respectively (Figure 5d), suggesting the presence of Au nanoparticles in CeAu. Interestingly, the binding energy of Au4f for 1.0% CeAu–CN is slightly lower than the CeAu sample, indicative of a strong interaction between Au and CN and CeO_2. To some extent, this further confirms that there is close contact between the three, which probably facilitates the transfer of photo-induced carriers, enhancing the photocatalytic performance for photocatalysts.

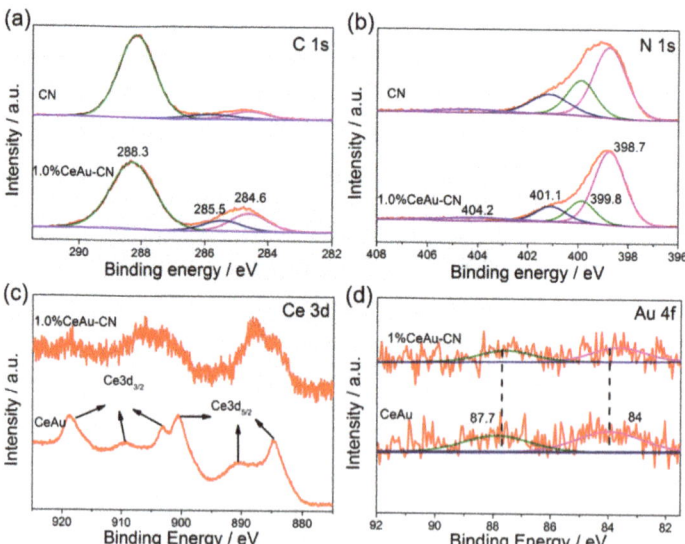

Figure 5. High-resolution XPS spectra of (**a**) C1s region and (**b**) N1s region of the CN and 1.0% CeAu–CN composite; (**c**) Ce3d region and (**d**) Au4f region of the 1.0% CeAu–CN and CeAu.

Visible-light-induced photocatalytic H_2 generation was then attempted by capitalizing on the prepared samples in the presence of 3 wt% Pt as a co-catalyst and 10 vol% triethanolamine (TEOA) as the sacrificial electron donor. As illustrated in Figure 6a, the CeAu has almost no photocatalytic activity of hydrogen production probably owing to the intrinsic property of CeO_2. Of note, compared with CN, the photocatalytic activity of CeCN is slightly improved, while the CeAu–CN composites display compelling photocatalytic performance. The 1.0% CeAu–CN sample shows the fastest rate of H_2 evolution driven by visible light ($\approx$80 µmol h^{-1}). As such, the materials of CeAu and CN just physically mixed (CeAu+CN) were selected as the reference and tested under the same reaction condition. Not surprisingly, there is no significant enhancement in catalytic activity. Therefore, the above experiments' results jointly confirm that the outstanding photocatalytic performance may be ascribed to the intimate contact between the host and guest materials and the existence of gold nanoparticles, both of which are indispensable. In addition, as shown in Figure S4, it is observed that 1.0% CeAu–CN still exhibits the H_2 evolution activity under extended wavelength irradiation, demonstrating that the CeAu strengthens the capture of visible light consistent with the analysis of DRS. In Figure 6b, the H_2 evolution rates of the CeAu–CN ternary photocatalysts sensitively depend on the added amount of CeAu. The photocatalytic water reduction for the H_2 evolution rate first increases and then decreases when the CeAu densities are varied. As the excessive accumulation of CeAu on the CN surface will build a light shielding effect, it may promote the recombination of photo-induced electrons and holes, thereby decreasing the photocatalytic efficiency. A comparison between the photocatalytic H_2 evolution of our 1.0% CeAu–CN and that of other reported catalysts is listed in Table 1. It is worth noting that, compared with some photocatalysts, 1.0% CeAu–CN showed better photocatalytic performance in hydrogen production under visible light irradiation.

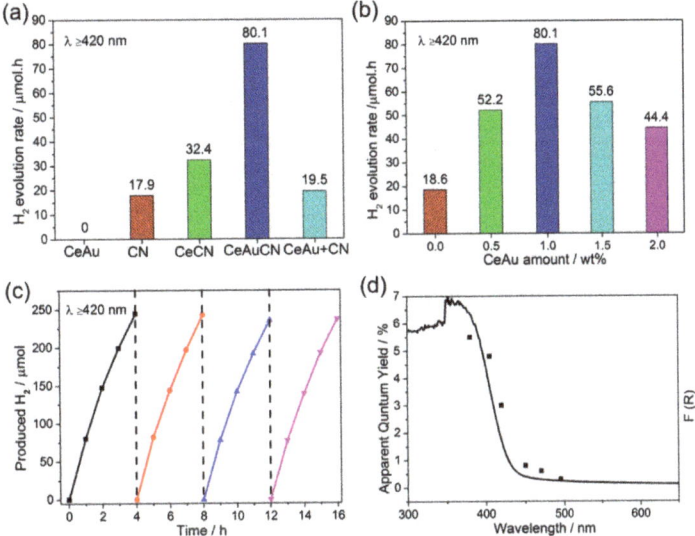

Figure 6. (**a**) Photocatalytic performance for H_2 production of CeAu, pure CN, CeCN, CeAuCN and CeAu+CN; (**b**) Photocatalytic activity for H_2 evolution rate of modified CN with different amounts of CeAu under visible light irradiation ($\lambda \geq$ 420 nm); (**c**) The photocatalytic stability of the 1.0% CeAu–CN; (**d**) Wavelength dependence of AQY on the H_2 evolution in the 1.0% CeAu–CN sample.

To evaluate the photostability of 1.0% CeAu–CN composites, the photocatalytic recycling performance of the sample was investigated under the same reaction conditions (Figure 6c). No obvious attenuation of hydrogen evolution rate (HER) is observed after the 16 h test, implying that 1.0% CeAu–CN ternary composition has excellent photostability. Moreover, the catalysts after the reaction were further subjected to characterization by XRD and FTIR (Figure S5), which reveals that there is no noticeable change in the crystal and chemical structure of the catalysts before and after the reaction. These observations explicitly elucidate the ternary CeAu–CN materials with the robust nature and high operation stability benefit for the photocatalysis of hydrogen evolution. Next, the best sample's apparent quantum yield (AQY) was measured under various monochromatic light irradiation. As shown in Figure 6d, the AQY of H_2 evolution matches well with the UV–Vis DRS spectra of the 1.0% CeAu–CN, confirming that the reaction is indeed driven by light irradiation, and the AQY value of 1.0% CeAu–CN is about 3.0% at 420 nm, which is higher than that of bulk CN previously reported.

Table 1. Comparison between the photocatalytic activity of 1.0% CeAu–CN and that of other reported catalysts for photocatalytic hydrogen evolution.

Photocatalysts	Reaction Conditions	Light Source	H_2 Production (μmol h^{-1} g^{-1})	Ref.
1.0% CeAu–CN	0.05 g catalyst, 3 wt%Pt, TEOA solution (10%)	300 W Xe lamp $\lambda > 420$ nm	1602	This work
CeO_2–g–C_3N_4	0.05 g catalyst, 3 wt%Pt, TEOA solution (10%)	300 W Xe lamp $\lambda > 420$ nm	1100	[25]
CeO_2–g–C_3N_4	0.05 g catalyst, 0.5 wt%Pt, lactic acid solution (20%)	300 W Xe lamp $\lambda > 420$ nm	73.12	[40]
N-CeO_x–g–C_3N_4	0.05 g catalyst, 1 wt%Pt, TEOA solution (10%)	300 W Xe lamp $\lambda > 420$ nm	292.5	[41]
Au–SnO_2–g–C_3N_4	0.1 g catalyst, methanol solution (20%)	300 W Xe lamp $\lambda > 400$ nm	770	[42]
g–C_3N_4–Au–C–TiO_2	0.01 g catalyst, TEOA solution (10%)	300 W Xe lamp $\lambda > 420$ nm	129	[43]

Considering that the photocatalytic performance of photocatalyst is closely related to the charge–carrier separation and transfer process, we performed room-temperature photoluminescence (PL) and photoelectrochemical characterization. The room-temperature photoluminescence was conducted using 380 nm as the exciting wavelength. From Figure 7a, it is clear that all the photocatalysts show similar broad peaks centered at around 480 nm due to the band–band PL phenomenon, which is consistent with the results of DRS. At the same time, the PL emission intensities gradually decrease as the loading content of CeAu increases, which reveals the prohibited recombination of light-excited charge by the construction of heterojunctions. Time-resolved PL spectroscopy was utilized to probe the charge carrier dynamics of pristine CN and 1.0% CeAu–CN composite (Figure 7b). According to the three radiative lifetimes with different percentages, we can obtain the average lifetimes of pure CN and 1.0% CeAu–CN composites as 7.67 ns and 6.27 ns, respectively. Compared with CN, the lifetime of 1.0% CeAu–CN composite is relatively short, probably due to the internal transmission path constructed by Au NPs at the interface or the close interface contact between CN and CeO_2, resulting in faster separation of photogenerated carriers. In addition, more charge carrier migration information was also reflected by the EPR analysis of CN and 1.0% CeAu–CN samples. In Figure S6, a single Lorentzian line centering at a g value of 2.003 is observed for CN and 1.0% CeAu–CN samples, indicating the presence of unpaired electrons on π-conjugated CN aromatic rings. Evidently, the EPR intensity of 1.0% CeAu–CN is greatly strengthened compared with CN. This strongly verifies that the modification of CeAu hybrids promotes the electron migration in the π-conjugated system of CN, possibly due to the intimated interface effects of CN and CeAu.

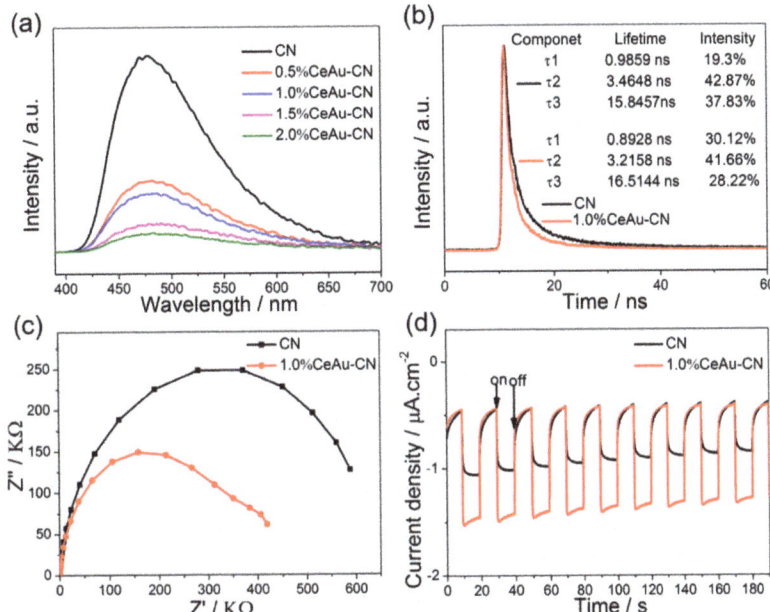

Figure 7. (a) PL spectra and (b) Time-resolved PL spectra of CN and x% CeAu–CN (x = 0.5, 1.0, 1.5, 2.0) samples; (c) EIS tests and (d) Photocurrent response of CN and 1.0% CeAu–CN.

The electron transfer behavior can also be further demonstrated in the following electrochemical experiments. Firstly, the interface charge transfer resistance of the electrons was implemented via an electrochemical impedance spectroscopy (EIS) test. The experimental results show that the 1.0% CeAu–CN photocatalyst possesses a smaller high-frequency semicircle than the pure CN (Figure 7c), meaningfully indicating the lower charge-transfer resistance in a hybrid that ensures faster electrons trainer. Next, the interface charge separation kinetics of the samples can also be reflected by the photocurrent density-time response plot. As displayed in Figure 7d, the photocurrent drastically increases and decreases, respectively, when the power supply is switched on and off. Only about 0.5 μA cm^{-2} of photocurrent density emerged for pure CN, whereas 1.2 μA cm^{-2} of photocurrent density is generated for 1.0% CeAu–CN electrode, implying a lower recombination rate of electron–hole pairs and good electrical conductivity for the 1.0% CeAu–CN electrode compared to the pristine one. Therefore, combined with the PL analysis and the results of photo/electrochemical studies, it is well established that the modification of CeAu successfully improved the separation and migration of light-generated charges compared with the pristine CN. The enhanced photogenerated charge separation of CeAu–CN is responsible for improving its photocatalytic activity.

In order to further investigate the transfer process of photogenerated charge, it is urgent to obtain the relative band position of CN and CeO$_2$. Therefore, in the next experiment, we carried out an electrochemical analysis of CN and CeO$_2$ to evaluate their electronic band structure. Figure 8a,b show the positive slope of the plot indicating both CN and CeO$_2$ are n-type semiconductors, and their corresponding flat potential is calculated to be −1.2 and −0.69 V reference the saturated calomel electrode (SCE), which are equivalent to −0.96 and −0.45 V versus the normal hydrogen electrode (NHE), respectively. As previously established, for the n-type semiconductor, the actual conduction band is 0~0.1 eV higher than the flat potential due to the electron effective mass and the carrier concentration [44–46]. Here, the voltage difference between the conduction band and the flat potential is set to 0.1 eV. According to the electrochemical results and the related calculations, the bottom conduction bands for CN and CeO$_2$ are −1.06 and −0.55 eV, while the corresponding valence bands are 1.59 and 2.35 eV, respectively.

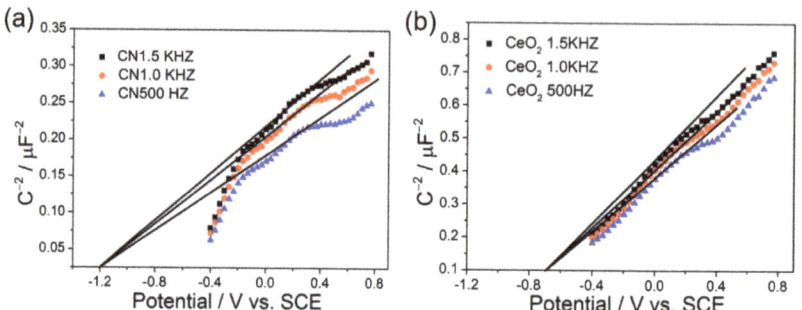

Figure 8. Mott–Schottky plots of (**a**) CN and (**b**) CeO$_2$ film electrodes at different frequency in 0.2 M Na$_2$SO$_4$ aqueous solution (pH = 6.8).

Based on the results and analysis of the previous characterizations and the above relative energy band levels of CN and CeO$_2$, a plausible mechanism schematic of photogenerated charge transfer and separation and photochemical reactions of CeAu–CN ternary composites driven by the light is proposed. As displayed in Figure 9, under visible light irradiation, it is acceptable that the photogenerated electrons of CN transfer to the CB of CeO$_2$ with the assistance of Au NPs. Because of its strong "electron sink" effect, Au NPs can be served as carrier conductors, providing an interior direct channel to facilitate the separation and transport of photo-induced carriers at the interface of type-II heterostructure [47–49]. The icing on the cake is that the tight interface between CN and CeO$_2$ also provides a good platform for charge transfer. As a result, the photogenerated charges of PCN are well separated and transferred for a more efficient reaction with target reactants, so its photocatalytic hydrogen production performance is greatly improved.

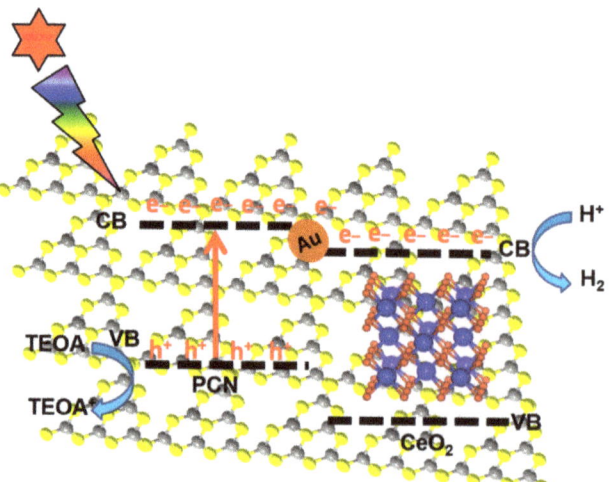

Figure 9. Schematic illustrations of the possible transfer channels of electron–hole pairs separation.

3. Materials and Methods

3.1. Materials

All chemicals used in the synthesis were of analytical grade without further purification. Cerium nitride hexahydrate (Ce(NO$_3$)$_3$·6H$_2$O), sodium hydroxide (NaOH), sodium sulfate (Na_2SO_4) and urea purchased from Sinopharm Chemical Reagent Co. Ltd., Shanghai, China. Chloroauric acid ($HAuCl_4·4H_2O$) and chloroplatinic acid($H_2PtCl_6·6H_2O$) were supplied by Alfa Aesar China Co., Ltd. (Tianjin, China).

3.2. Synthesis

3.2.1. Synthesis of CeO_2

As previously reported, CeO_2 nanomaterials were synthesized by the hydrothermal method using cerium nitride hexahydrate ($Ce(NO_3)_3·6H_2O$) as the precursor [50]. To describe the experiment briefly, first, 4 mmol of $Ce(NO_3)_3·6H_2O$ was dissolved in 10 mL water, subsequently, 70 mL of 6 M NaOH solution was added to the above solution dropwise and continuously stirred at room temperature for 2 h. Then, the gray mud was transferred into a 100 mL Teflon-lined stainless-steel autoclave and heated at 120 °C for 24 h under autogenous pressure. After cooling to room temperature, the gray precipitates were collected by centrifugation with deionized water and ethanol many times, followed by drying at 80 °C overnight in an oven, resulting in CeO_2.

3.2.2. Synthesis of Au–CeO_2

Au-loaded CeO_2 was prepared via a simple impregnation method using aqueous $HAuCl_4·4H_2O$ solution [51]. The detailed preparation procedure employed is described below. Firstly, 0.3 g CeO_2 was dispersed in 15 mL of deionized water to obtain uniform suspension by stirring for 0.5 h, then added a certain amount of $HAuCl_4·4H_2O$ solution, and the mixture was stirred for another 1 h. The lavender powder was formed after dried at 80 °C for 16 h under vigorous magnetic stirring. In this paper, the Au–CeO_2 sample with theoretical 1.5 wt% Au was obtained and denoted as CeAu.

3.2.3. Synthesis of CeAu–CN, Ce–CN and CN

The CeAu–CN photocatalysts were obtained through a modified thermal polymerization of urea molecules with certain amount of CeAu similar to the experimental procedure described by our group [52]. Typically, 10 g urea mixed with different amount of CeAu in 20 mL deionized water with vigorous magnetic stirring at room temperature for 2 h and then stirring at 85 °C to remove water. Afterwards the resultant solids were ground and calcined at 550 °C for 2 h in N_2 with the speed of 4.6 °C min^{-1} to obtain the final samples. There were noted as x% CeAu–CN, where x (x = 0.5, 1, 1.5, 2.0) is the percentage weight content of CeAu to urea. The Ce–CN composite was prepared by the same procedure of CeAu–CN, only with the CeAu replaced by CeO_2. The pure PCN (denoted as CN) was prepared by the same method without adding the CeAu.

3.3. Characterizations

The crystal structure of the samples was determined with the support of Powder X-ray diffraction (XRD) which was executed on Bruker D8 Advance focus using Cu Kα1 radiation and recorded in the 2θ rang 10–80°. Fourier transform infrared (FTIR) spectra were measured on a Nicolet-6700 in the frequency range of 4000–400 cm^{-1}. The morphology of the as-made samples was investigated by field emission scanning electron microscopy (SEM) (JSM-6700F) and FEI Tecnai20 transmission electron microscopy (TEM). Brunauer–Emmett–Teller (BET) specific surface area and porosity of photocatalysts were carried out by sorption using Micromeritics ASAP 2010 instrument at 77 K. The compositions and chemical valence states of as-synthesis samples were collected from XPS spectra analyzer using a Thermo ESCALAB250 instrument with a monochromatized Al Kα line source (200 W). The optical properties of the materials can be demonstrated by the ultraviolet-visible diffuse reflectance spectra (UV–Vis DRS) measured on a Varian Cary 500 Scan UV–Vis system using $BaSO_4$ as a reference. In addition, photoluminescence (PL) spectra obtained on an Edinburgh FI/FSTCSPC 920 spectrophotometer using 380 nm as the exciting wavelength. Electron paramagnetic resonance (EPR) measurements were performed on a Bruker model A300 spectrometer. The BioLogic VSP-300 electrochemical system was used to measure the electrochemical performance of samples in a traditional three-electrode cell with a Pt plate and saturated calomel electrode (SCE) as counter electrode and reference electrode, respectively.

3.4. Photocatalytic Activity Test

The photocatalytic hydrogen evolution experiment was carried out in an online reaction system. The schematic of photocatalytic water splitting reaction is shown in Figure S7, which mainly includes three parts: photocatalytic water splitting reaction system, vacuum circulation system and analysis and testing system. The detailed experimental process is as follows: the powder photocatalyst (50 mg) was suspended in an aqueous solution (100 mL) containing 10 mL triethanolamine as the holes' sacrificial reagent. The 3 wt% Pt co-catalyst was loaded onto the surface of photocatalysts by in situ photo-deposition method using H_2PtCl_6 as precursor during the reaction. The reaction temperature was always maintained at 12 °C by the circulating condensing equipment. Subsequently, the reaction system was sealed and evacuated many times to completely remove the air, which was then irradiated under a 300 W Xe lamp using a 420 nm cutoff filter. The different wavelength experiments were similar to previous ones, except that the cutoff filters were changed to a different cutoff wavelength. The generated hydrogen was determined by a gas chromatograph equipped with a thermal conductive detector (TCD) and a 5 Å molecular sieve column with high-purity argon as the carrier gas.

The apparent quantum yield (AQY) for H_2 evolution was measured using monochromatic LED lamps with band pass filter of 380, 405, 420, 450, 470, 495 nm. The irradiation area was controlled as 3×3 cm^2. The AQY was calculated based on the following formula: AQY = Ne/Np $\times$ 100 = 2MNAhc/SPtλ $\times$ 100%, where Ne is the amount of electrons involved in the reaction, Np is the amount of incident photons, M is the quantity of hydrogen molecules produced by reaction, NA is Avogadro constant, h is the Planck constant, c is the speed of light, S is the irradiation area, P is the intensity of the irradiation, t is the photo-irradiation time, and λ is the wavelength of the monochromatic light.

4. Conclusions

In summary, a three-component CeAu–CN heterostructure has been successfully established via a modified thermal polymerization method. Owing to the interfacial interaction between PCN and CeO_2 via N-O or C-O bands and the interior electronic transmission channel constructed by the decorated of Au NPs at the interface, CeAu–CN has been confirmed to be highly efficient in the separation and transfer of photogenerated carriers, and greatly enhanced photocatalytic activity. The amount-optimized 1.0% CeAu–CN nanocomposite exhibits the highest H_2 evolution rate of 80.1 µmol h^{-1} under visible light irradiation ($\lambda > 420$ nm), and the photocatalytic activity of CeAu–CN still remains unchanged for four runs in 4 h. More importantly, the present study not only discloses the importance of interfacial effects on photocatalytic activity, but also opens a promising avenue for the rational design and fabrication of heterogeneous interface-containing metals for highly effective solar water splitting.

Supplementary Materials: The following are available online at https://www.mdpi.com/article/10.3390/molecules27217489/s1, Figure S1: SEM images of (a,b) CN, and (c,d) 1.0% CeAu–CN; Figure S2: (a) Nitrogen adsorption–desorption isotherms of CN and 1.0% CeAu–CN samples at 77 K; Figure S3: (a) Typical XPS survey spectra of 1.0% CeAu–CN and high resolution spectra of (b) O 1s of CeAu and 1.0% CeAu–CN; Figure S4: Room-temperature EPR spectra of CN and 1.0% CeAu–CN; Figure S5: Photocatalytic activity for H_2 evolution rate of 1.0% CeAu–CN under different wavelength irradiation; Figure S6: (a) XRD and (b) FTIR spectra of 1.0% CeAu–CN samples before and after photochemical reaction. Figure S7: Schematic of photocatalytic water splitting system.

Author Contributions: Conceptualization, R.L. and G.Y.; Investigation, L.Z., L.C., Y.X., Z.L. and R.H.; Writing—Review and editing, L.Z. All authors have read and agreed to the published version of the manuscript.

Funding: This work was supported by the Research Project of Ningde Normal University (No. 2020Y016), Natural Science Foundation of Fujian province (No. 2022J05269 and 2020J05224), Education & Research Project for Young and Middle-aged Teachers of Fujian province (No. JAT200698), Scientific Research Project for Young and middle-aged teachers of Ningde Normal University (No. 2021Q107). Moreover, we are also grateful to the Program of IRTSTFJ for the financial support.

Institutional Review Board Statement: Not applicable.

Informed Consent Statement: Not applicable.

Data Availability Statement: Not applicable.

Conflicts of Interest: The authors declare no conflict of interest.

References

1. Kudo, A.; Miseki, Y. Heterogeneous photocatalyst materials for water splitting. *Chem. Soc. Rev.* **2009**, *38*, 253–278. [CrossRef] [PubMed]
2. Hisatomi, T.; Domen, K. Introductory lecture: Sunlight-driven water splitting and carbon dioxide reduction by heterogeneous semiconductor systems as key processes in artificial photosynthesis. *Faraday Discuss.* **2017**, *198*, 11–35. [CrossRef] [PubMed]
3. Fujishima, A.; Honda, K. Electrochemical photolysis of water at a semiconductor electrode. *Nature* **1972**, *238*, 37–38. [CrossRef] [PubMed]
4. Liu, Y.; Mao, J.; Huang, Y.; Qian, Q.; Luo, Y.; Xue, H.; Yang, S. Pt-Chitosan-TiO$_2$ for Efficient Photocatalytic Hydrogen Evolution via Ligand-to-Metal Charge Transfer Mechanism under Visible Light. *Molecules* **2022**, *27*, 4673. [CrossRef] [PubMed]
5. Chen, S.; Takata, T.; Domen, K. Particulate photocatalysts for overall water splitting. *Nat. Rev.Mater.* **2017**, *2*, 17050. [CrossRef]
6. Wang, X.; Maeda, K.; Thomas, A.; Takanabe, K.; Xin, G.; Carlsson, J.M.; Domen, K.; Antonietti, M. A metal-free polymeric photocatalyst for hydrogen production from water under visible light. *Nat. Mater.* **2009**, *8*, 76–80. [CrossRef]
7. Yang, C.; Wang, B.; Zhang, L.; Yin, L.; Wang, X. Synthesis of Layered Carbonitrides from Biotic Molecules for Photoredox Transformations. *Angew. Chem. Int. Ed.* **2017**, *56*, 6627–6631. [CrossRef]
8. Kessler, F.K.; Zheng, Y.; Schwarz, D.; Merschjann, C.; Schnick, W.; Wang, X.; Bojdys, M.J. Functional carbon nitride materials—design strategies for electrochemical devices. *Nat. Rev.Mater.* **2017**, *2*, 17030. [CrossRef]
9. Ong, W.J.; Tan, L.L.; Ng, Y.H.; Yong, S.T.; Chai, S.P. Graphitic Carbon Nitride (g-C$_3$N$_4$)-Based Photocatalysts for Artificial Photosynthesis and Environmental Remediation: Are We a Step Closer To Achieving Sustainability? *Chem. Rev.* **2016**, *116*, 7159–7329. [CrossRef]
10. Fang, Y.X.; Wang, X.C. Photocatalytic CO$_2$ conversion by polymeric carbon nitrides. *Chem. Commun.* **2018**, *54*, 5674–5687. [CrossRef]
11. Ou, H.; Chen, X.; Lin, L.; Fang, Y.; Wang, X. Biomimetic Donor-Acceptor Motifs in Conjugated Polymers for Promoting Exciton Splitting and Charge Separation. *Angew. Chem. Int. Ed.* **2018**, *57*, 8729–8733. [CrossRef]
12. Fu, J.; Yu, J.; Jiang, C.; Cheng, B. g-C$_3$N$_4$-Based Heterostructured Photocatalysts. *Adv. Energy. Mater.* **2018**, *8*, 1701503. [CrossRef]
13. Ruan, X.; Cui, X.; Jia, G.; Wu, J.; Zhao, J.; Singh, D.J.; Liu, Y.; Zhang, H.; Zhang, L.; Zheng, W. Intramolecular heterostructured carbon nitride with heptazine-triazine for enhanced photocatalytic hydrogen evolution. *Chem. Eng. J.* **2022**, *428*, 132579. [CrossRef]
14. Lu, S.-S.; Zhang, L.-M.; Fan, K.; Xie, J.-Y.; Shang, X.; Zhang, J.-Q.; Chi, J.-Q.; Yang, X.-L.; Wang, L.; Chai, Y.-M.; et al. In Situ formation of ultrathin C$_3$N$_4$ layers on metallic WO$_2$ nanorods for efficient hydrogen evolution. *Appl. Surf. Sci.* **2019**, *487*, 945–950. [CrossRef]
15. Moussa, H.; Chouchene, B.; Gries, T.; Balan, L.; Mozet, K.; Medjahdi, G.; Schneider, R. Growth of ZnO Nanorods on Graphitic Carbon Nitride gCN Sheets for the Preparation of Photocatalysts with High Visible-Light Activity. *ChemCatChem* **2018**, *10*, 4987–4997. [CrossRef]
16. Zhou, W.; Jia, T.; Shi, H.; Yu, D.; Hong, W.; Chen, X. Conjugated polymer dots/graphitic carbon nitride nanosheet heterojunctions for metal-free hydrogen evolution photocatalysis. *J. Mater. Chem. A* **2019**, *7*, 303–311. [CrossRef]
17. Lu, X.J.; Jin, Y.L.; Zhang, X.Y.; Xu, G.Q.; Wang, D.M.; Lv, J.; Zheng, Z.X.; Wu, Y.C. Controllable synthesis of graphitic C$_3$N$_4$/ultrathin MoS$_2$ nanosheet hybrid nanostructures with enhanced photocatalytic performance. *Dalton Trans.* **2016**, *45*, 15406–15414. [CrossRef]
18. Liang, Z.Y.; Chen, F.; Huang, R.K.; Huang, W.J.; Wang, Y.; Liang, R.W.; Yan, G.Y. CdS Nanocubes Adorned by Graphitic C$_3$N$_4$ Nanoparticles for Hydrogenating Nitroaromatics: A Route of Visible-Light-Induced Heterogeneous Hollow Structural Photocatalysis. *Molecules* **2022**, *27*, 5438. [CrossRef]
19. Wang, Y.; Wang, H.; Li, Y.; Zhang, M.; Zheng, Y. Designing a 0D/1D S-Scheme Heterojunction of Cadmium Selenide and Polymeric Carbon Nitride for Photocatalytic Water Splitting and Carbon Dioxide Reduction. *Molecules* **2022**, *27*, 6286. [CrossRef]
20. Xie, S.; Wang, Z.; Cheng, F.; Zhang, P.; Mai, W.; Tong, Y. Ceria and ceria-based nanostructured materials for photoenergy applications. *Nano Energy* **2017**, *34*, 313–337. [CrossRef]
21. Tran, D.P.H.; Pham, M.T.; Bui, X.T.; Wang, Y.F.; You, S.J. CeO$_2$ as a photocatalytic material for CO$_2$ conversion: A review. *Sol. Energy* **2022**, *240*, 443–466. [CrossRef]

22. Montini, T.; Melchionna, M.; Monai, M.; Fornasiero, P. Fundamentals and Catalytic Applications of CeO_2-Based Materials. *Chem. Rev.* **2016**, *116*, 5987–6041. [CrossRef] [PubMed]
23. Feng, T.; Ding, J.; Li, H.; Wang, W.; Dong, B.; Cao, L. Amorphous $Fe(OH)_3$ Passivating CeO_2 Nanorods: A Noble-Metal-Free Photocatalyst for Water Oxidation. *ChemSusChem* **2021**, *14*, 3382–3390. [CrossRef] [PubMed]
24. Zhu, C.; Wang, Y.; Jiang, Z.; Xu, F.; Xian, Q.; Sun, C.; Tong, Q.; Zou, W.; Duan, X.; Wang, S. CeO_2 nanocrystal-modified layered MoS_2/g-C_3N_4 as 0D/2D ternary composite for visible-light photocatalytic hydrogen evolution: Interfacial consecutive multi-step electron transfer and enhanced H_2O reactant adsorption. *Appl. Catal. B* **2019**, *259*, 118072. [CrossRef]
25. Zou, W.; Deng, B.; Hu, X.; Zhou, Y.; Pu, Y.; Yu, S.; Ma, K.; Sun, J.; Wan, H.; Dong, L. Crystal-plane-dependent metal oxide-support interaction in CeO_2/g-C_3N_4 for photocatalytic hydrogen evolution. *Appl. Catal. B* **2018**, *238*, 111–118. [CrossRef]
26. Wei, X.; Wang, X.; Pu, Y.; Liu, A.; Chen, C.; Zou, W.; Zheng, Y.; Huang, J.; Zhang, Y.; Yang, Y.; et al. Facile ball-milling synthesis of CeO_2/g-C_3N_4 Z-scheme heterojunction for synergistic adsorption and photodegradation of methylene blue: Characteristics, kinetics, models, and mechanisms. *Chem. Eng. J.* **2021**, *420*, 127719. [CrossRef]
27. Song, S.; Li, K.; Pan, J.; Wang, F.; Li, J.; Feng, J.; Yao, S.; Ge, X.; Wang, X.; Zhang, H. Achieving the Trade-Off between Selectivity and Activity in Semihydrogenation of Alkynes by Fabrication of (Asymmetrical Pd@Ag Core)@(CeO_2 Shell) Nanocatalysts via Autoredox Reaction. *Adv. Mater.* **2017**, *29*, 1605332. [CrossRef]
28. Yoon, S.; Ha, H.; Kim, J.; Nam, E.; Yoo, M.; Jeong, B.; Kim, H.Y.; An, K. Influence of the Pt size and CeO_2 morphology at the Pt–CeO_2 interface in CO oxidation. *J. Mater. Chem. A* **2021**, *9*, 26381–26390. [CrossRef]
29. Tanaka, A.; Hashimoto, K.; Kominami, H. Preparation of Au/CeO_2 Exhibiting Strong Surface Plasmon Resonance Effective for Selective or Chemoselective Oxidation of Alcohols to Aldehydes or Ketones in Aqueous Suspensions under Irradiation by Green Light. *J. Am. Chem. Soc.* **2012**, *134*, 14526–14533. [CrossRef]
30. Primo, A.; Marino, T.; Corma, A.; Molinari, R.; Garcia, H. Efficient Visible-Light Photocatalytic Water Splitting by Minute Amounts of Gold Supported on Nanoparticulate CeO_2 Obtained by a Biopolymer Templating Method. *J. Am. Chem. Soc.* **2012**, *134*, 1892. [CrossRef]
31. Baranchikov, A.E.; Razumov, M.I.; Kameneva, S.V.; Sozarukova, M.M.; Beshkareva, T.S.; Filippova, A.D.; Kozlov, D.A.; Ivanova, O.S.; Shcherbakov, A.B.; Ivanov, V.K. Facile Synthesis of Stable Cerium Dioxide Sols in Nonpolar Solvents. *Molecules* **2022**, *27*, 5028. [CrossRef]
32. Khan, M.M.; Ansari, S.A.; Ansari, M.O.; Min, B.K.; Lee, J.; Cho, M.H. Biogenic Fabrication of Au@CeO_2 Nanocomposite with Enhanced Visible Light Activity. *J. Phys. Chem. C* **2014**, *118*, 9477–9484. [CrossRef]
33. Kang, Y.; Yang, Y.; Yin, L.-C.; Kang, X.; Wang, L.; Liu, G.; Cheng, H.-M. Selective Breaking of Hydrogen Bonds of Layered Carbon Nitride for Visible Light Photocatalysis. *Adv. Mater.* **2016**, *28*, 6471–6477. [CrossRef]
34. Kurtikyan, T.S.; Eksuzyan, S.R.; Hayrapetyan, V.A.; Martirosyan, G.G.; Hovhannisyan, G.S.; Goodwin, J.A. Nitric Oxide Dioxygenation Reaction by Oxy-Cobogloblin Models: In-situ Low-Temperature FTIR Characterization of Coordinated Peroxynitrite. *J. Am. Chem. Soc.* **2012**, *134*, 13861–13870. [CrossRef]
35. Patterson, E.M.; Shelden, C.E.; Stockton, B.H. Kubelka-Munk optical properties of a barium sulfate white reflectance standard. *Appl. Opt.* **1977**, *16*, 729–732. [CrossRef]
36. Lin, Z.; Wang, X. Ionic Liquid Promoted Synthesis of Conjugated Carbon Nitride Photocatalysts from Urea. *ChemSusChem* **2014**, *7*, 1547–1550. [CrossRef]
37. Xiao, Y.T.; Chen, Y.J.; Xie, Y.; Tian, G.H.; Guo, S.E.; Han, T.R.; Fu, H.G. Hydrogenated CeO_2-xSx mesoporous hollow spheres for enhanced solar driven water oxidation. *Chem. Commun.* **2016**, *52*, 2521–2524. [CrossRef]
38. Li, W.; Jin, L.; Gao, F.; Wan, H.; Pu, Y.; Wei, X.; Chen, C.; Zou, W.; Zhu, C.; Dong, L. Advantageous roles of phosphate decorated octahedral CeO_2 {111}/g-C_3N_4 in boosting photocatalytic CO_2 reduction: Charge transfer bridge and Lewis basic site. *Appl. Catal. B* **2021**, *294*, 120257. [CrossRef]
39. Wang, Y.; Wang, H.; Chen, F.; Cao, F.; Zhao, X.; Meng, S.; Cui, Y. Facile synthesis of oxygen doped carbon nitride hollow microsphere for photocatalysis. *Appl. Catal. B* **2017**, *206*, 417–425. [CrossRef]
40. Tian, N.; Huang, H.; Liu, C.; Dong, F.; Zhang, T.; Du, X.; Yu, S.; Zhang, Y. In situ co-pyrolysis fabrication of CeO_2/g-C_3N_4 n-n type heterojunction for synchronously promoting photo-induced oxidation and reduction properties. *J. Mater. Chem. A* **2015**, *3*, 17120–17129. [CrossRef]
41. Chen, J.; Shen, S.; Wu, P.; Guo, L. Nitrogen-doped CeOx nanoparticles modified graphitic carbon nitride for enhanced photocatalytic hydrogen production. *Green Chem.* **2015**, *17*, 509–517. [CrossRef]
42. Zada, A.; Humayun, M.; Raziq, F.; Zhang, X.; Qu, Y.; Bai, L.; Qin, C.; Jing, L.; Fu, H. Exceptional Visible-Light-Driven Cocatalyst-Free Photocatalytic Activity of g-C_3N_4 by Well Designed Nanocomposites with Plasmonic Au and SnO_2. *Adv. Energy Mater.* **2016**, *6*, 1601190. [CrossRef]
43. Zou, Y.; Shi, J.-W.; Ma, D.; Fan, Z.; Niu, C.; Wang, L. Fabrication of g-C_3N_4/Au/C-TiO_2 Hollow Structures as Visible-Light-Driven Z-Scheme Photocatalysts with Enhanced Photocatalytic H_2 Evolution. *ChemCatChem* **2017**, *9*, 3752–3761. [CrossRef]
44. Ishikawa, A.; Takata, T.; Kondo, J.N.; Hara, M.; Kobayashi, H.; Domen, K. Oxysulfide Sm2Ti2S2O5 as a stable photocatalyst for water oxidation and reduction under visible light irradiation (lambda <= 650 nm). *J. Am. Chem. Soc.* **2002**, *124*, 13547–13553. [CrossRef] [PubMed]
45. Wang, J.; Yu, Y.; Zhang, L. Highly efficient photocatalytic removal of sodium pentachlorophenate with Bi_3O_4Br under visible light. *Appl. Catal. B* **2013**, *136–137*, 112–121. [CrossRef]

46. Luo, W.; Li, Z.; Jiang, X.; Yu, T.; Liu, L.; Chen, X.; Ye, J.; Zou, Z. Correlation between the band positions of $(SrTiO_3)_{1-x} \cdot (LaTiO_2N)_x$ solid solutions and photocatalytic properties under visible light irradiation. *Phys. Chem. Chem. Phys.* **2008**, *10*, 6717–6723. [CrossRef]
47. Ye, W.; Long, R.; Huang, H.; Xiong, Y. Plasmonic nanostructures in solar energy conversion. *J. Mater. Chem. C* **2017**, *5*, 1008–1021. [CrossRef]
48. Zheng, D.; Pang, C.; Wang, X. The function-led design of Z-scheme photocatalytic systems based on hollow carbon nitride semiconductors. *Chem. Commun.* **2015**, *51*, 17467–17470. [CrossRef]
49. Zhang, L.; Feng, W.; Wang, B.; Wang, K.; Gao, F.; Zhao, Y.; Liu, P. Construction of dual-channel for optimizing Z-scheme photocatalytic system. *Appl. Catal. B* **2017**, *212*, 80–88. [CrossRef]
50. Xu, Q.; Lei, W.; Li, X.; Qi, X.; Yu, J.; Liu, G.; Wang, J.; Zhang, P. Efficient Removal of Formaldehyde by Nanosized Gold on Well-Defined CeO_2 Nanorods at Room Temperature. *Environ. Sci. Technol.* **2014**, *48*, 9702–9708. [CrossRef]
51. Kesavan, L.; Tiruvalam, R.; Ab Rahim, M.H.; bin Saiman, M.I.; Enache, D.I.; Jenkins, R.L.; Dimitratos, N.; Lopez-Sanchez, J.A.; Taylor, S.H.; Knight, D.W.; et al. Solvent-Free Oxidation of Primary Carbon-Hydrogen Bonds in Toluene Using Au-Pd Alloy Nanoparticles. *Science* **2011**, *331*, 195–199. [CrossRef] [PubMed]
52. Zhang, G.; Huang, C.; Wang, X. Dispersing Molecular Cobalt in Graphitic Carbon Nitride Frameworks for Photocatalytic Water Oxidation. *Small* **2015**, *11*, 1215–1221. [CrossRef]

Review

Photocatalytic Water Splitting: How Far Away Are We from Being Able to Industrially Produce Solar Hydrogen?

Parnapalle Ravi [1] and Jinseo Noh [2,*]

[1] Bionano Research Institute, Gachon University, 1342 Seongnamdaero, Sujeong-gu, Seongnam-si 13120, Gyeonggi-do, Korea
[2] Department of Physics, Gachon University, 1342 Seongnamdaero, Sujeong-gu, Seongnam-si 13120, Gyeonggi-do, Korea
* Correspondence: jinseonoh@gachon.ac.kr; Tel.: +82-317505611

Abstract: Solar water splitting (SWS) has been researched for about five decades, but despite successes there has not been a big breakthrough advancement. While the three fundamental steps, light absorption, charge carrier separation and diffusion, and charge utilization at redox sites are given a great deal of attention either separately or simultaneously, practical considerations that can help to increase efficiency are rarely discussed or put into practice. Nevertheless, it is possible to increase the generation of solar hydrogen by making a few little but important adjustments. In this review, we talk about various methods for photocatalytic water splitting that have been documented in the literature and importance of the thin film approach to move closer to the large-scale photocatalytic hydrogen production. For instance, when comparing the film form of the identical catalyst to the particulate form, it was found that the solar hydrogen production increased by up to two orders of magnitude. The major topic of this review with thin-film forms is, discussion on several methods of increased hydrogen generation under direct solar and one-sun circumstances. The advantages and disadvantages of thin film and particle technologies are extensively discussed. In the current assessment, potential approaches and scalable success factors are also covered. As demonstrated by a film-based approach, the local charge utilization at a zero applied potential is an appealing characteristic for SWS. Furthermore, we compare the PEC-WS and SWS for solar hydrogen generation and discuss how far we are from producing solar hydrogen on an industrial scale. We believe that the currently employed variety of attempts may be condensed to fewer strategies such as film-based evaluation, which will create a path to address the SWS issue and achieve sustainable solar hydrogen generation.

Keywords: solar energy; photocatalytic water splitting; hydrogen production; thin films; large scale evolution

Citation: Ravi, P.; Noh, J. Photocatalytic Water Splitting: How Far Away Are We from Being Able to Industrially Produce Solar Hydrogen?. *Molecules* 2022, 27, 7176. https://doi.org/10.3390/molecules27217176

Academic Editor: Ruowen Liang

Received: 29 September 2022
Accepted: 21 October 2022
Published: 23 October 2022

Publisher's Note: MDPI stays neutral with regard to jurisdictional claims in published maps and institutional affiliations.

Copyright: © 2022 by the authors. Licensee MDPI, Basel, Switzerland. This article is an open access article distributed under the terms and conditions of the Creative Commons Attribution (CC BY) license (https://creativecommons.org/licenses/by/4.0/).

1. Introduction

A quick survey was conducted in Scopus to find out how many research projects on water splitting (both with and without sacrificial agents) had been conducted internationally, and a few selected efficient photocatalysts are summarized in Table 1. Despite the rise in reports of solar hydrogen generation over the past two decades, we think the field is still in its infancy and that intense efforts are required to achieve the breakthrough stage. So far, the greatest hydrogen generation efficiency recorded was obtained from a 6 L reactor exposed to UV radiation [1]. The reason why the efficiency of solar light-driven photocatalytic water splitting remained unchanged since the proof-of-concept work published in 1972 is crucial to understand [2]. There are numerous scientific explanations given in the reviews that have recently been published, including an increase in the absorption of light from various solar wavelength regions, manipulation of the band-edge positions of photocatalysts, an extension of the lifetime of photogenerated charge carriers, and the incorporation of photocatalysts with appropriate co-catalysts such as graphene, metals, or metal oxides [3–22]. The current review focuses primarily on several methods for photocatalytic water splitting that have been described in the literature, particularly with a view to increasing hydrogen output and scaling to greater regions when exposed to direct sunlight or one sun.

Table 1. A few highly efficient reported photocatalytic systems for better hydrogen efficacy.

Photocatalyst	Cocatalyst	Efficiency	Reference
TiO_2	Pt/RuO_2	QE: 30 ± 10% at 310 nm	[23]
$La_2Ti_2O_7$:Ba	NiO_x	QE: 35% at (<360 nm)	[24]
$Sr_2Nb_2O_7$	Ni	QE: 23% at (<300 nm)	[25]
$NaTaO_3$:La	NiO	AQE: 56% at 270 nm	[26]
Ga_2O_3:Zn	$Rh_{2-y}Cr_yO_3$	AQY: 71% at 254 nm	[27]
GaN:Mg/InGaN:Mg	Rh/Cr_2O_3	AQE: 12.3% at 400–475 nm	[28]
CDots-C_3N_4		AQE: 16% at 420 nm	[29]

A viable energy source, hydrogen has the advantages of clean energy, high conversion efficiency, and environmental friendliness. In the previous era, natural gas (or methane) steam reforming has been used to make hydrogen at a low cost, although this process actually produces additional greenhouse gases such as CO_2 as a byproduct. The 196-nation Paris Agreement, which was signed in April 2016, calls for a decrease in global warming through reduced emissions, which lowers the danger of climate change. As a result, it is critical to switch to using resources that are abundant in nature to generate green and clean fuels. One of the possible routes for the generation of hydrogen in this context is the SWS reaction under direct sunlight. To succeed on a commercial basis, though, there remains a very long way to go. Domen et al. recently outlined one such scenario [30]. It is suggested that 10,000 solar plants, each with a 25 km^2 surface area and a solar energy conversion efficiency of 10% under AM 1.5G irradiation, would be adequate to provide one-third of the predicted energy needs of the world's population by 2050. The solar power plants are anticipated to create 570 tons of H_2 per day from a total projected area of 250,000 km^2, which is equivalent to 1% of the desert area on Earth. It is essential to mention that the photocatalytic process creating hydrogen mimics natural photosynthesis, in which chlorophyll is known to act as a light absorber. Over the past few decades, numerous photocatalyst materials have been discovered and exploited by researchers in their efforts to generate solar hydrogen efficiently. Throughout the whole literature, the hydrogen evolution rate (HER) of photocatalyst materials in powder form is reported to range from a few hundred mmol h^{-1}·g^{-1} to 100 mmol h^{-1}·g^{-1}. Following are some key findings from a thorough analysis of existing literature reports: (a) When a photocatalyst is suspended, more light scatters than is absorbed, although even low-quality thin films appear to have substantially reduced this issue [31–36], thus, making films able to generate additional charge carriers from the first stage of light absorption. (b) Thin films, like solar photovoltaic panels, have a strong potential for scaling up to bigger size panels (supposedly, ~m^2) and will thus be favored for large-scale hydrogen generation. To create photocatalyst panels, a simplistic painting technique might be used [35,36]. (c) Using certain design features, stirring problems can be resolved with a thin-film/panel technique, which saves energy and is anticipated to reduce material disintegration because the catalyst films will not be subjected to mechanical strain. In this regard, the majority of issues with powder suspension may be solved using films for increased hydrogen production. Based on the various photocatalytic hydrogen generation methods described in the earlier, the present review goes into great detail about the drawbacks of the conventional method of conducting photocatalysis experiments, the benefits of thin film and panel forms of photocatalysts which help to reach commercial hydrogen production, and the underlying science.

2. Fundamentals of Water-Splitting Reaction

The water-splitting process entails several photophysical and photochemical stages. Initially, the first process is started by photon absorption, which creates an electron–hole (e^--h^+) pair. The electrons will move to the conduction band and the holes remain at the valance band. The position of the CB minimum (CBM) should be more negative than the water reduction potential of H^+/H_2 (0 V vs. RHE), while the VB maximum (VBM) should be more positive than the water oxidation potential of H_2O/O_2 (1.23 V). In order to generate more electron–hole pairs, effective light absorption with minimal light scattering is the crucial factor in addition to bandgap engineering of the material [35,37,38]. Moreover, it is the fact that particle catalysts have a considerable light scattering issue rather than an absorption one. The electron–hole pair should be separated and diffuse towards the respective redox co-catalyst sites present in the photocatalyst. Due to (a) charge carriers' short lifespan and (b) chemical processes' lengthy time scales, a significant portion of charge carriers (>90%) undergo recombination; this leads to poor net redox reaction, which is one of the major issues of photocatalysis. It should be stressed that the photocatalyst's function is essential for efficient light harvesting.

General redox reactions that take place to produce oxygen and hydrogen as follows:

$$2H^+ + 2e^- \rightarrow H_2 \quad (1)$$

$$2H_2O + 4h^+ \rightarrow 4H^+ + O_2 \quad (2)$$

The main factors increasing SWS efficiency are electron–hole pair separation and diffusion. It should be emphasized once more that charge recombination occurs more quickly than charge usage. To address this issue, the scientific community has created a number of materials and techniques [16].

2.1. Hydrogen Generation Using Scavengers/Sacrificial Agents

As previously mentioned, it is essential to separate the electron-hole pairs and diffuse them towards the redox co-catalyst sites to reduce the amount of recombination. In this context, use of sacrificial agents is one of the strategies involved to improve the efficiency. In order to devour the holes, organic molecules are often utilized as the sacrificial agents and involved in photooxidation. In the literature, a plethora of sacrificial agents such as alcohols, amines, and sulphides are used for better HERs [39]. The choice of the sacrificial agent must be carefully considered in order to achieve effective HER. To facilitate the simple electron/hole transfer between light-harvesting photocatalysts and sacrificial reagents, the energy levels (VBM and CBM) of the two materials must match for the sacrificial agent to be chosen. In this context, Na_2S/Na_2SO_3 is used for high HERs in the majority of chalcogenide-based photocatalysts, including CdS, PbS, ZnS, and MoS_2, which are also known to be visible light-active materials. However, for improved hole utilization and thus superior HERs, TiO_2 and titania-based composites employ alcohol molecules including methanol, ethanol, and glycerol. However, triethanol amine (TEOA) is used as a sacrificial agent for enhanced HERs in the instance of graphitic carbon nitride (g-C_3N_4), which is recognized to be one of the intriguing materials that functions as a visible light-active material with a bandgap of 2.4 eV [8]. However, the quantity of hydrogen produced from the sacrificial agent and water is still up for discussion along with oxidized byproducts such as CO_2.

Here is the detailed hole utilization mechanism of categorized sacrificial reagents such as alcohols, amines, and sulphides as follows [39,40].

2.1.1. Alcohols (Glycerol):

Glycerol adsorbed on the surface of photocatalyst will interact with holes and forms radicals as shown in Equation (5). Once glycerol engages the holes, the excited electrons on the conduction band consequently produce H_2 gas as shown Equations (3)–(9). The hydroxyl glycerol in the second stage is unstable, and it interacts with holes, resulting in the formation of glyceraldehyde as shown in Equation (6). Glyceraldehyde dissociates further producing an unstable CO radical that combines with a few holes and eventually converts to CO_2 as shown in Equations (7) and (8).

$$Metal\ oxide\ NP \xrightarrow{h\nu} h^+ + e^- \qquad (3)$$

$$3H_2O + 3e^- \rightarrow 3H^+ + 3OH^* \qquad (4)$$

$$C_3H_8O_3 + 3h^+ \rightarrow C_3H_5O_3^* + 3H^+ \qquad (5)$$

$$C_3H_5O_3^* + 3h^+ \rightarrow C_3H_2O_3^* + 3H^+ \qquad (6)$$

$$C_3H_2O_3^* + 2h^+ \rightarrow 3CO^* + 2H^+ \qquad (7)$$

$$3CO^* + 3OH^* \rightarrow 3CO_2 + 3H^+ \qquad (8)$$

Overall net reaction

$$14H^+ + 14e^- \rightarrow 7H_2 \qquad (9)$$

2.1.2. Triethanolamine (TEOA):

In the case of TEOA as sacrificial reagent, TEOA becomes formaldehyde and then to hydrogen as shown in Equations (14) and (15).

$$Photocatalyst \xrightarrow{h\nu} h^+ + e^- \qquad (10)$$

$$2H^+ + 2e^- \rightarrow H_2 \qquad (11)$$

$$h^+ + H_2O \rightarrow OH^* + H^+ \qquad (12)$$

$$h^+ + OH^- \rightarrow OH^* \qquad (13)$$

$$TEOA + OH^* \rightarrow HCHO + NH_3 \qquad (14)$$

$$HCHO + OH^* \rightarrow H_2 + by\ products(CO_2) \qquad (15)$$

2.1.3. Sodium Sulphide and Sodium Sulphite Mixture (Na_2S and Na_2SO_3):

$$Photocatalyst \xrightarrow{2h\nu} 2h^+ + 2e^- \qquad (16)$$

$$At_{CB}(2e^-) + 2H_2O \rightarrow H_2 + 2OH^- \qquad (17)$$

$$At_{VB}(2h^+) + SO_3^{2-} + 2OH^- \rightarrow SO_4^{2-} + H_2O \qquad (18)$$

$$2SO_3^{2-} + 2h^+ \rightarrow S_2O_6^{2-} \qquad (19)$$

$$2S^{2-} + 2h^+ \rightarrow S_2^{2-} \qquad (20)$$

$$S_2^{2-} + SO_3^{2-} \rightarrow S_2O_3^{2-} + S^{2-} \qquad (21)$$

In this type of systems, sulphide (S^{2-}) and sulphite (SO_3^{2-}) can act as sacrificial inorganic reagents for the photocatalytic hydrogen generation because they are very efficient hole acceptors, enabling the effective separation of the charge carriers. The oxidation of S^{2-} and SO_3^{2-} can occur either by a two-electron transfer process (16) to (18) or one-electron oxidation (19) to (21).

2.2. Overall Water Splitting

Overall water splitting (OWS) is an act of producing hydrogen that involves utilizing water as the lone reactant at a pH of 7 with the right photocatalyst under full sunshine. Even if HER and OER happen as shown in Equations (1) and (2), the rate of HER is much lower than it would be with a sacrificial agent. As a matter of fact, the HER values reported from OWS are at least 2–3 orders of magnitude less than those reported using sacrificial agents. The OER (Equation (2)) is a four electron process that cause sluggish kinetics and lowers the total rate of OWS. In this regard, several catalyst materials have been published in the literature with the aim of enhancing the overall kinetics of OWS and so achieving efficient hydrogen production. In this scenario, Domen and colleagues as well as several others have made contributions using a Z-scheme method to effective hydrogen production [17,29,41–46]. In which $BiVO_4$:Mo and $SrTiO_3$:La,Rh used as photocatalyst-I (PS-I) and photocatalyst-II (PS-II), and Au used as electron mediator. The CB of PS-I is closer to the VB of PS-II via Au metallic electron mediator to reduce the resistance to the migration of electrons and holes. Whenever the composite got excited with light energy, both PS-I and PS-II generates photoexcitons and the electrons at CB of PS-I will combined with the holes at VB of PS-II via Au mediator. The terminal electrons and holes will take place in redox reaction. However, compared to the anticipated STH (solar to hydrogen) conversion efficiency values (10% and higher), the efficiency (1%) that has been recorded so far is quite low for commercial-scale hydrogen generation. We think there is a chance to raise the total rate of OWS by employing alternate strategy such as the thin film technique. The efficiency of converting solar energy into hydrogen is often measured using one of two techniques as follows.

$$\text{Apparent quantum yield (AQY)} = 2 \times \frac{\text{no of hydrogen molecules}}{\text{no of photons}} \times 100 \qquad (22)$$

From the above equation, the number of incident photons can be measured using a radiant power energy meter, which considers the following equation (Equation (23)):

$$E = nh\nu = nh\frac{C}{\lambda} \qquad (23)$$

where E stands for incoming light energy, n stands for photon number, h for Planck's constant $(6.634 \times 10^{-34}\ j\ S)$, c for light velocity, and λ for incident light wavelength.

2.3. Factors for Achieving the High Efficiency of SWS

It is practically impossible for a single semiconductor material to function as an effective SWS photocatalyst and to carry out all three of the key processes of (a) light absorption from the broad spectrum of sunlight, (b) charge separation and diffusion to redox sites, and (c) actual redox reactions.

Habitually, a composite photocatalyst made up of at least two (for example, Pt and TiO_2), three (for example, Au, rGO, and TiO_2), or more (for Z-scheme photocatalysts) components are effective to move closer to overcome the above-mentioned key processes. Very few researchers have tried to combine the three basic photocatalytic processes using effective synthetic techniques [37,47,48]. Furthermore, a charge diffusion is one approach that might be used to improve the activity by seamlessly combining the charge generating sites and charge usage (or redox) sites [16]. To accomplish this, designing a highly integrated single composite material for effective hydrogen generation from SWS is required. A systematic increase in the complexity of the photocatalyst design of Au-gC_3N_4/TiO_2 (221 mmol $h^{-1} \cdot g^{-1}$) outperforms SWS kinetics from g-C_3N_4/TiO_2 (91 mmol $h^{-1} \cdot g^{-1}$) [49]. When an integrated Au–N–TiO_2–graphene composite was used instead of titania alone, the HER of 525 mmol $h^{-1} \cdot g^{-1}$ was found [50]. The improvement in activity with 1275 mmol $h^{-1} \cdot g^{-1}$ and sustained activity for 125 h were attributed to the electronic integration of an Au–Pt bimetal cluster with titania [51]. On titania, a similar observation was achieved using Au–

Ag or Au–nanorod [52,53]. According to a recent study, a single titania nanotube material with heterojunctions between native and nonnative structures, such as the presence of the anatase, rutile, and brookite phases, and Pt as the co-catalyst exhibited higher HER activity from SWS under one sun conditions than anatase-rutile (2.5 mmol h^{-1}·g^{-1}) or bare anatase (0.15 mmol h^{-1}·g^{-1}) [54,55]. The difference in activity is attributable to the existence of bulk heterojunctions in a composite material as opposed to a single junction (anatase-rutile) or bare anatase nanotubes, which can aid in effective charge transfer between the particles. Independent of the photocatalyst systems, the outcomes given here suggest that SWS activity increased from a negligible value to 7.6 mmol h^{-1}·g^{-1} [47–54]. Furthermore, the HER was measured using a photocatalyst made of Au–Pd/rGO/TiO$_2$ both in particulate and thin-film forms by easy casting on a glass plate. The measured HERs from thin-film and particulate versions are 21.5 and 0.50 mmol h^{-1}·g^{-1}, respectively. It is specifically due to the difference of light absorption. Despite having all of the necessary photocatalyst components, the HER values are incredibly low for particulate photocatalyst systems and the evaluation of HER in the particulate form of the photocatalyst composite acts as a unifying factor among all these findings. Simple comparisons between photocatalyst suspension and the identical photocatalyst preserved in an aqueous methanol solution as a thin film demonstrate (Figure 1) that the former exhibits low light penetration while the later exhibits high light penetration [56]. Regardless of the angle of light incidence, light penetration would be inadequate with photocatalyst suspension.

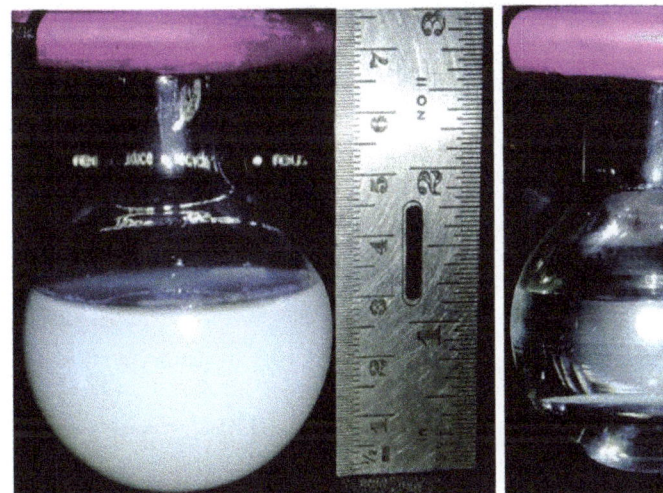
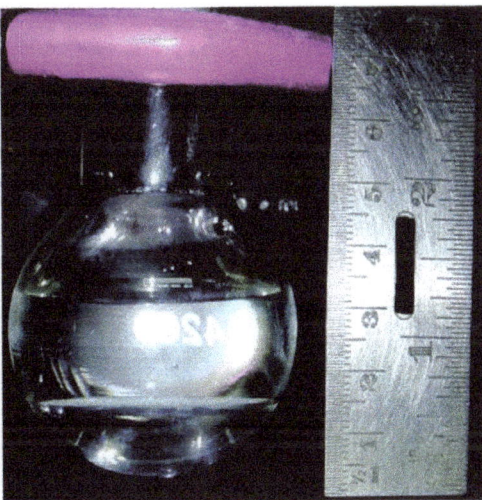

Figure 1. A digital image of a quartz reactor with a capacity of 70 mL (56 mm in diameter at the center) filled with 40 mL of solution and 25 mg of titania-based catalyst powder while it was spinning (left), and in static circumstances, using thin films produced with 1 mg of the same photocatalyst (right). Replicated from ref. [56].

Although there are multiple reasons that prevent SWS from operating at a high efficiency, major SWS limiting issues and some potential solutions are provided as listed below:

(a) The primary limiting element is the very different time scales between photophysical and photochemical processes and how to connect them. Structural and electronic components integration is crucial to be optimized for efficient diffusion of charge carriers to redox sites and their exploitation for redox reactions. Quantum dots (QDs) and 2D-layered materials might be used to overcome the aforementioned issue. It is also important to note that the nanoscience, which only involves photophysical processes, has made a significant contribution to the rapid expansion of applications involving light emission. It is very desirable to use synthesis methods that would result in bulk heterojunctions in a photocatalyst composite. For instance, the assembly of QDs in the pores of wide bandgap materials using techniques such as SILAR (successive ionic layer adsorption and reaction) results in bulk heterojunctions.

(b) In general, scaling up the catalyst quantity, even from 10 to 100 to 1000 mg at a laboratory level, substantially reduces the effectiveness of any photocatalyst system. This is largely because of poor light absorption combined with excessive charge recombination [57]. The current review tackles this issue in depth, through the use of a photocatalyst thin film or a panel with increased hydrogen generation.

(c) Since OWS is often carried out in extremely acidic environments [58], it is essential to be able to conduct the tests at a pH close to neutral (pH = 7). There is definitely a need for greater study in this area.

(d) To the best of our knowledge, noble metals are frequently used, which is not a cost-effective solution, and there is no reasonable consideration given to the selection of a certain co-catalyst for a specific semiconductor [58]. More research must be done to examine more affordable and plentiful co-catalysts, with a focus on rational selection.

(e) It is necessary to switch to using environmentally beneficial and/or biomass-derived materials such as glycerol and cellulose instead of sacrificial ones such as methanol. Sacrificial agent use in water splitting may be a temporary fix, and OWS will be the long-term fix [39].

(f) Despite the fact that CdS, PbS, and other chalcogenide QDs have very strong visible light absorption qualities and the capacity to control the bandgap, they are also vulnerable to photo-corrosion and are unfriendly to the environment [59,60]. The oxide-based QDs should be the focus of additional efforts.

Until now, a lot of work has gone into creating effective systems for the SWS process, which produces hydrogen. However, none of the photocatalysts can produce hydrogen in a practical manner. Indeed, under sunshine or one-sun circumstances, none of the powder catalyst systems have been studied at the gram scale. Table 2 lists some of the top hydrogen production activity values obtained using several powder-based catalysts.

Table 2. The highest recorded solar hydrogen production using several photocatalyst systems that have been thoroughly studied in powder form.

Photocatalyst	Co-Catalyst	Sacrificial Agent	Light Source	H_2 Yield (mmol $h^{-1} \cdot g^{-1}$)	Catalyst wt. (mg)	Stability (h)	Ref
ZnS–In$_2$S$_3$–CuS$_2$	—	0.1 M Na$_2$S + 1.2 M Na$_2$SO$_3$	300 W xenon lamp (UV-cut off filter, λ > 420 nm)	360	10	9	[61]
TiO$_2$/CdS	Pt	Lactic acid	300 W xenon lamp (λ > 400 nm)	128.3	50	15	[62]
CdS nanowires	MoS$_2$	Lactic acid	300 W xenon lamp (UV-cut off filter, λ > 420 nm)	95.7	20	24	[63]
CdS@TiO$_2$	Pt	0.1 M Na$_2$S + 0.02 M Na$_2$SO$_3$	Sun light	44.8	10	24	[64]
CdS	MoS$_2$	Ethanol	300 W Xe arc lamp (λ > 420 nm)	140	10	150	[65]
CdS	CoP	Lactic acid	300 W Xe arc lamp (λ > 420 nm)	106	20	18	[66]
CdS	MoS$_2$	Lactic acid	300 W Xe arc lamp (λ > 420 nm)	49.8	200	24	[67]
CdS–titanate	Ni	Ethanol	Sun light	31.82	100	15	[68]
ZnS–In$_2$S$_3$–Ag$_2$S	—	0.6 M Na$_2$SO$_3$–0.1 M Na$_2$S	300 W Xe arc lamp (λ > 420 nm)	220	15	16	[69]
TiO$_2$	Au-Pt	Methanol	One sun condition	6	20	125	[70]

The current analysis concentrates on the additional crucial factors that contribute to improving the overall effectiveness of solar light-driven water splitting. The current study stresses a thin film-based technique. It is advantageous to boost efficiency as opposed to the particulate-based research that is commonly used by many researchers worldwide.

2.3.1. Mechanical Stirring Is Unfavorable

It is difficult to analyze the photocatalyst powder on a big scale, and continual stirring of the powder solution is necessary to enhance light harvesting. It is an energy expensive when mechanical churning is used at such huge loads. Following the conclusion of reaction investigations utilizing photocatalyst powders, the collection of the catalyst using centrifugation and filtering is another time-consuming and energy-intensive step [71]. Above all, the essential notion of better light absorption does not appear to happen with an increase in the photocatalyst quantity in the suspension. Additionally, when the catalyst concentration rises, light cannot reach all areas of the solution due to the turbidity of the solution. Moreover, the price of hydrogen generated using traditional steam reforming techniques is $3–4 per kg, therefore any new approach needs to at least be competitive with that to be taken into consideration. The typical thin film method of producing solar cells would be advantageous for better light absorption, and many benefits and drawbacks of photocatalysts in their thin-film and particulate forms, respectively, will be examined in this context.

2.3.2. Loading Effect

It is anticipated that more catalyst will need to be loaded into the solution in order to produce solar hydrogen on a wide scale. In this context, Maeda et al. examined the impact of loading on the quantum yield of hydrogen for a Z-scheme photocatalyst using $Pt/ZrO_2/TaON$ as the hydrogen evolution catalyst and Pt/WO_3 as the oxygen evolution catalyst [57]. It was discovered that the AQY drops from 6.3 to 2.7% when the total catalyst dosage rises from 75 to 150 mg. In contrast to the predicted rise in light absorption with an increase in the quantity of powder catalyst loading, only a portion of the particles are exposed to photons at a given moment, and the remainder particles are inactive. However, when there is only a tiny quantity of catalyst present in the test solution, the majority of the particles are exposed to photons and form a greater number of charge carriers, which helps to achieve high efficiency. This is directly corroborated by the research done by Nalajala et al., who found that using 25 mg of powder Pd/TiO_2 in 40 mL of aqueous methanol produced less H_2 (9 mmol $h^{-1} \cdot g^{-1}$) than using 1 mg of the same catalyst under the same circumstances (32 mmol $h^{-1} \cdot g^{-1}$) [34]. The hydrogen yield (HY) per gram of catalyst in water splitting studies with a tiny quantity of catalyst (1 mg) would appear to be quite high (Table 3). Reporting the hydrogen evolution rate for a few different weights for a certain volume of water or aqueous solution is the preferred method, since it enables other labs to replicate the findings. However, this raises the unavoidable question of what is the best method to use for handling massive quantities of catalyst for SWS.

2.3.3. Scale-Up and Disintegration Issues of Photocatalysts

Under reaction circumstances, the catalyst system must remain intact in order to produce hydrogen from SWS sustainably. The catalyst system in particular is anticipated to be stable for prolonged exposure to light and a wet environment. The development of photocatalyst materials in the particulate form is now the focus of several initiatives. The greatest recorded hydrogen production activity using various catalyst systems are shown in Table 2; However, they were selected because (a) they could demonstrate sustainability for at least 10 h in one day of sunlight and (b) the catalyst quantity used in real studies needed to be at least 10 mg. Whatever the provided catalysts were confirmed to be active, only a small number of investigations have demonstrated their stability for 100 h or more [65,70]. A contrasting truth may be found by simply comparing the findings provided in Tables 2 and 3. Higher HER values were seen with 1 mg of catalyst from the actual data (Table 3) compared to those shown in Table 2, which use 10–200 mg of catalyst. As shown in Figure 1, a suspension form of a catalyst cannot increase activity linearly with an increase in catalyst amount. A few studies also indicate a decline in activity within 10 h after the reaction, raising the likelihood of catalyst breakdown [61]. The activity and stability of the thin film were established by Schroder et al. [35], during the 30 days active light period, and Goto et al. investigated the films for 1000 h [36]. A stable photocatalyst would sustain HER activity for a very long time.

Table 3. Highest recorded solar hydrogen production using several photocatalysts that were tested in powder form at low concentrations.

Photocatalyst	Co-Catalyst	Sacrificial Agent	Light Source	Hydrogen Yield (mmol h^{-1}·g^{-1})	Mass (mg)	Stability (h)	Ref.
CdS nanorods	Ni$_2$P	Na$_2$S–Na$_2$SO$_3$	300 W Xe lamp (λ > 420 nm)	1200	1	12	[72]
CdS nanorods	Co$_x$P	Na$_2$S–Na$_2$SO$_3$	300 W Xe lamp (λ > 420 nm)	500	3	25	[73]
CdS/ZnS	—	0.5 M Na$_2$SO$_4$	300 W Xe lamp (λ > 400 nm)	239	1	12	[74]
CdS nanorods	FeP	Lactic acid	LED: 30 × 3 W, λ > 420 nm	202	5	100	[75]
CdS nanorods	Cu$_3$P	Na$_2$S–Na$_2$SO$_3$	300 W Xe lamp (λ > 420 nm)	200	1	12	[76]
CdS nanorods	WS$_2$	Lactic acid	300 W Xe lamp (λ > 420 nm)	185.8	1	50	[77]
CdS nanorods	Co$_3$N	Na$_2$S–Na$_2$SO$_3$	300 W Xe lamp (λ > 420 nm)	137.3	1	48	[78]
CdS nanorods	PdPt	Lactic acid	150 W Xe lamp	130.3	1	20	[79]
CdS/MoS$_2$	—	Lactic acid	Sunlight	174	1	25	[80]
MoS$_2$-RGO-CoP/CdS	MoS$_2$/CoP	Lactic acid	Sunlight	83.9	1	20	[81]

3. Light Absorption and Scattering

While there are several semiconductor materials with adequate band gaps and band edge locations for SWS, effective light absorption in their powder state is a significant problem. According to the Beer–Lambert equation, the absorption coefficient that depends on wavelength and material thickness determines how much light is attenuated. However, it is typically believed that enough light absorption in the normal UV-Vis absorption spectrum is required for its effective light absorption properties. In actuality, smooth, high-quality solid surfaces absorb solar light more effectively than rough, poorly absorbent particle surfaces.

As seen in Figure 2, light is primarily dispersed from all sides of particles (in suspension) as opposed to thin films, which typically absorb light due to their comparatively flat surfaces and high absorption coefficients. Thin films can actually absorb more light with a 50–500 nm uneven surface structure than with a pristine smooth surface, which is undesirable since it would reflect more light, and an anti-reflective coating used in solar cells would be necessary for very flat surfaces. It is anticipated that the thin film-based photocatalyst would yield more charge carriers than the particulate catalyst [6]. When compared to the intensity at the material's surface, light penetration depth (also known as skin depth) is described as a drop in light intensity to 37% (or 13%). The same reasoning holds true here as they did in the case of dye-sensitized solar cells (DSSCs), where this feature has been well investigated. For maximal incoming light absorption, a thickness range of 8–12 µm is recommended. The decreased thickness of the material in a film with a thickness of less than 8 µm results in less light absorption. Whereas, due to the additional layers of materials present at the bottom FTO/ITO plate for solar cell applications, films thicker than 12 µm result in increased recombination as the charge carriers must diffuse across these layers.

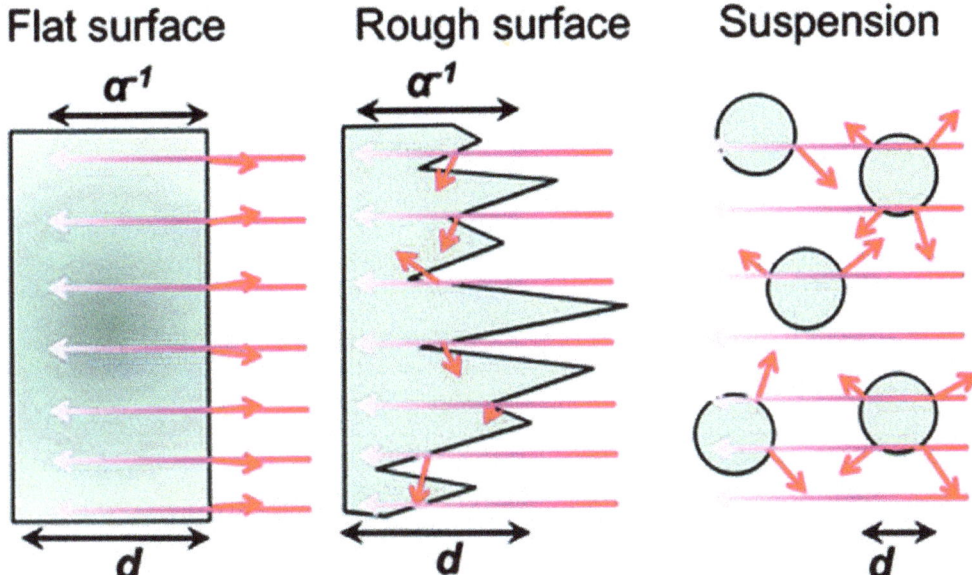

Figure 2. Dispersion of light in nanostructured and flat films as well as in suspensions of particles. A α^{-1} is the optical penetration depth, and d is the film or particle thickness. Short arrows denote light that has been reflected or dispersed. Replicated from ref. [13].

4. Towards Enhanced Hydrogen Production with a Thin-Film Approach

The advantages, production methods for large-scale thin films, their photocatalytic activities, stability factors, and the re-absorption of emitted light for better activity are all covered in this part along with other procedures that are successful for SWS.

4.1. Thin Film Approach

By placing active photocatalysts as films over a fixed substrate, a considerable number of problems with the particulate form of the catalyst may be resolved. Better light absorption may be anticipated than with equivalents in the solution that are suspended powders since the catalyst is attached to the substrate with the specified thickness. Use of centrifugation for gathering the suspended powders or mechanical stirring for better dispersion will be avoided. Small volumes of liquid may be used to operate thin films for reactions, which is unquestionably a procedure that uses less time and energy. Furthermore, this method may be an economically feasible choice for sustainable hydrogen generation due to the minimal infrastructure (little number of materials and no expensive equipment needed) and simplicity of scaling up.

Advantages of Thin Films over Particles for Photocatalysis of Hydrogen Production

It is vital to draw attention to a number of benefits that a thin film form has over a particle form. The following details are noteworthy:

(i) Light absorption: As illustrated in Figures 1 and 2, films rather than particle suspensions are more commonly used in the initial stage of effective light absorption in photocatalysis. Thin films produce a significant number of charge carriers by efficient and consistent light absorption. Thin films allow for the generation of many charge carriers since the photocatalyst is fully and evenly exposed to the photon source during the whole time period. This is clear from the video depiction of solar hydrogen that is discussed in the literature [34].

(ii) Maximal activity, little input, and continuous process: It should be stressed that there is a significant decrease in hydrogen production in photocatalysis studies with a particulate suspension as catalyst concentration rises. High hydrogen yields are produced using a thin film that is the ideal thickness (8–12 μm). The maximal activity with the least amount of material is shown by up to an order of magnitude and greater activity recorded with the same quantity of the substance in thin film form as opposed to powder. Additionally, as opposed to a batch procedure using suspension, the thin film makes the process continuous.

(iii) Energy saving: Due to the lack of mechanical stirring, which may be replaced by centrifugation for catalyst removal in the following batch or recycling, it is anticipated that thin film-based solar panels would have much lower operating costs. However, they will involve suspension, thus there is a cost/energy consideration.

(iv) Running cost: It should be noted that solar hydrogen production with thin films is also possible with water (or solution) layers as thin as 1 mm, and hydrogen bubbles are discharged smoothly. However, it is possible to further reduce the water layer thickness to a few nanometers. This will be crucial for using extremely effective catalysts, which would continuously and without resistance create hydrogen bubbles. The smooth and immediate release of bubbles is made possible by the thinness of the water layer. It should be noted that delayed bubble development reduces the catalyst's ability to absorb light since bubbles have a strong tendency to disperse light.

(v) Coalescing diverse material components: The artificial leaf and QuAL ideas propose to combine many elements, such as co-catalysts for reduction and oxidation processes and various light-absorbing quantum dots in one device. Therefore, this method takes care of the (structural and maybe electronic) integration of various material components, whereas it is challenging to evenly include all of the aforementioned elements throughout the particle bulk catalyst.

(vi) Mass transfer issues: The greater engineering problem is distributing the reactant(s) equally across the film/panel device. However, since there is no need for pressure and hydrogen may be produced with or without scavengers in merely a millimeter of liquid water thickness, a simplistic tilting mechanism may be used to circulate the solution using gravity. In addition, tilting and sun tracking might be properly integrated to increase hydrogen generation.

4.2. Scaling Up and Thin Film Preparation Techniques

Thin film technology is well recognized to have many uses in diverse science and technology fields, including solar cells, optics, electronics, and many more. Low material usage and the frequent use of flexible substrates are benefits of thin-film devices that are difficult to obtain when applying bulk materials directly. It should be noted that thin film characteristics and activity are frequently influenced by the type of deposition technique. Because of uncomplicated, inexpensive, and convenient features for depositing a wide range of materials, chemical techniques are often effective for the deposition of large-area thin films. Although there are several photocatalysts available to produce the solar hydrogen, the discoveries made in the lab must be scaled up. In this context, a few techniques, including spray coating, slot-die coating, and screen printing, are mentioned in the literature to create films of various diameters [82]. For the film preparation using particle precursor photocatalysts, there are just a few reports accessible at this time. The following section contains some literature reports on various photocatalysts prepared as thin films using drop-cast and particle transfer techniques.

4.2.1. Drop-Cast Method

A thin layer of a $Rh_{2-y}Cr_yO_3/(Ga_{1-x} - Zn_x)(N_{1-x}O_x)$ photocatalyst was created by Xiong et al. using a drop casting approach on a flat piece of frosted glass [83]. The researchers added silica powder with different particle sizes (from nanometers to micrometers) to the photocatalyst in order to control the porosity and hydrophilic character of thin films. To create a homogeneous suspension, the photocatalyst powder (20 mg) was sonicated in 200 mL of deionized water. A quantity of 20 mg of silica powder was added with the aforementioned combination and held for sonication in order to better understand how the photocatalyst's porosity and hydrophilic nature affected its performance. The suspension was then dropped onto a clean, frosted glass plate and held there to dry while it was heated to 323 K. In order to create a film with a consistent thickness, this operation was done ten times.

In a different study by Schroder et al., a Pt@mp–CN (Pt cocatalyst on mesoporous carbon nitride) photocatalyst was immobilized on a stainless-steel plate by a drop casting methodology over different substrate dimensions, such as $3.5 \times 3.5 \times 0.25$ cm^3 for the laboratory reactor and $30 \times 28 \times 0.1$ cm^3 for the demonstration reactor, as shown in Figure 3I [35]. Under direct sunlight, the photocatalyst panels were employed to produce solar hydrogen. It is important to note that the negative zeta potentials of both Pt and carbon nitride caused Pt nanoparticles to be manufactured separately by a microemulsion process [84].

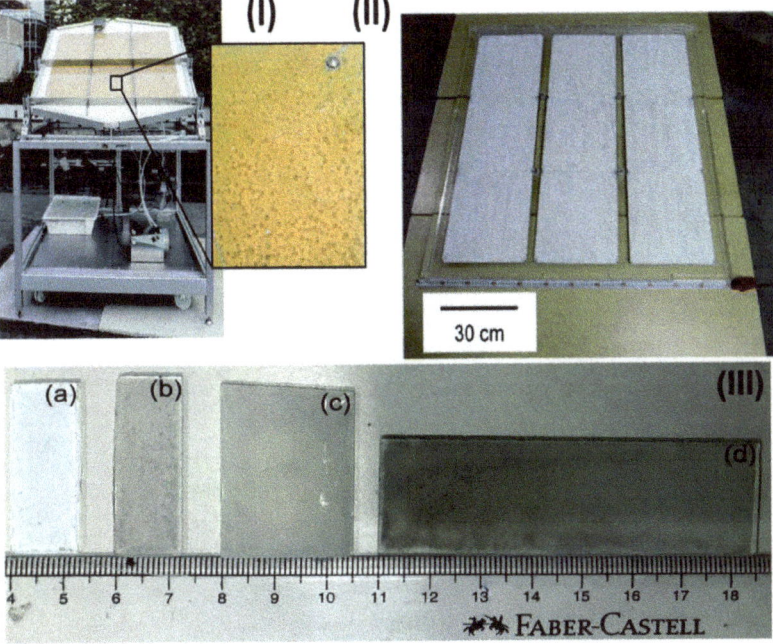

Figure 3. Images of panels and thin sheets in various sizes: (**I**) picture of the large-scale hydrogen generation system for photocatalysis illuminated by the sunshine. Nine stainless steel plates, each measuring $28 \times 30 \times 0.1$ cm^3, make up a photocatalyst (Pt@mp–CN) panel. The mesoporous carbon nitride photocatalyst emitting hydrogen bubbles can be seen when magnified. (**II**) Nine 33×33 cm^2 sheets make up a 1×1 m^2 SrTiO$_3$:Al photocatalyst panel. (**III**) Sizes of thin films (**a**) 1.25×3.75 of P25 (**b**) 1.25×3.85, (**c**) 2.5×3.75, and (**d**) 2.5×7.5 cm^2 of Pd/P25. Replicated from ref. [36].

By using a similar approach to Schroder et al., Goto et al. developed a 1×1 m^2 size SrTiO$_3$-Al panel (Figure 3II) for large-scale hydrogen production [36]. The study focused on a number of factors to enhance the performance of panel-type catalysts for water splitting reactions, including tilting angle (10–20°) and water layer thickness (1–5 and 5 mm). These factors are crucial to reduce liquid weight and, consequently, pressure, in order to achieve the smooth release of gas bubbles. Without induced convection, a layer of 1 mm thick water might release hydrogen bubbles.

Prominently, this study showed that crack development did not impair hydrogen production in contrast to PEC cells and PV-based electrolyzers, where photocatalyst sheets do not form a series circuit in the panels [36]. The constructed photocatalyst panels were examined for H_2 production from pure water under both artificial and natural light sources.

In a recent study, Nalajala et al. developed thin films of titania-based photocatalysts (Pd/P25) on glass plates and assessed their hydrogen generation capabilities in the presence of direct sunlight. The creation of diverse sized and shaped thin films was accomplished using a drop-cast process (Figure 3III). They employed methanol as a sacrificial agent, Pd as a co-catalyst, and P25 (TiO_2) as a light-harvesting material to create hydrogen in their experiment. Similar to the previous research by Goto et al., the thin films showed micron-sized fracture development, although this did not impair the thin film's functionality.

4.2.2. Particle Transfer Method

For the creation of thin films from powder catalyst suspensions, the particle transfer (PT) method was developed. In this procedure, a combination of photocatalyst granules was dispersed in isopropanol and then applied to clean glass plates measuring 3 × 3 cm^2. A thin layer of catalyst film was created over a glass plate by quickly drying the catalyst suspension after it had been deposited (Figure 4). The contact layer that will make contact with semiconductor particles was created using a vacuum evaporation process. As a contact layer, any of the metals (Au, Ag, Al, Ni, or Rh) can be employed. Then, using a vacuum evaporation process, an additional Au layer was deposited over the previously developed contact layer of the film in order to provide the prepared film with the necessary conductivity and mechanical stability. Taking off the first or primary glass plate, the metal film that makes up the particle photocatalysts was adhered to a second glass plate using carbon tape. To remove any extra particles that had accumulated on the photocatalyst layer, the resultant glass plate underwent a brief ultrasonic treatment.

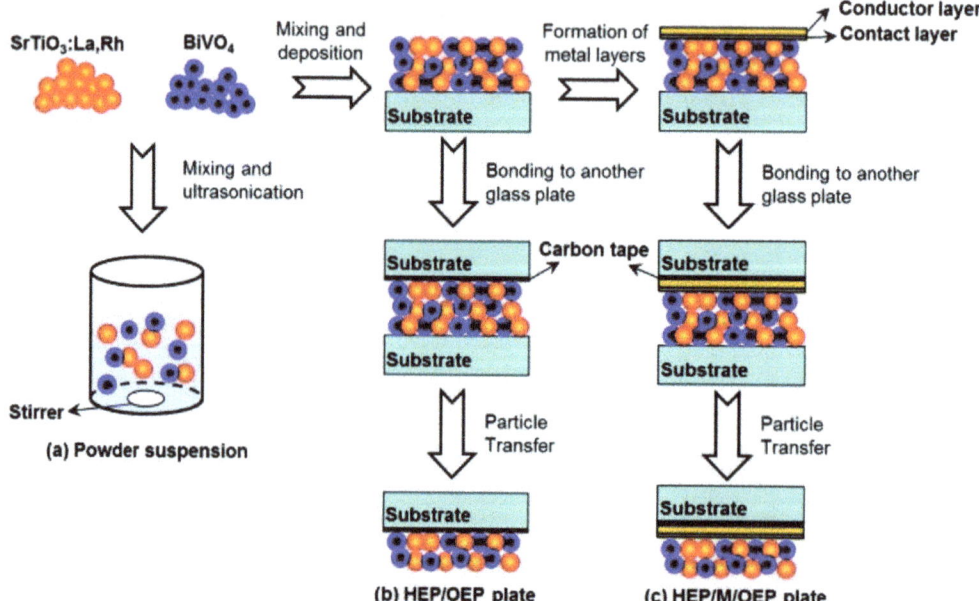

Figure 4. An illustration showing how photocatalyst particles are transferred across glass substrates to create photocatalyst sheets: (a) preparation for powder suspension, (b) preparation of HEP/OEP plates on carbon tape and without a metal layer, and (c) preparation of HEP/M/OEP plates with sandwiched metal layers on carbon tape. Replicated from ref. [41].

Domen and coworkers used this PT approach extensively and published their findings for a range of particle photocatalyst materials [29,41–45,84–87]. The possibility of destroying thin films when removing (lifting off) the main glass plate is a major worry with this method of preparation. In order to facilitate the conductivity between the HEP and OEP particles, it is also important to take into account that a complicated equipment needed to generate the metal contact layer.

4.2.3. Screen Printing

One of the reliable methods that offers large-scale thin films with maximal material usage is screen printing. The drying and/or sintering processes were crucial to dictate the production capacity of screen printing rather than the screen-printing method itself. Hu et al. showed that the screen printing (Figure 5a) can give the cells of 10×10 cm^2 size with an efficiency of 10% and exhibited its stability under light for 1000 h, stability outside for 30 days, and stability over a year of storage [88].

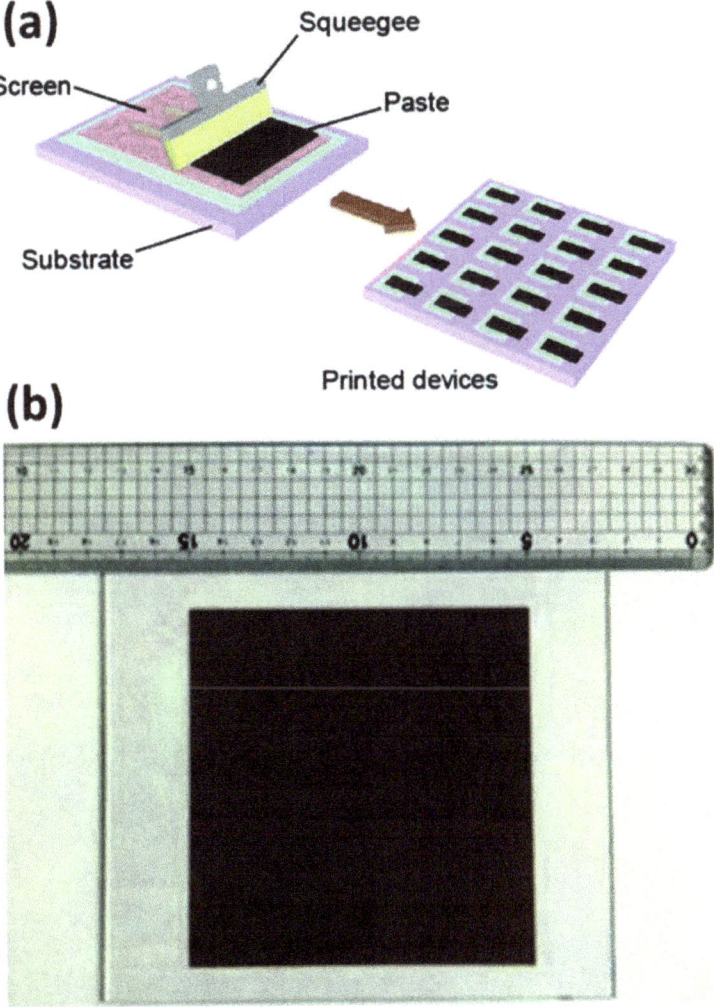

Figure 5. (a) Schematic for screen printing method, and (b) photo image of a 10×10 cm^2 screen printed device. Replicated from ref. [29,82].

This method was also developed by Wang et al., who used a powder photocatalyst to generate large-scale thin films [29]. In their investigation, printing ink was made from a combination of organic binders, catalyst powder (SrTiO$_3$:La, RhBiVO$_4$:Mo), and Au colloid solution in ethanol. The paste was screen printed on a 3 × 3 cm^2 glass substrate after the ethanol had evaporated. Furthermore, this method may also be used to create a 10 × 10 cm^2 photocatalyst sheet (Figure 5b) with various Au concentrations (up to 40 wt.%). As a result, this method also showed that systems for large-scale hydrogen synthesis in the presence of direct sunlight are possible. It should be noted that large-area solar cells, particularly for DSSCs, were produced using the screen-printing technique [89].

4.2.4. Doctor Blade Method

The doctor blade approach is a well-known and popular technique for creating a homogeneous covering of broad bandgap semiconductor layers for dye-sensitized solar cells (DSSCs). A well-mixed slurry of target materials and additives suspended in a solvent was applied on an appropriate substrate (often an FTO or ITO plate). A doctor blade was then continuously moved across the substrate. Then the slurry spreads across the substrate and a high-quality thin layer is created after drying (at 400 °C for TiO$_2$ films). Different micron-thicknesses of thin films can be produced according to the quantity of materials used, the size of the substrate, and the speed of the doctor blade. High quality thin films are likely to be produced when the right additives and solvents are used in a slurry under the best possible circumstances. For modest size, up to 10 cm^2, this method of preparation is effective at a laboratory level. However, as already noted, it may be expanded to a bigger extent, perhaps in conjunction with other techniques, so it is worthwhile to further investigate. N-doped mesoporous titania sheets (TiO$_{2-X}$N$_X$) production for DSSC applications was described by Sivaranjani et al. [6]. Patra et al. [60,90] described a quasi-artificial leaf device that was made using AuTiO$_2$ thin films employing the doctor blade approach and was then sensitized with chalcogenides (CdS, PbS, and ZnS) using the SILAR method. We believe the SILAR approach should be expanded to build active catalytic components on porous supports, which is not feasible with any current techniques.

4.3. Effect of Binders in Thin Films

For making thin films, organic and inorganic binders are frequently employed in order to maintain the thin films in place on the surface of the substrate and avoid peeling. However, employing binders might sometimes result in undesirable characteristics or actions. The following is a list of potential outcomes: (a) Due to high temperature calcination, organic binders employed in the doctor blade process, such as cellulose, are often eliminated [40,60,90]. The porosity of the thin films is preserved when binders are removed at high temperatures. However, if an organic binder is left in place, the presence of reactive oxygen species in application circumstances may cause them to oxidize [91]. In order to create stable and uniform thin films, Nafion was utilized as a binder [35]; in fact, Nafion serves to prevent the separation of meso porous carbon nitride (mp-CN) from the stainless steel substrates. Furthermore, it should be noted that Nafion possesses effective proton conductivity, which actually reduces the mass transfer resistance inside the catalytically active layer. (b) SrTiO$_3$–Al thin films were created on glass substrates by using inorganic binders, 20 nm SiO$_2$ nanoparticles [36]. Silica nanoparticles' primary function was to create thin coatings with excellent adhesion to the substrate. The catalyst and silica (1:2) solution was coated, dried, and then calcined at 623 K to produce films that were 10–20 microns thick. (c) It must be emphasized that thin films with good stability may still be created without binders and can be tested for hydrogen production. In fact, we advise to use this technique to assess the performance of any photocatalyst in thin film forms. The stability of the film should be improved by the inclusion of binders, which is crucial for HER research. Binders should be carefully selected such that their role does not hinder the underlying activity but rather enhances it. If the binder is included in the final photocatalyst thin film, it must be chemically inert and incapable of reacting with charge carriers or with water in an aqueous media when light is present. The amount of binder applied must also be tuned for every photocatalyst system because one binder does not necessarily work for all semiconductors or composites.

4.4. Photocatalytic Activity of Sheets and Films

Recently, Xiong et al. demonstrated the photocatalyst panels of $Rh_{2-y}Cr_yO_3/(Ga_{1-x}Zn_x)(N_{1-x}O_x)$ consisting of $(Ga_{1-x}Zn_x)(N_{1-x}O_x)$ main catalyst and $Rh_{2-y}Cr_yO_3$ cocatalyst [86]. In this investigation, using several methods including squeegee and drop-casting processes, the catalyst and silica (micron size) combination was applied to 25 cm^2 frosted glass plates. The OWS activity of the panel-type catalyst yielded a determined HER (9 mmol h^{-1}·mg^{-1}) that is equivalent to HER rate (11 mmol h^{-1}·mg^{-1}) from the standard technique of assessing photocatalysts in powder form. It was also discovered that the problem of transport of water and product gases via the interparticle void spaces is avoided by the inclusion of micrometer-sized hydrophilic SiO$_2$ particles. It should be noted that photocatalyst particles range in size from 200 to 400 nm, and that the absence of silica in the photocatalyst layers makes water diffusion challenging. From the above, the ability to scale up thin films and a reduction in mass transfer issues are the key benefits.

Wang et al. showed high activity using semiconductors immobilized as thin films on metal layers for Z-scheme water splitting (Figure 4); particulate BiVO$_4$ (OEP) and SrTiO$_3$:La,Rh (HEP) semiconductors [41]. In the absence of redox mediators, the activity of the photocatalyst in thin film form with an Au layer between them is discovered to be 4.5 mmol h^{-1}·cm^2, which is 6 and 20 times greater than that of their comparable powder counterparts (0.8 mmol h^{-1}·cm^2) and without metal layers (0.2 mmol h^{-1}·cm^2), respectively. The constructed system's remarkable performance is related to the ease with which electrons may move between BiVO$_4$ and SrTiO$_3$:La,Rh through the metal layer (Au). Since both HEP and OEP are presented close to one another, a drop in H$^+$ and OH$^-$ overpotentials has also been noted. As a result, the study focused on how one might increase activity using film-based systems by facilitating efficient charge transfer via metal layer between the photocatalysts of the HEP and OEP, which in reality give correct contact as well as retain the particles intact with one another. For better activity of photocatalyst sheets linked to metal layers, Wang et al. conducted a research on Mo- and La-doped BiVO$_4$ (BiVO$_4$:Mo) particles encased in a gold (Au) layer [29]. The above Z-scheme catalyst complexes demonstrate 1.1% STH conversion efficiency with an AQY of 30% using monochromatic light of 419 nm and pure water of pH = 6.9. Furthermore, it has been demonstrated that the surface modification using Cr$_2$O$_3$ and annealing the system at 573 K for 20 min may be used to optimize the electron transport of SrTiO$_3$:La, Rh/Au/BiVO$_4$:Mo and reduce side reactions, respectively. Since the production of H$_2$ and O$_2$ occurs near together, a significant backward reaction results, which needs to be reduced for the developed photocatalyst sheet systems to function more effectively. By employing carbon as a conducting layer instead of Au, these issues may be solved. Another study used rGO as a conductive binder to increase the photocatalyst sheets' ability to split water [84].

Due to the ability to absorb visible light and improve solar hydrogen evolution, oxysulfides are considered to be a viable HEP component of the Z-scheme photocatalyst. In this scenario, Sun et al. recently showed a photocatalyst sheet using the compounds of La$_5$Ti$_2$CuS$_5$O$_7$ (LTC) and BiVO$_4$ as the HEP and OEP, respectively [43]. In this work, p-type doping (Ga^{3+}, Al^{3+}, Sc^{3+}, and Mg^{2+}) at the Ti sites and the production of La$_5$Ti$_2$Cu$_{0.9}$Ag$_{0.1}$S$_5$O$_7$ (LTCA) solid solution for the OWS reaction greatly increased the activity of this system. When LTC/Au/BiVO$_4$ was in powder form in pure water, no activity was seen. However, when LTC/Au/BiVO$_4$ was in sheet form and exposed to visible light, substantial activity (0.47 mmol h^{-1}·mg^{-1}) was seen. In pure water under visible light, Ga-LTCA/Au/BiVO$_4$ had more activity (2.2 mmol h^{-1}·mg^{-1}) than LTCA/Au/BiVO$_4$. Ga^{3+} is demonstrated to be a superior dopant to increase the activity of the sheet among the p-type doping elements.

A panel-type reactor was created by Goto et al., in which the Al-doped SrTiO$_3$ catalyst was applied as a thin layer on a glass plate using a drop casting process and tested for OWS reaction. Doping Al^{3+} into the starting material is crucial for improving SrTiO$_3$ activity and controlling particle development. Two significant parameters that were examined during the investigation are hydrophobicity and hydrophilicity of the inner surface of the window that was covered over the reactor setup (Figure 6). It has been discovered that the hydrophobic properties of the window allow gas bubbles to develop and proliferate (Figure 6), although they have no effect on the panel's overall functionality. In order to prevent the buildup of potentially explosive H$_2$ and O$_2$ gas combination bubbles inside the reactor, it is advised that the hydrophilic nature of the window be used. Under natural sunlight on clean water, the 1×1 m^2 flat photocatalyst panel displayed the STH of 0.4 percent. High rates of H$_2$ and O$_2$ were found at 5.6 mL. h^{-1}·cm^{-2}, which corresponds to a smooth release of bubbles from the panel and a STH of 10%. Here, it should be noted that elevating light output was required to reach the STH of 10%. The overall view of comparison between thin-film systems and powder form systems are given in the Table 4.

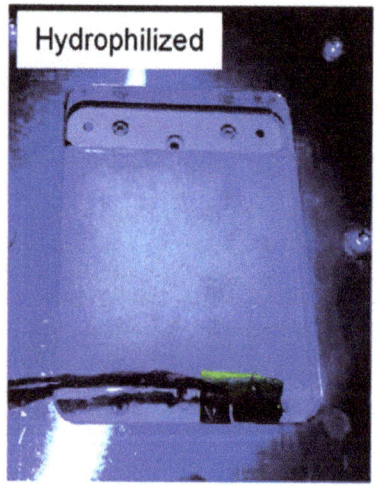 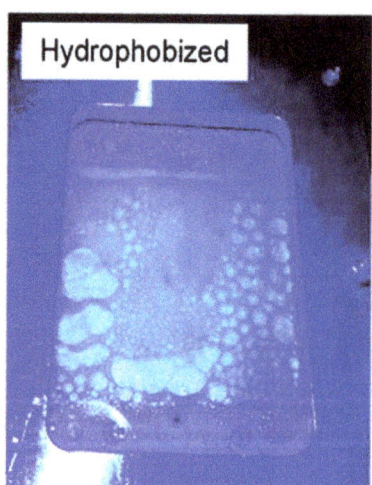

Figure 6. Photos taken while being illuminated via hydrophilic and hydrophobic windows. Replicated from ref. [36].

Table 4. Overview of photocatalytic activity existed in thin film as well as powder form.

Photocatalyst	H$_2$ Evolution-mmol h^{-1}·g^{-1}.		Reference
	Powder Form	Thin Film Form	
Au–Pd/rGO/TiO$_2$	0.50	21.5	[56]
BiVO$_4$ and SrTiO$_3$:La,Rh	0.8	4.5	[41]
(RhCrO$_x$/LaMg$_{1/3}$Ta$_{2/3}$O$_2$N/(Au,RGO)/BiVO$_4$:Mo)	0.11	0.45	[84]
LTC/Au/BiVO$_4$	Nil	0.47	[43]
Ga-LTCA/Au/BiVO$_4$	Nil	2.2	[43]
Al-SrTiO$_3$/RhCrO$_x$	Nil	5.6	[36]
Pt@mp–gC$_3$N$_4$	0.08	0.6	[35]
Pd/TiO$_2$	9.1	30	[34]
Cu–Ni (1:1)/TiO$_2$	1.75	41.7	[92]
NEC/WS$_2$/CF	4.9	64.85	[93]
Au–Pd/C/TiO$_2$	0.48	6.42	[56]

4.5. Quasi Artificial Leaf Concept with a Novel Design of Solar Cell Materials and Devices

The idea of an artificial leaf is widely recognized for dividing water. Trevisan et al. and Patra et al. used the quasi-artificial leaf (QuAL) concept to produce hydrogen [59,60,93]. Here, certain charge carriers such as electrons are collected and used at the Pt-co-catalyst sites, and holes are directly introduced into the solution to oxidize sacrificial agents. Based on the QuAL theory, porous titania (also known as Au–TiO$_2$) and Pt were used to create a solar cell, and a SILAR approach was used to build PbS and CdS quantum dots in situ within the pores of titania. The FESEM-EDX chemical mapping was used to show the distribution of PdS and CdS quantum dots in the solar cell from the top to bottom layers of titania (Figure 7). The key benefit of employing the SILAR technique (assembly QDs) in this system is the production of many bulk heterojunctions between chalcogenide quantum dots and a porous Au–TiO$_2$ matrix. Despite having a total composition of only 3%, chalcogenide is evenly distributed throughout the photoanode's porous matrix, which greatly aids in the efficient conversion of light into chemical energy.

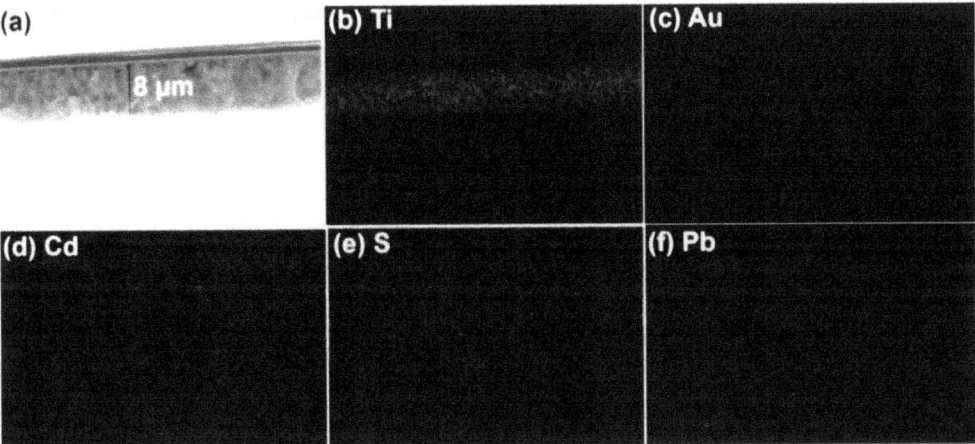

Figure 7. Bulk heterojunctions are demonstrated throughout the thin film employing (**a**) FESEM, and EDX chemical mapping analysis of (**b**) Ti, (**c**) Au, (**d**) Cd, (**e**) S, and (**f**) Pb. EDX result shows the uniform distribution of chalcogenides. Replicated from ref. [60].

The following is two key benefits of the QuAL solar cell: (a) The whole region of UV and visible light as well as a little portion of the NIR range is completely absorbed by a combination of distinct band gap semiconductors and nano gold. (b) The QuAL device operates at zero applied potential and doesn't require any potential. In this instance, 1 cm^2 was covered with 2 mg of photoanode material (AuTiO$_2$/PbS/CdS), which produced 12 mL.h^{-1} H$_2$ under one sun circumstances. The results might be linearly extrapolated to large photoanodes, such as 23 × 23 cm^2 and 46 × 46 cm^2 result in 6 and 24 L.h^{-1} H$_2$, respectively.

Patra et al. employed a cutting-edge concept to significantly expand the QuAL concept's ability to generate more hydrogen. Due to the extensive charge recombination, the photocatalyst's efficiency is inevitably low and frequently results in radiative emission, which is visible as fluorescence. In light emitting diode applications, self-absorption of the light is a difficult problem since it reduces efficiency. However, to produce more charge carriers for water splitting and therefore more hydrogen, this issue was used as a benefit. In the QuAL technique, Mn-doped CdS quantum dots are put together in the titania mesopores using the SILAR method. Pt or Ni–Cu alloy was employed as a cocatalyst. In this instance, 16 mL (10.5 mL) H$_2$ was seen under one sun circumstances. It should be emphasized that the device lacks expensive gold and PbS, making HER more advanced than their prior work with the Au–TiO$_2$/PbS/CdS-based device [93]. The photoluminescence study provided conclusive confirmation, and Figure 8 compares its results using UV-Vis absorption spectra. A tiny and large grey color triangle displayed in Figure 8 for TiO$_2$/CdS and TiO$_2$/Mn-CdS, respectively, demonstrates the same for the considerable overlap in absorption and emission characteristics (recorded with 420 nm excitation source). When the excitation source used was 400 nm, white light-like emission was seen for TiO$_2$/Mn–CdS. (Figure 8 inset). Figure 8 presents the possibility for self-absorption of emitted light as a secondary light source to generate additional charge carriers and therefore greater H$_2$ evolution. Considering that the solid lattice is at the source of the emission, in situ light absorption is known to be beneficial and can be compared to the field effect caused by surface plasmons.

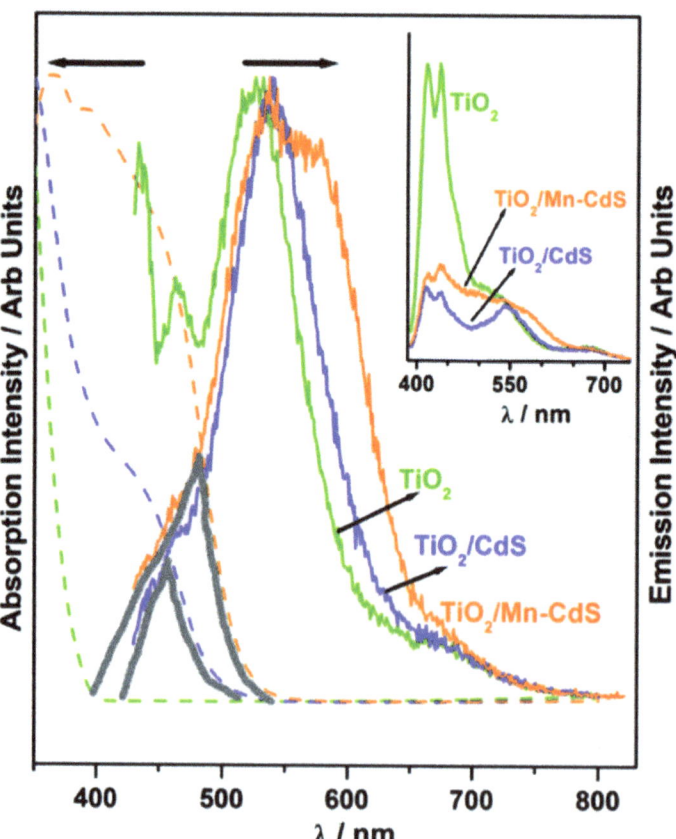

Figure 8. Photoluminescence spectra of TiO$_2$, TiO$_2$/CdS, and TiO$_2$/Mn–CdS obtained with a 420 nm excitation source; emission characteristics are normalized. Additionally, photoluminescence is compared to the UV-visible absorption spectra that were obtained for all three materials (shown as a dashed line in the same color). In the picture, a solid triangle area emphasizing the secondary light source accessible for chalcogenide absorption highlights the wavelength range where the absorption and emission spectra coincide. The photoluminescence spectra of TiO$_2$, TiO$_2$/CdS, and TiO$_2$/Mn-CdS, which were obtained at a wavelength of 370 nm, are displayed in the inset. Replicated from ref. [93].

5. Commercial Feasibility of Solar Hydrogen

It is anticipated that the cost of producing hydrogen utilizing PEC technology will be higher (USD 10.4 per kg) than the cost of producing hydrogen from particle suspension (USD 1.6 per kg) for water splitting [94]. The high capital cost (USD 154.95 per m^2) is blamed for the high cost hydrogen generation from PEC, whereas it is only 1.5% of PEC (USD 2.21 per m^2) for particle suspension for a plant size of 1 ton per day (TPD) hydrogen net output. The aforementioned figures may be reviewed due to the vast and inconsistent light absorption with films/panels and powder/suspension, respectively. However, it should be noted that just 19.8% of the total capital cost is contributed by particles, compared to 89.6% for the production of PEC cells. It is well known that PEC technology is more costly than particle suspension since the former consists of many layers of various materials such as semiconductor layer(s) for light absorption, co-catalyst for charge usage to produce products, interface layers to provide contact between semiconductor and co-catalyst, and then properly integrating them for effective water oxidation. Additionally, the preparation procedure for the photoelectrode requires the use of sophisticated/expensive methods such as atomic layer deposition (ALD) and molecular beam epitaxy (MBE) in order to produce the layers of photoelectrode materials with correct contact among them. Although particle suspension can produce hydrogen at a low cost, there are still several obstacles to overcome: (i) lack of knowledge about how solar flux is used by the particles due to light scattering and poor light absorption coefficient; (ii) plant functioning is questionable due to safety concerns brought up by the cogeneration of H_2/O_2 gases, and (iii) Poor solar light penetration into the deeper particles in the bed causes the system to work poorly. It is advised to use particle catalysts in the form of films for efficient light use in order to solve several of the issues mentioned above. Furthermore, the creation of thin films of active particles does not require a sophisticated infrastructure, therefore producing hydrogen would be much less expensive attributable to minimal capital costs.

Even if the PEC technique for solar hydrogen generation with 10% STH may be competitive with traditional methane reforming, PEC-WS must first clear a number of scientific and technological hurdles before it can be considered for large-scale operation [95]. Below are a few of them:

(i) It is necessary to increase the light absorption by using materials with narrow band gaps and by covering the photoelectrode with one more light-absorbing layer; (ii) antireflective coating must be used to decrease light reflection from the photoelectrode; (iii) in order to decrease recombination, it is necessary to tailor the nanostructural characteristics and doping levels; (iv) surface reconstruction is required, and passivation layers must be designed to reduce surface recombination; (v) the selection of the proper materials with the required energetics for effective charge separation; (vi) in order to solve the stability problems, designing protective layers, adjusting electrolyte compositions, and tuning semiconductor architectures are all important; (vii) Due to a sizable number of extremely regulated experimental settings and strict vacuum requirements, the procedures for producing photoelectrodes on a commercial scale are exceedingly difficult. It is important to note that, with the exception of the antireflective coating (ii) and the fabrication of the electrodes (vii), most of these barriers (i, iii, iv, and v) are also relevant to thin film photocatalysis. Additionally, the economically advantageous SILAR approach provides the opportunity to construct light-absorbing components in the pores of wide band gap semiconductor, decreases charge recombination, and only partially answers point (iv).

According to Jacobsson et al., the PEC of water splitting may soon become obsolete, since significant advancements in silicon PV technology have rendered the development of PEC-WS systems no longer worthwhile [91]. The cost of producing hydrogen is significantly higher even with a 10% STH conversion efficiency compared to solar PV technology with a 15% conversion, which is a key argument against the PEC-WS. Finding a photo-absorber with all the required characteristics, such as a proper bandgap, band edge locations, and stability, which drive all the functions, from light absorption to catalytic reaction, is the optimum situation for a functional PEC-WS technology. Finding a photo-absorber with stable hydrogen generation is exceedingly challenging in practice. It should be noted that, due to lower material and equipment costs compared to the latter, thin film photocatalysis, which has an efficiency of 3–4%, may be economically comparable with PEC-WS, which has an efficiency of 10% for STH conversion. Although PEC uses a thin sheet approach, the utilization of bias and complex manufacturing techniques raises the price of such devices. It must be stressed that a simplistic strategy such as artificial leaf is likely to produce superior outcomes with no bias imposed and uncomplicated (thin film) production techniques. Despite the fact that Jacobssen's point of view is fascinating, Si-PV manufacture is a well-known, extremely polluting sector. Recycling of Si-PV panels will soon become a problem for the environment. In essence, many people recognize that the Si-PV technology's "green energy" label is false. As a result, we must continue searching for true green solar technologies. For operational reasons, it is also advantageous to have rival energy conversion technologies, even those with varying price tags.

6. Conclusions and Prospects

The current review focused on the importance and producing economy of hydrogen as fuel. The progress based on particulate photocatalytic systems and their drawbacks also discussed. In order to move towards the economic scenario, we discussed about the advantages of thin film-based techniques and covered a wide range of options with improved solar hydrogen generation. We explored a number of strategies for enhancing photocatalytic water splitting activity. Apart from activity, thin-film (or panel) versions of particles have previously been shown to be scalable and would undoubtedly be a better form of the catalyst. In comparison to their powder counterparts, the films have the ability to produce hydrogen at levels up to two orders of magnitude greater. The following are the causes: (a) efficient light absorption with minimal light scattering; (b) the use of charge carriers on a small scale in a few hundred square nanometers for water splitting to produce hydrogen. The charge carriers produced in solar cells and DSSCs must travel across a distance of several microns (8–14 μm) in order to produce current. In this regard, it is important to note that the large-sized photocatalyst panels discussed in the current research may offer a different and sustainable method of producing hydrogen. The fabrication of huge photocatalyst panels is inexpensive and has the unique benefit of local charge carrier usage, unlike photoelectrodes which need highly expensive technology. It should be observed that the H_2 generation between OWS and with sacrificial agent differs by 3–4 orders of magnitude, with the latter showing the high rate of H_2 production. Glycerol is a plentiful chemical that may be employed for short-term purposes since transesterification produces biodiesel. Nevertheless, OWS and improving its STH effectiveness should be the main priority. It is wise to take advantage of this characteristic to speed up the pace of H_2 generation because experiments carried out in direct sunlight raise the temperature of the catalyst and solution. In any scenario, research on water splitting should ultimately be performed in full sunshine. Utilizing the 4–5% UV light that is found in sunshine is also crucial. Overall price reduction is ultimately being driven by increased device efficiency. Therefore, we advise photocatalysis experts to conduct their upcoming studies using thin films under direct sunlight. Although nanoscience and materials science have advanced significantly over the last several decades, it is still a challenge to employ the findings in a way that is appropriate for solar hydrogen production by photocatalysis. In this overview, certain achievements are emphasized, including the in situ assembly of light-absorbing semiconductor quantum dots in substrate pores, which results in effective charge separation via heterojunctions. We believe that by using nanoscience and nanotechnology wisely, the enormous gap in time scales between photophysical and photochemical processes might potentially be resolved. Therefore, coordinated efforts are necessary, and it is probable that they will result in a solution for economically effective and scalable water splitting to produce significant amounts of hydrogen.

Author Contributions: Conceptualization and writing—original draft preparation, P.R.; Editing, supervision and project administration, J.N. All authors have read and agreed to the published version of the manuscript.

Funding: This research was funded by [MSIT] grant number [No. 2019R1A2C1008746].

Institutional Review Board Statement: Not applicable.

Informed Consent Statement: Not applicable.

Data Availability Statement: Not applicable.

Acknowledgments: This work was supported by the National Research Foundation of Korea (NRF) grant funded by the Korea government (MSIT) (No. 2019R1A2C1008746).

Conflicts of Interest: There are no conflict to declare.

References

1. Escobedo Salas, S.; Serrano Rosales, B.; de Lasa, H. Quantum Yield with Platinum Modified TiO_2 Photocatalyst for Hydrogen Production. *Appl. Catal. B Environ.* **2013**, *140–141*, 523–536. [CrossRef]
2. Abe, R. Recent Progress on Photocatalytic and Photoelectrochemical Water Splitting under Visible Light Irradiation. *J. Photochem. Photobiol. C Photochem. Rev.* **2010**, *11*, 179–209. [CrossRef]
3. Kudo, A.; Miseki, Y. Heterogeneous Photocatalyst Materials for Water Splitting. *Chem. Soc. Rev.* **2009**, *38*, 253–278. [CrossRef]
4. Rahman, M.Z.; Davey, K.; Qiao, S.Z. Carbon, Nitrogen and Phosphorus Containing Metal-Free Photocatalysts for Hydrogen Production: Progress and Challenges. *J. Mater. Chem. A* **2018**, *6*, 1305–1322. [CrossRef]
5. Sivaranjani, K.; Gopinath, C.S. Porosity Driven Photocatalytic Activity of Wormhole Mesoporous $TiO_{2-x}N_x$ in Direct Sunlight. *J. Mater. Chem.* **2011**, *21*, 2639–2647. [CrossRef]
6. Sivaranjani, K.; Agarkar, S.; Ogale, S.B.; Gopinath, C.S. Toward a Quantitative Correlation between Microstructure and DSSC Efficiency: A Case Study of $TiO_{2-x}N_x$ Nanoparticles in a Disordered Mesoporous Framework. *J. Phys. Chem. C* **2012**, *116*, 2581–2587. [CrossRef]
7. Devaraji, P.; Sathu, N.K.; Gopinath, C.S. Ambient Oxidation of Benzene to Phenol by Photocatalysis on Au/Ti0.98V0.02O2: Role of Holes. *ACS Catal.* **2014**, *4*, 2844–2853. [CrossRef]
8. Bellamkonda, S.; Shanmugam, R.; Gangavarapu, R.R. Extending the π-Electron Conjugation in 2D Planar Graphitic Carbon Nitride: Efficient Charge Separation for Overall Water Splitting. *J. Mater. Chem. A* **2019**, *7*, 3757–3771. [CrossRef]
9. Yao, T.; An, X.; Han, H.; Chen, J.Q.; Li, C. Photoelectrocatalytic Materials for Solar Water Splitting. *Adv. Energy Mater.* **2018**, *8*, 1800210. [CrossRef]
10. Liu, G.; Du, K.; Xu, J.; Chen, G.; Gu, M.; Yang, C.; Wang, K.; Jakobsen, H. Plasmon-Dominated Photoelectrodes for Solar Water Splitting. *J. Mater. Chem. A* **2017**, *5*, 4233–4253. [CrossRef]
11. Kment, S.; Riboni, F.; Pausova, S.; Wang, L.; Wang, L.; Han, H.; Hubicka, Z.; Krysa, J.; Schmuki, P.; Zboril, R. Photoanodes Based on TiO_2 and α-Fe_2O_3 for Solar Water Splitting-Superior Role of 1D Nanoarchitectures and of Combined Heterostructures. *Chem. Soc. Rev.* **2017**, *46*, 3716–3769. [CrossRef]
12. Su, J.; Wei, Y.; Vayssieres, L. Stability and Performance of Sulfide-, Nitride-, and Phosphide-Based Electrodes for Photocatalytic Solar Water Splitting. *J. Phys. Chem. Lett.* **2017**, *8*, 5228–5238. [CrossRef] [PubMed]
13. Osterloh, F.E. Inorganic Nanostructures for Photoelectrochemical and Photocatalytic Water Splitting. *Chem. Soc. Rev.* **2013**, *42*, 2294–2320. [CrossRef] [PubMed]
14. Devi, L.G.; Kavitha, R. A Review on Non Metal Ion Doped Titania for the Photocatalytic Degradation of Organic Pollutants under UV/Solar Light: Role of Photogenerated Charge Carrier Dynamics in Enhancing the Activity. *Appl. Catal. B Environ.* **2013**, *140–141*, 559–587. [CrossRef]
15. Yi, S.S.; Zhang, X.B.; Wulan, B.R.; Yan, J.M.; Jiang, Q. Non-Noble Metals Applied to Solar Water Splitting. *Energy Environ. Sci.* **2018**, *11*, 3128–3156. [CrossRef]
16. Kim, J.H.; Hansora, D.; Sharma, P.; Jang, J.W.; Lee, J.S. Toward Practical Solar Hydrogen Production-an Artificial Photosynthetic Leaf-to-Farm Challenge. *Chem. Soc. Rev.* **2019**, *48*, 1908–1971. [CrossRef]
17. Wang, Y.; Suzuki, H.; Xie, J.; Tomita, O.; Martin, D.J.; Higashi, M.; Kong, D.; Abe, R.; Tang, J. Mimicking Natural Photosynthesis: Solar to Renewable H_2 Fuel Synthesis by Z-Scheme Water Splitting Systems. *Chem. Rev.* **2018**, *118*, 5201–5241. [CrossRef]
18. RAJAAMBAL, S.; SIVARANJANI, K.; GOPINATH, C.S. Recent Developments in Solar H_2 Generation from Water Splitting. *J. Chem. Sci.* **2015**, *127*, 33–47. [CrossRef]
19. Chen, S.; Takata, T.; Domen, K. Particulate Photocatalysts for Overall Water Splitting. *Nat. Rev. Mater.* **2017**, *2*, 17050. [CrossRef]
20. Fabian, D.M.; Hu, S.; Singh, N.; Houle, F.A.; Hisatomi, T.; Domen, K.; Osterloh, F.E.; Ardo, S. Particle Suspension Reactors and Materials for Solar-Driven Water Splitting. *Energy Environ. Sci.* **2015**, *8*, 2825–2850. [CrossRef]
21. Natarajan, T.S.; Thampi, K.R.; Tayade, R.J. Visible Light Driven Redox-Mediator-Free Dual Semiconductor Photocatalytic Systems for Pollutant Degradation and the Ambiguity in Applying Z-Scheme Concept. *Appl. Catal. B Environ.* **2018**, *227*, 296–311. [CrossRef]

22. Bowker, M. Sustainable Hydrogen Production by the Application of Ambient Temperature Photocatalysis. *Green Chem.* **2011**, *13*, 2235–2246. [CrossRef]
23. Zhang, G.; Lan, Z.A.; Lin, L.; Lin, S.; Wang, X. Overall Water Splitting by Pt/g-C_3N_4 Photocatalysts without Using Sacrificial Agents. *Chem. Sci.* **2016**, *7*, 3062–3066. [CrossRef] [PubMed]
24. Lin, L.; Lin, Z.; Zhang, J.; Cai, X.; Lin, W.; Yu, Z.; Wang, X. Molecular-Level Insights on the Reactive Facet of Carbon Nitride Single Crystals Photocatalysing Overall Water Splitting. *Nat. Catal.* **2020**, *3*, 649–655. [CrossRef]
25. Tao, X.; Zhao, Y.; Wang, S.; Li, C.; Li, R. Recent Advances and Perspectives for Solar-Driven Water Splitting Using Particulate Photocatalysts. *Chem. Soc. Rev.* **2022**, *51*, 3561–3608. [CrossRef]
26. Shi, M.; Li, R.; Li, C. Halide Perovskites for Light Emission and Artificial Photosynthesis: Opportunities, Challenges, and Perspectives. *EcoMat* **2021**, *3*, e12074. [CrossRef]
27. Maeda, K. Z-Scheme Water Splitting Using Two Different Semiconductor Photocatalysts. *ACS Catal.* **2013**, *3*, 1486–1503. [CrossRef]
28. Sayama, K.; Mukasa, K.; Abe, R.; Abe, Y.; Arakawa, H. Stoichiometric Water Splitting into H_2 and O_2 Using a Mixture of Two Different Photocatalysts and an IO_3^-/I^- Shuttle Redox Mediator under Visible Light Irradiation. *Chem. Commun.* **2001**, *1*, 2416–2417. [CrossRef]
29. Wang, Q.; Hisatomi, T.; Jia, Q.; Tokudome, H.; Zhong, M.; Wang, C.; Pan, Z.; Takata, T.; Nakabayashi, M.; Shibata, N.; et al. Scalable Water Splitting on Particulate Photocatalyst Sheets with a Solar-to-Hydrogen Energy Conversion Efficiency Exceeding 1%. *Nat. Mater.* **2016**, *15*, 611–615. [CrossRef] [PubMed]
30. Chao, Y.; Zheng, J.; Chen, J.; Wang, Z.; Jia, S.; Zhang, H.; Zhu, Z. Highly Efficient Visible Light-Driven Hydrogen Production of Precious Metal-Free Hybrid Photocatalyst: CdS@NiMoS Core-Shell Nanorods. *Catal. Sci. Technol.* **2017**, *7*, 2798–2804. [CrossRef]
31. Chauhan, M.; Soni, K.; Karthik, P.E.; Reddy, K.P.; Gopinath, C.S.; Deka, S. Promising Visible-Light Driven Hydrogen Production from Water on a Highly Efficient $CuCo_2S_4$ Nanosheet Photocatalyst. *J. Mater. Chem. A* **2019**, *7*, 6985–6994. [CrossRef]
32. Chao, Y.; Zheng, J.; Zhang, H.; Li, F.; Yan, F.; Tan, Y.; Zhu, Z. Oxygen-Incorporation in Co_2P as a Non-Noble Metal Cocatalyst to Enhance Photocatalysis for Reducing Water to H_2 under Visible Light. *Chem. Eng. J.* **2018**, *346*, 281–288. [CrossRef]
33. Moon, S.Y.; Gwag, E.H.; Park, J.Y. Hydrogen Generation on Metal/Mesoporous Oxides: The Effects of Hierarchical Structure, Doping, and Co-Catalysts. *Energy Technol.* **2018**, *6*, 459–469. [CrossRef]
34. Nalajala, N.; Patra, K.K.; Bharad, P.A.; Gopinath, C.S. Why the Thin Film Form of a Photocatalyst Is Better than the Particulate Form for Direct Solar-to-Hydrogen Conversion: A Poor Man's Approach. *RSC Adv.* **2019**, *9*, 6094–6100. [CrossRef] [PubMed]
35. Schröder, M.; Kailasam, K.; Borgmeyer, J.; Neumann, M.; Thomas, A.; Schomäcker, R.; Schwarze, M. Hydrogen Evolution Reaction in a Large-Scale Reactor Using a Carbon Nitride Photocatalyst under Natural Sunlight Irradiation. *Energy Technol.* **2015**, *3*, 1014–1017. [CrossRef]
36. Goto, Y.; Hisatomi, T.; Wang, Q.; Higashi, T.; Ishikiriyama, K.; Maeda, T.; Sakata, Y.; Okunaka, S.; Tokudome, H.; Katayama, M.; et al. A Particulate Photocatalyst Water-Splitting Panel for Large-Scale Solar Hydrogen Generation. *Joule* **2018**, *2*, 509–520. [CrossRef]
37. Mapa, M.; Gopinath, C.S. Combustion Synthesis of Triangular and Multifunctional ZnO 1-XNx (x = 0.15) Materials. *Chem. Mater.* **2009**, *21*, 351–359. [CrossRef]
38. Devaraji, P.; Mapa, M.; Hakkeem, H.M.A.; Sudhakar, V.; Krishnamoorthy, K.; Gopinath, C.S. ZnO-ZnS Heterojunctions: A Potential Candidate for Optoelectronics Applications and Mineralization of Endocrine Disruptors in Direct Sunlight. *ACS Omega* **2017**, *2*, 6768–6781. [CrossRef] [PubMed]
39. Kumaravel, V.; Imam, M.; Badreldin, A.; Chava, R.; Do, J.; Kang, M.; Abdel-Wahab, A. Photocatalytic Hydrogen Production: Role of Sacrificial Reagents on the Activity of Oxide, Carbon, and Sulfide Catalysts. *Catalysts* **2019**, *9*, 276. [CrossRef]
40. Schneider, J.; Bahnemann, D.W. Undesired Role of Sacrificial Reagents in Photocatalysis. *J. Phys. Chem. Lett.* **2013**, *4*, 3479–3483. [CrossRef]
41. Wang, Q.; Li, Y.; Hisatomi, T.; Nakabayashi, M.; Shibata, N.; Kubota, J.; Domen, K. Z-Scheme Water Splitting Using Particulate Semiconductors Immobilized onto Metal Layers for Efficient Electron Relay. *J. Catal.* **2015**, *328*, 308–315. [CrossRef]
42. Wang, Q.; Hisatomi, T.; Suzuki, Y.; Pan, Z.; Seo, J.; Katayama, M.; Minegishi, T.; Nishiyama, H.; Takata, T.; Seki, K.; et al. Particulate Photocatalyst Sheets Based on Carbon Conductor Layer for Efficient Z-Scheme Pure-Water Splitting at Ambient Pressure. *J. Am. Chem. Soc.* **2017**, *139*, 1675–1683. [CrossRef]
43. Sun, S.; Hisatomi, T.; Wang, Q.; Chen, S.; Ma, G.; Liu, J.; Nandy, S.; Minegishi, T.; Katayama, M.; Domen, K. Efficient Redox-Mediator-Free Z-Scheme Water Splitting Employing Oxysulfide Photocatalysts under Visible Light. *ACS Catal.* **2018**, *8*, 1690–1696. [CrossRef]
44. Hisatomi, T.; Yamamoto, T.; Wang, Q.; Nakanishi, T.; Higashi, T.; Katayama, M.; Minegishi, T.; Domen, K. Particulate Photocatalyst Sheets Based on Non-Oxide Semiconductor Materials for Water Splitting under Visible Light Irradiation. *Catal. Sci. Technol.* **2018**, *8*, 3918–3925. [CrossRef]
45. Pan, Z.; Hisatomi, T.; Wang, Q.; Chen, S.; Nakabayashi, M.; Shibata, N.; Pan, C.; Takata, T.; Katayama, M.; Minegishi, T.; et al. Photocatalyst Sheets Composed of Particulate $LaMg_{1/3}Ta_{2/3}O_2N$ and Mo-Doped $BiVO_4$ for Z-Scheme Water Splitting under Visible Light. *ACS Catal.* **2016**, *6*, 7188–7196. [CrossRef]
46. Xia, X.; Song, M.; Wang, H.; Zhang, X.; Sui, N.; Zhang, Q.; Colvin, V.L.; Yu, W.W. Latest Progress in Constructing Solid-State Z Scheme Photocatalysts for Water Splitting. *Nanoscale* **2019**, *11*, 11071–11082. [CrossRef]
47. Bard, A.J. Photoelectrochemistry and Heterogeneous Photo-Catalysis at Semiconductors. *J. Photochem.* **1979**, *10*, 59–75. [CrossRef]

48. Shwetharani, R.; Sakar, M.; Fernando, C.A.N.; Binas, V.; Balakrishna, R.G. Recent Advances and Strategies to Tailor the Energy Levels, Active Sites and Electron Mobility in Titania and Its Doped/Composite Analogues for Hydrogen Evolution in Sunlight. *Catal. Sci. Technol.* **2019**, *9*, 12–46. [CrossRef]
49. Devaraji, P.; Gopinath, C.S. Pt–g-C_3N_4–(Au/TiO_2): Electronically Integrated Nanocomposite for Solar Hydrogen Generation. *Int. J. Hydrogen Energy* **2018**, *43*, 601–613. [CrossRef]
50. Bharad, P.A.; Sivaranjani, K.; Gopinath, C.S. A Rational Approach towards Enhancing Solar Water Splitting: A Case Study of Au–RGO/N-RGO–TiO_2. *Nanoscale* **2015**, *7*, 11206–11215. [CrossRef]
51. Melvin, A.A.; Bharad, P.A.; Illath, K.; Lawrence, M.P.; Gopinath, C.S. Is There Any Real Effect of Low Dimensional Morphologies towards Light Harvesting? A Case Study of Au-RGO-TiO_2 Nanocomposites. *ChemistrySelect* **2016**, *1*, 917–923. [CrossRef]
52. Patra, K.K.; Gopinath, C.S. Harnessing Visible-Light and Limited Near-IR Photons through Plasmon Effect of Gold Nanorod with AgTiO_2. *J. Phys. Chem. C* **2018**, *122*, 1206–1214. [CrossRef]
53. Patra, K.K.; Gopinath, C.S. Bimetallic and Plasmonic Ag-Au on TiO_2 for Solar Water Splitting: An Active Nanocomposite for Entire Visible-Light-Region Absorption. *ChemCatChem* **2016**, *8*, 3294–3311. [CrossRef]
54. Preethi, L.K.; Mathews, T.; Nand, M.; Jha, S.N.; Gopinath, C.S.; Dash, S. Band Alignment and Charge Transfer Pathway in Three Phase Anatase-Rutile-Brookite TiO_2 Nanotubes: An Efficient Photocatalyst for Water Splitting. *Appl. Catal. B Environ.* **2017**, *218*, 9–19. [CrossRef]
55. Preethi, L.K.; Antony, R.P.; Mathews, T.; Walczak, L.; Gopinath, C.S. A Study on Doped Heterojunctions in TiO_2 Nanotubes: An Efficient Photocatalyst for Solar Water Splitting. *Sci. Rep.* **2017**, *7*, 14314. [CrossRef]
56. Tudu, B.; Nalajala, N.P.; Reddy, K.; Saikia, P.; Gopinath, C.S. Electronic Integration and Thin Film Aspects of Au–Pd/RGO/TiO_2 for Improved Solar Hydrogen Generation. *ACS Appl. Mater. Interfaces* **2019**, *11*, 32869–32878. [CrossRef]
57. Maeda, K.; Higashi, M.; Lu, D.; Abe, R.; Domen, K. Efficient Nonsacrificial Water Splitting through Two-Step Photoexcitation by Visible Light Using a Modified Oxynitride as a Hydrogen Evolution Photocatalyst. *J. Am. Chem. Soc.* **2010**, *132*, 5858–5868. [CrossRef]
58. Wang, H.; Chen, Z.; Wu, D.; Cao, M.; Sun, F.; Zhang, H.; You, H.; Zhuang, W.; Cao, R. Significantly Enhanced Overall Water Splitting Performance by Partial Oxidation of Ir through Au Modification in Core–Shell Alloy Structure. *J. Am. Chem. Soc.* **2021**, *143*, 4639–4645. [CrossRef]
59. Trevisan, R.; Rodenas, P.; Gonzalez-Pedro, V.; Sima, C.; Sanchez, R.S.; Barea, E.M.; Mora-Sero, I.; Fabregat-Santiago, F.; Gimenez, S. Harnessing Infrared Photons for Photoelectrochemical Hydrogen Generation. A PbS Quantum Dot Based "Quasi-Artificial Leaf". *J. Phys. Chem. Lett.* **2013**, *4*, 141–146. [CrossRef]
60. Patra, K.K.; Bhuskute, B.D.; Gopinath, C.S. Possibly Scalable Solar Hydrogen Generation with Quasi-Artificial Leaf Approach. *Sci. Rep.* **2017**, *7*, 6515. [CrossRef]
61. Li, Y.; Chen, G.; Wang, Q.; Wang, X.; Zhou, A.; Shen, Z. Hierarchical ZnS-In_2S_3-CuS Nanospheres with Nanoporous Structure: Facile Synthesis, Growth Mechanism, and Excellent Photocatalytic Activity. *Adv. Funct. Mater.* **2010**, *20*, 3390–3398. [CrossRef]
62. Wang, J.; Wang, Z.; Qu, P.; Xu, Q.; Zheng, J.; Jia, S.; Chen, J.; Zhu, Z. A 2D/1D TiO_2 Nanosheet/CdS Nanorods Heterostructure with Enhanced Photocatalytic Water Splitting Performance for H_2 Evolution. *Int. J. Hydrogen Energy* **2018**, *43*, 7388–7396. [CrossRef]
63. He, J.; Chen, L.; Wang, F.; Liu, Y.; Chen, P.; Au, C.-T.; Yin, S.-F. CdS Nanowires Decorated with Ultrathin MoS_2 Nanosheets as an Efficient Photocatalyst for Hydrogen Evolution. *ChemSusChem* **2016**, *9*, 624–630. [CrossRef] [PubMed]
64. El-Maghrabi, H.H.; Barhoum, A.; Nada, A.A.; Moustafa, Y.M.; Seliman, S.M.; Youssef, A.M.; Bechelany, M. Synthesis of Mesoporous Core-Shell CdS@TiO_2 (0D and 1D) Photocatalysts for Solar-Driven Hydrogen Fuel Production. *J. Photochem. Photobiol. A Chem.* **2018**, *351*, 261–270. [CrossRef]
65. Iqbal, S.; Pan, Z.; Zhou, K. Enhanced Photocatalytic Hydrogen Evolution from in Situ Formation of Few-Layered MoS_2/CdS Nanosheet-Based van Der Waals Heterostructures. *Nanoscale* **2017**, *9*, 6638–6642. [CrossRef] [PubMed]
66. Zhao, D.; Sun, B.; Li, X.; Qin, L.; Kang, S.; Wang, D. Promoting Visible Light-Driven Hydrogen Evolution over CdS Nanorods Using Earth-Abundant CoP as a Cocatalyst. *RSC Adv.* **2016**, *6*, 33120–33125. [CrossRef]
67. Yin, X.-L.; Li, L.-L.; Jiang, W.-J.; Zhang, Y.; Zhang, X.; Wan, L.-J.; Hu, J.-S. MoS_2/CdS Nanosheets-on-Nanorod Heterostructure for Highly Efficient Photocatalytic H_2 Generation under Visible Light Irradiation. *ACS Appl. Mater. Interfaces* **2016**, *8*, 15258–15266. [CrossRef]
68. Dinh, C.-T.; Pham, M.-H.; Kleitz, F.; Do, T.-O. Design of Water-Soluble CdS–Titanate–Nickel Nanocomposites for Photocatalytic Hydrogen Production under Sunlight. *J. Mater. Chem. A* **2013**, *1*, 13308. [CrossRef]
69. Li, Y.; Chen, G.; Zhou, C.; Sun, J. A Simple Template-Free Synthesis of Nanoporous ZnS–In_2S_3–Ag_2S Solid Solutions for Highly Efficient Photocatalytic H_2 Evolution under Visible Light. *Chem. Commun.* **2009**, *15*, 2020. [CrossRef]
70. Melvin, A.A.; Illath, K.; Das, T.; Raja, T.; Bhattacharyya, S.; Gopinath, C.S. M–Au/TiO_2 (M = Ag, Pd, and Pt) Nanophotocatalyst for Overall Solar Water Splitting: Role of Interfaces. *Nanoscale* **2015**, *7*, 13477–13488. [CrossRef]
71. Shaner, M.R.; Atwater, H.A.; Lewis, N.S.; McFarland, E.W. A Comparative Technoeconomic Analysis of Renewable Hydrogen Production Using Solar Energy. *Energy Environ. Sci.* **2016**, *9*, 2354–2371. [CrossRef]
72. Sun, Z.; Zheng, H.; Li, J.; Du, P. Extraordinarily Efficient Photocatalytic Hydrogen Evolution in Water Using Semiconductor Nanorods Integrated with Crystalline Ni_2P Cocatalysts. *Energy Environ. Sci.* **2015**, *8*, 2668–2676. [CrossRef]

73. Dong, Y.; Kong, L.; Wang, G.; Jiang, P.; Zhao, N.; Zhang, H. Photochemical Synthesis of CoxP as Cocatalyst for Boosting Photocatalytic H_2 Production via Spatial Charge Separation. *Appl. Catal. B Environ.* **2017**, *211*, 245–251. [CrossRef]
74. Jiang, D.; Sun, Z.; Jia, H.; Lu, D.; Du, P. A Cocatalyst-Free CdS Nanorod/ZnS Nanoparticle Composite for High-Performance Visible-Light-Driven Hydrogen Production from Water. *J. Mater. Chem. A* **2016**, *4*, 675–683. [CrossRef]
75. Cheng, H.; Lv, X.-J.; Cao, S.; Zhao, Z.-Y.; Chen, Y.; Fu, W.-F. Robustly Photogenerating H_2 in Water Using FeP/CdS Catalyst under Solar Irradiation. *Sci. Rep.* **2016**, *6*, 19846. [CrossRef]
76. Sun, Z.; Yue, Q.; Li, J.; Xu, J.; Zheng, H.; Du, P. Copper Phosphide Modified Cadmium Sulfide Nanorods as a Novel p–n Heterojunction for Highly Efficient Visible-Light-Driven Hydrogen Production in Water. *J. Mater. Chem. A* **2015**, *3*, 10243–10247. [CrossRef]
77. Gopannagari, M.; Kumar, D.P.; Reddy, D.A.; Hong, S.; Song, M.I.; Kim, T.K. In Situ Preparation of Few-Layered WS_2 Nanosheets and Exfoliation into Bilayers on CdS Nanorods for Ultrafast Charge Carrier Migrations toward Enhanced Photocatalytic Hydrogen Production. *J. Catal.* **2017**, *351*, 153–160. [CrossRef]
78. Chen, H.; Jiang, D.; Sun, Z.; Irfan, R.M.; Zhang, L.; Du, P. Cobalt Nitride as an Efficient Cocatalyst on CdS Nanorods for Enhanced Photocatalytic Hydrogen Production in Water. *Catal. Sci. Technol.* **2017**, *7*, 1515–1522. [CrossRef]
79. Park, H.; Reddy, D.A.; Kim, Y.; Lee, S.; Ma, R.; Kim, T.K. Synthesis of Ultra-Small Palladium Nanoparticles Deposited on CdS Nanorods by Pulsed Laser Ablation in Liquid: Role of Metal Nanocrystal Size in the Photocatalytic Hydrogen Production. *Chem. A Eur. J.* **2017**, *23*, 13112–13119. [CrossRef]
80. Kumar, D.P.; Hong, S.; Reddy, D.A.; Kim, T.K. Noble Metal-Free Ultrathin MoS_2 Nanosheet-Decorated CdS Nanorods as an Efficient Photocatalyst for Spectacular Hydrogen Evolution under Solar Light Irradiation. *J. Mater. Chem. A* **2016**, *4*, 18551–18558. [CrossRef]
81. Reddy, D.A.; Choi, J.; Lee, S.; Kim, Y.; Hong, S.; Kumar, D.P.; Kim, T.K. Hierarchical Dandelion-Flower-like Cobalt-Phosphide Modified CdS/Reduced Graphene Oxide-MoS_2 Nanocomposites as a Noble-Metal-Free Catalyst for Efficient Hydrogen Evolution from Water. *Catal. Sci. Technol.* **2016**, *6*, 6197–6206. [CrossRef]
82. Rong, Y.; Ming, Y.; Ji, W.; Li, D.; Mei, A.; Hu, Y.; Han, H. Toward Industrial-Scale Production of Perovskite Solar Cells: Screen Printing, Slot-Die Coating, and Emerging Techniques. *J. Phys. Chem. Lett.* **2018**, *9*, 2707–2713. [CrossRef] [PubMed]
83. Xiong, A.; Ma, G.; Maeda, K.; Takata, T.; Hisatomi, T.; Setoyama, T.; Kubota, J.; Domen, K. Fabrication of Photocatalyst Panels and the Factors Determining Their Activity for Water Splitting. *Catal. Sci. Technol.* **2014**, *4*, 325–328. [CrossRef]
84. Pan, Z.; Hisatomi, T.; Wang, Q.; Chen, S.; Iwase, A.; Nakabayashi, M.; Shibata, N.; Takata, T.; Katayama, M.; Minegishi, T.; et al. Photoreduced Graphene Oxide as a Conductive Binder to Improve the Water Splitting Activity of Photocatalyst Sheets. *Adv. Funct. Mater.* **2016**, *26*, 7011–7019. [CrossRef]
85. Schröder, M.; Kailasam, K.; Rudi, S.; Fündling, K.; Rieß, J.; Lublow, M.; Thomas, A.; Schomäcker, R.; Schwarze, M. Applying Thermo-Destabilization of Microemulsions as a New Method for Co-Catalyst Loading on Mesoporous Polymeric Carbon Nitride—Towards Large Scale Applications. *RSC Adv.* **2014**, *4*, 50017–50026. [CrossRef]
86. Pan, Z.; Hisatomi, T.; Wang, Q.; Nakabayashi, M.; Shibata, N.; Pan, C.; Takata, T.; Domen, K. Application of $LaMg_{1/3}Ta_{2/3}O_2N$ as a Hydrogen Evolution Photocatalyst of a Photocatalyst Sheet for Z-Scheme Water Splitting. *Appl. Catal. A Gen.* **2016**, *521*, 26–33. [CrossRef]
87. Minegishi, T.; Nishimura, N.; Kubota, J.; Domen, K. Photoelectrochemical Properties of $LaTiO_2N$ Electrodes Prepared by Particle Transfer for Sunlight-Driven Water Splitting. *Chem. Sci.* **2013**, *4*, 1120. [CrossRef]
88. Hu, Y.; Si, S.; Mei, A.; Rong, Y.; Liu, H.; Li, X.; Han, H. Stable Large-Area (10×10 cm^2) Printable Mesoscopic Perovskite Module Exceeding 10% Efficiency. *Sol. RRL* **2017**, *1*, 1600019. [CrossRef]
89. Liu, J.; Li, Y.; Arumugam, S.; Tudor, J.; Beeby, S. Screen Printed Dye-Sensitized Solar Cells (DSSCs) on Woven Polyester Cotton Fabric for Wearable Energy Harvesting Applications. *Mater. Today Proc.* **2018**, *5*, 13753–13758. [CrossRef]
90. Patra, K.K.; Bharad, P.A.; Jain, V.; Gopinath, C.S. Direct Solar-to-Hydrogen Generation by Quasi-Artificial Leaf Approach: Possibly Scalable and Economical Device. *J. Mater. Chem. A* **2019**, *7*, 3179–3189. [CrossRef]
91. Heller, A.; Pishko, M.V.; Heller, E. Photocatalyst-binder compositions. US Patent US005854169A, 29 December 1998.
92. Tudu, B.; Nalajala, N.; Saikia, P.; Gopinath, C.S. Cu–Ni Bimetal Integrated TiO_2 Thin Film for Enhanced Solar Hydrogen Generation. *Sol. RRL* **2020**, *4*, 1900557. [CrossRef]
93. Lin, Z.; Li, J.; Zheng, Z.; Li, L.; Yu, L.; Wang, C.; Yang, G. A Floating Sheet for Efficient Photocatalytic Water Splitting. *Adv. Energy Mater.* **2016**, *6*, 1600510. [CrossRef]
94. Jacobsson, T.J. Photoelectrochemical Water Splitting: An Idea Heading towards Obsolescence? *Energy Environ. Sci.* **2018**, *11*, 1977–1979. [CrossRef]
95. Pinaud, B.A.; Benck, J.D.; Seitz, L.C.; Forman, A.J.; Chen, Z.; Deutsch, T.G.; James, B.D.; Baum, K.N.; Baum, G.N.; Ardo, S.; et al. Technical and Economic Feasibility of Centralized Facilities for Solar Hydrogen Production via Photocatalysis and Photoelectrochemistry. *Energy Environ. Sci.* **2013**, *6*, 1983. [CrossRef]

Article

Designing a 0D/1D S-Scheme Heterojunction of Cadmium Selenide and Polymeric Carbon Nitride for Photocatalytic Water Splitting and Carbon Dioxide Reduction

Yayun Wang [1], Haotian Wang [1], Yuke Li [2], Mingwen Zhang [3,*] and Yun Zheng [1,*]

[1] Xiamen Key Laboratory of Optoelectronic Materials and Advanced Manufacturing, College of Materials Science and Engineering, Huaqiao University, Xiamen 361021, China
[2] Department of Chemistry and Centre for Scientific Modeling and Computation, Chinese University of Hong Kong, Shatin, Hong Kong, China
[3] Fujian Provincial Key Lab of Coastal Basin Environment, School of Materials and Environment Engineering, Fujian Polytechnic Normal University, Fuzhou 350300, China
* Correspondence: mwzhang1989@163.com (M.Z.); zheng-yun@hqu.edu.cn (Y.Z.)

Abstract: Constructing photocatalysts to promote hydrogen evolution and carbon dioxide photoreduction into solar fuels is of vital importance. The design and establishment of an S-scheme heterojunction system is one of the most feasible approaches to facilitate the separation and transfer of photogenerated charge carriers and obtain powerful photoredox capabilities for boosting photocatalytic performance. Herein, a zero-dimensional/one-dimensional S-scheme heterojunction composed of CdSe quantum dots and polymeric carbon nitride nanorods (CdSe/CN) is created and constructed via a linker-assisted hybridization approach. The CdSe/CN composites exhibit superior photocatalytic activity in water splitting and promoted carbon dioxide conversion performance compared with CN nanorods and CdSe quantum dots. The best efficiency in photocatalytic water splitting (10.2% apparent quantum yield at 420 nm irradiation, 20.1 mmol g^{-1} h^{-1} hydrogen evolution rate) and CO_2 reduction (0.77 mmol g^{-1} h^{-1} CO production rate) was achieved by 5%CdSe/CN composites. The significantly improved photocatalytic reactivity of CdSe/CN composites primarily originates from the emergence of an internal electric field in the zero-dimensional/one-dimensional S-scheme heterojunction, which could greatly improve the photoinduced charge-carrier separation. This work underlines the possibility of employing polymeric carbon nitride nanostructures as appropriate platforms to establish highly active S-scheme heterojunction photocatalysts for solar fuel production.

Keywords: photocatalysis; water splitting; CO_2 reduction; S-scheme heterojunction; carbon nitride; quantum dot

1. Introduction

During past 50 years, the level of carbon dioxide (CO_2) in the atmosphere has increased significantly as a result of excessive combustion of fuel [1,2]. The development of photocatalytic technology to reduce water to hydrogen (H_2) and recycle CO_2 into value-added hydrocarbons will help decrease the level of CO_2 in the atmosphere and partially meet future energy requirements [3–5]. However, the photocatalytic efficiencies of most unitary photocatalysts can hardly meet the practical requirements primarily ascribed to the high electron-hole recombination rate. Designing efficient heterojunction photocatalysts with the boosted separation of photoinduced electron-holes remains as a great challenge in this field [6–8].

Polymeric carbon nitride (CN) materials have been shown to act as promising photocatalysts for multifunctional photoredox reactions such as water decomposition, CO_2 conversion, selective organic transformation, pollutant removal, nitrogen fixation, and bacterial inactivation [9–16]. Wang et al. synthesized one-dimensional polymeric carbon nitride nanorods by using chiral mesoporous silica nanorods as a hard-template, and showed that CN nanorods exhibited stronger photocatalytic reactivity than bulk CN in water splitting and CO_2 conversion [17]. However, the light-harvesting ability and photocatalytic activity of pristine CN and its related nanostructures remains limited [18–24]. The photocatalytic reactivity of CN can be further enhanced by heterostructure design to accelerate charge-carrier separation and optimize the visible light harvesting capability [25–27]. So far, CN have been hybridized with different metals or semiconductors to construct heterojunction photocatalysts for pollutant removal, CO_2 conversion, and water splitting [28–32]. Although better charge separation has been achieved in these heterojunction systems, most previously reported heterojunction systems are based on Schottky and type II heterojunctions at the expense of photogenerated electron reduction power.

Semiconductor quantum dots (QDs) have stimulated widespread research interest in photocatalysis, which can be attributed to the unique properties of quantum size effect and multiexciton generation effect [33–36]. In particular, CdSe quantum dots (CdSe QDs) have stimulated considerable interest in photocatalytic H_2 production due to the high surface volume ratio, size-dependent light absorption capability, and the ability to induce multiple electron and hole production via single photon absorption [37–40]. Nevertheless, the agglomeration and photocorrison issues of CdSe QDs result in decreased surface area and a stronger recombination rate of photogenerated electrons and holes.

An S-scheme heterojunction, which is composed of two n-type semiconductors with the "S" shape transfer path of photogenerated charge carriers at the interface, has been reported to possess the highest redox capacity of heterojunction with boosted photocatalytic activity for photoredox reactions [41–46]. However, there have been few reports on the construction of zero-dimensional/one-dimensional (0D/1D) S-scheme heterojunctions for photocatalytic water splitting and CO_2 reduction. The different work function of CdSe QDs and CN nanorods is highly likely to form S-scheme heterojunctions with accelerated charge-carrier separation efficiency and promote redox activity for photoredox reactions. Furthermore, in the 0D/1D heteronanostructure, CN nanorods possess small nanoparticles and nanosheets with abundant voids and rough surfaces, and tend to form loose networks via randomly stacking, which provides an ideal host for immobilizing CdSe QDs, offers abundant active sites, and promotes the adsorption, desorption, and transportation of reactants and products.

In this paper, we describe a 0D/1D S-scheme heterojunction photocatalyst constructed by electrostatic self-assembly of CN nanorods and CdSe QD to promote water splitting and CO_2 reduction. Both experimental studies and density functional theory (DFT) calculations confirmed the existence of an internal electric field (IEF) in the CdSe/CN heterojunction, which can more efficiently separate photoinduced charge carriers and result in stronger redox ability. The S-scheme CdSe/CN heterojunctions showed excellent activity in water splitting and reducing CO_2 to solar fuel. This study provides a view of CN-based photocatalysts for efficient water splitting and CO_2 photoreduction following the S-scheme electron transfer pathways.

2. Results and Discussion

2.1. Preparation of Photocatalysts

The synthetic process of CdSe/CN hybrids is shown in Figure 1. A nanocasting method was utilized to fabricate CN nanorods by using chiral mesoporous silica hard-template, and then use a linker-assisted hybridization approach to prepare CdSe QDs-modified CN nanorods. Since water-soluble CdSe QDs were covered by mercaptoacetic acid, sulfhydryl groups (-SH) and carboxylic groups (-COOH) were conjugated on the surface of CdSe QDs and ionized in water, respectively. The amino groups ($-NH_2$, $=NH$) on the surface of CN nanorods shows a strong affinity for carboxylic acid groups (-COOH) of CdSe QDs, forming the resultant CdSe/CN hybrid materials.

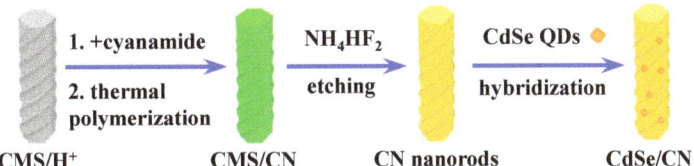

Figure 1. Schematic illustration of the synthetic process of CdSe/CN hybrids.

2.2. Morphological Characterization

The morphology and nanostructure of CdSe QDs, CN nanorods, and 5% CdSe/CN were studied by scanning emission microscopy (SEM) and transmission electron microscopy (TEM). The SEM images of CN nanorods and 5% CdSe/CN both showed a uniform rod-like morphology with an outer diameter of ca. 0.15 μm and a length of ca. 2 μm (Figures 2a, S1 and S2). The TEM and HRTEM images of 5% CdSe/CN presents that small nanoparticles with the size of ca. 5 nm are attached onto the surface of nanorods, confirming the formation of 0D/1D heteronanostructure (Figure 2b,c and S3). The lattice spacings of the CdSe QDs were 0.215 and 0.351 nm, ascribed to (220) and (111) faces of cubic CdSe (JCPDS19-0191), respectively (Figure 2d). The high-angle annular dark-field scanning transmission electron microscopy (HAADF-STEM) images, elemental mapping images, and energy dispersive X-ray (EDX) spectrum validated the existence of C, N, Cd, and Se elements for 5% CdSe/CN composite (Figures 2e–i and S4).

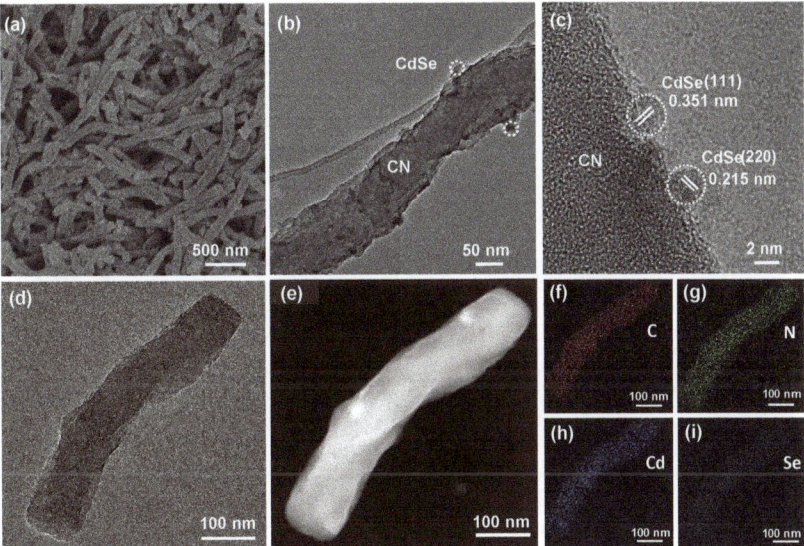

Figure 2. (**a**) SEM, (**b–d**) TEM and HRTEM images, (**e**) HAADF-STEM images, and TEM element mapping images of (**f**) C, (**g**) N, (**h**) Cd, and (**i**) Se of 5% CdSe/CN.

2.3. Structural Characterization

The X-ray diffraction (XRD) patterns of CdSe/CN, CN nanorods, and CdSe QDs samples are demonstrated in Figure 3a. Concerning the XRD pattern of CN nanorods, the two diffraction peaks at 13.0° and 27.4° are indexed to be the (100) reflection of the continuous heptazine framework with an in-plane repetition period of 0.685 nm and the (002) reflection of the graphitic structure with d value 0.326 nm, respectively [47]. CdSe QDs have a face-centered cubic CdSe crystal structure (JCPDS19-0191). The diffraction peaks at 25.4°, 42.0°, and 49.7° correspond to the (111), (220), and (311) crystal planes of

CdSe QDs. The XRD patterns of CdSe/CN hybrids exhibit diffraction peaks corresponding to both CN nanorods and CdSe QDs, indicating the presence of two phases. As the amount of CdSe QDs increases, the diffraction peaks of the CdSe/CN composites at 13.0° decrease gradually, while the diffraction peaks at 42.0° and 49.7° became increasingly more obvious. These results prove that CdSe QDs were indeed incorporated with CN nanorods.

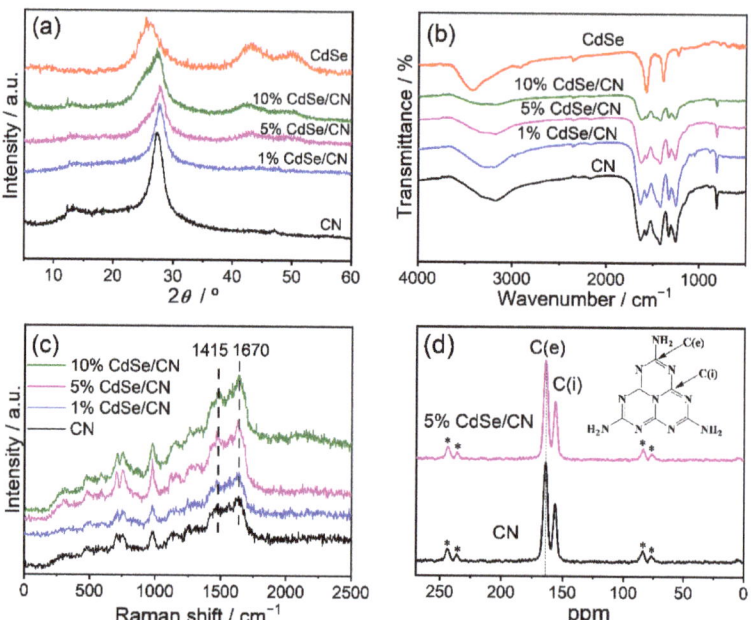

Figure 3. (**a**) XRD patterns, (**b**) FTIR spectra, (**c**) UV–Raman spectra, and (**d**) solid-state ^{13}C NMR spectra of CdSe/CN composite and CN nanorods. The stars (*) correspond to the spinning sidebands.

The Fourier transform infrared (FTIR) spectra of pristine CN nanorods, CdSe QDs, and CdSe/CN hybrids are shown in Figure 3b. For CN nanorods, the stretching mode of the carbon and nitrogen heterocycle and breathing mode of the *s*-triazine unit are presented as the characteristic band in the regions of 1200–1600 and 810 cm^{-1}, respectively. FTIR spectra of mercaptoacetic acid-coated CdSe QDs showed characteristic peaks at 1220, 1390, and 1580 cm^{-1}, corresponding to the vibrations of hydroxyl and carboxyl groups because the ligands are attached to the nanoparticles. Characteristic bands of CN nanorods and CdSe QDs both appear in the FTIR spectra of CdSe/CN composites, confirming the emergence of composite photocatalysts. Additionally, the broadband in the region of 3000–3800 and 2349 cm^{-1} are assigned to the absorption of H$_2$O and CO$_2$ on the catalysts from the atmosphere.

The Raman spectra were acquired to investigate the chemical structure of CdSe/CN (Figure 3c). There are not any bands (in the region of 2000–2500 cm^{-1}) assigned to triple C≡N units or N=C=N groups of CN structure. The bands in the range 1200–1700, 690, and 980 cm^{-1} are assigned to the C-N tensile vibration of disordered graphitic carbon-based materials, double degenerate mode of bending vibration in the plane of heptazine, and the symmetric N-breathing mode of the heptazine unit, respectively. The peaks at ca. 1415 and 1620 cm^{-1} are assigned to the D (disorder) and G (graphitic) bands of CN, related to structurally disordered graphitic carbons and other materials containing layered carbon and nitrogen. These features were observed for all CN nanorods and CdSe/CN composite catalysts.

Additionally, solid-state ^{13}C NMR spectra showed that the heptazine units were presented for both CN nanorods and 5% CdSe/CN (Figure 3d). The peaks at ca. 164.3 and

155.6 ppm correspond to the C (e) atom of [CN$_2$(NH$_x$)] and C(i) atoms of melem (CN$_3$) of poly (heptazine) structures. Raman spectra and ^{13}C NMR spectra showed that there was a graphitic structure comprising heptazine heterocycles in the CdSe/CN composites.

The chemical state of the CdSe/CN hybrid was measured by X-photoelectron spectroscopy (XPS). Six elements (C, N, Cd, Se, O, and S) were determined for the XPS survey spectra of 5% CdSe/CN (Figure S5). In comparison with CdSe, an additional N peak at the shoulder next to the Cd peaks is observed for CdSe/CN hybrid, which confirms the presence of additional CN in the composite. Apart from Cd and Se, other elements of C, O, and S for CdSe/CN hybrid originate from the mercaptoacetic acid ligand that encapsulated the CdSe QDs. The two peaks centered at 284.8 and 288.3 eV for the C1s spectrum belong to sp^2 C-C and sp^2-hybridized carbon in the N aromatic ring (N-C=N), respectively (Figure 4). Three peaks centered at 398.5, 400.0, and 401.2 eV for the N 1s spectrum are ascribed to be the sp^2-hybridized nitrogen in the triazine ring (C-N=C), the tertiary nitrogen N-(C)$_3$ group, and amino functions (C-N-H) due to incomplete polymerization of poly (tri-s-triazine) structures. The sp^2-hybridized nitrogen in the triazine ring (C-N=C, 398.5 eV), the tertiary nitrogen group (N-(C)$_3$, 400.0 eV), and sp^2 hybrid carbon (N-C=N, 288.0eV) comprise heptazine heterocyclic ring units of CN polymers. The two peaks at 405.0 and 412.0 eV for the Cd 3d spectrum correspond to Cd 3d$_{5/2}$ and Cd 3d$_{3/2}$, respectively. The two peaks at 54.0 and 63.5 eV are assigned to Se 3d and selenium oxide (formed by the partial oxidation of CdSe QDs in the air), respectively. Based on the XPS spectra of CN nanorods, CdSe QDs, and 5% CdSe/CN, it can be concluded that CdSe QDs are successfully hybridized with CN nanorods. In particular, the binding energies of C 1s and N 1s of 5% CdSe/CN were shifted negatively by 0.2 eV compared with the original CN, and binding energy Cd 3d and Se 3d of 5% CdSe/CN was more positive compared to the pristine CN, implying the existence of charge transfer pathways between CdSe QDs and CN nanorods.

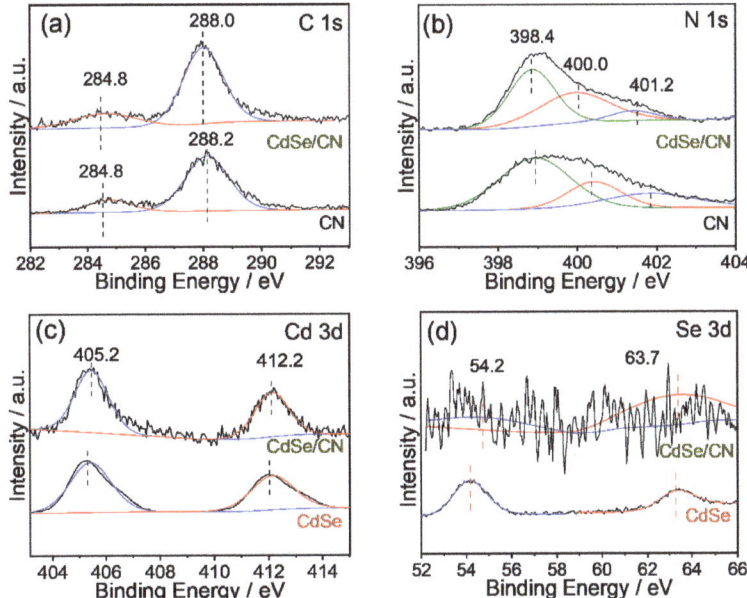

Figure 4. XPS spectra of 5% CdSe/CN: (**a**) C 1s, (**b**) N 1s, (**c**) Cd 3d, and (**d**) Se 3d.

2.4. Photochemical Properties and Band Structure

The electronic properties and light-harvesting ability were explored via UV–vis diffuse reflectance spectroscopy (DRS). The CdSe QDs exhibits obvious visible-light absorption with a band edge of 521 nm (Figure 5a). The pristine CN nanorods sample presents its basic absorption edge at 452 nm. All CdSe/CN hybrids exhibit stronger visible-light absorption ability than pristine CN nanorods. As the CdSe QDs content increases, the coverage spectrum of the composite sample becomes wider and the color of the sample becomes redder. Based on the Tauc plots, the bandgap values of CN nanorods, 5% CdSe/CN, and CdSe QDs are 2.74, 2.67, and 2.38 eV, respectively (Figure S6).

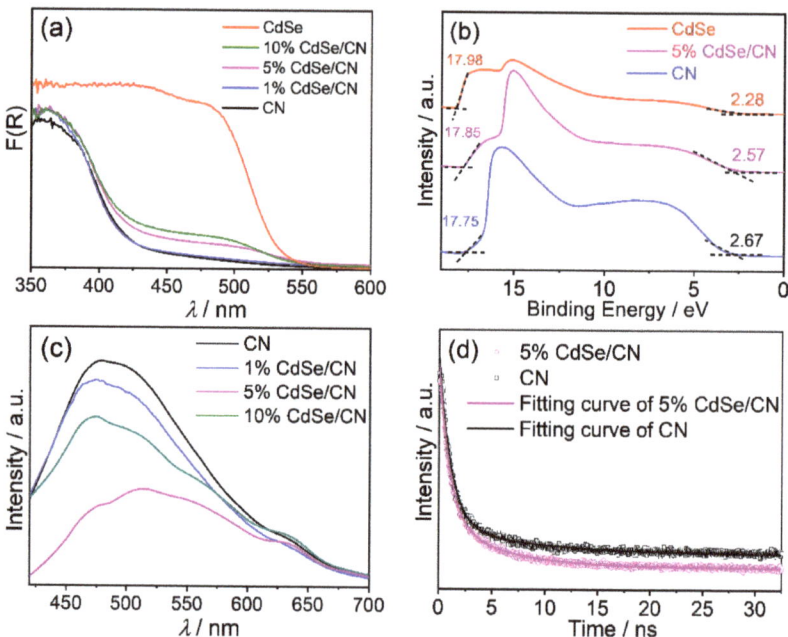

Figure 5. Optical properties of CdSe/CN: (**a**) UV–vis DRS spectra; (**b**) UPS valence band (VB) spectra of CN, 5% CdSe/CN, and CdSe; (**c**) steady-state photoluminescence spectra; and (**d**) time-resolved photoluminescence spectra of CdSe/CN composites at room temperature.

The work function (Φ) and valance band potential (E_{VB}) of CN, CdSe/CN, and CdSe were monitored by ultraviolet photoelectron spectroscopy (UPS) measurement (Figure 5b). The work function (Φ) is ascertained by the difference between the photon energy (21.22 eV) and the binding energy of the secondary cutoff edge. The secondary cutoff edge values of CN, CdSe/CN, and CdSe were 17.75, 17.85, and 17.98 eV, respectively. The work function of CN, CdSe/CN, and CdSe were 3.47, 3.37, and 3.24 eV vs. vacuum level, respectively. Thus, the Fermi energy level of CN, CdSe/CN, and CdSe are determined to be -0.97, -1.07, and -1.20 V vs. reversible hydrogen electrode (RHE).

The UPS widths (ΔE) of CN, CdSe/CN, and CdSe are 15.11, 15.28, and 15.70 eV, respectively. The E_{VB} of the catalysts are determined according to Equation (1):

$$E_{VB} = \Delta E - 21.22 \text{ eV} \tag{1}$$

The E_{VB} values of CN, CdSe/CN, and CdSe are estimated to be 6.11, 5.94, and 5.52 eV (vs. vacuum level). Since the reference standard 0 V vs. RHE (reversible hydrogen electrode) equals to -4.44 eV vs. vacuum level, the calculated value in eV is converted to potentials in volts. The E_{VB} values of CN, CdSe/CN, and CdSe correspond to 1.67, 1.50, and 1.08 V vs. RHE, respectively.

Based on the E_{VB} and E_g of the photocatalyst, the conduction band potential (E_{CB}) of the photocatalysts are calculated based on Equation (2):

$$E_{CB} = E_{VB} - E_g \quad (2)$$

Thus, the E_{CB} of CN, 5% CdSe/CN, and CdSe QDs are -1.07, -1.17, and -1.30 V vs. RHE, respectively.

The photoluminescence (PL) spectra of the CdSe/CN hybrids were tested with light excitation of 400 nm (Figure 5c). The primary emission band of CN nanorods is centered at ca. 480 nm. The photoluminescence intensity of CN nanorods is the largest among these samples, indicating that the original CN nanorods have the highest exciton energy and electron-hole recombination rate among these samples. This energy-wasteful process can be greatly suppressed by constructing an ideal heterostructure system on the surface of CN nanorods via integration with CdSe QDs. The photoluminescence intensity of CdSe/CN samples is remarkably reduced in comparison with that of CN nanorods. As the CdSe QDs content rises, the photoluminescence intensity of the CdSe/CN composite gradually decreases. Coating CN nanorods with CdSe QDs restricts the recombination of photoinduced charge carriers.

The transient PL spectra of CN and CdSe/CN are shown in Figure 5d. The short lifetime (τ_1) reflects radiative processes such as the recombination of the photogenerated charge carriers resulting in fluorescent emission, and the long lifetime (τ_2) reveals non-radiative energy transfer processes. The short lifetimes (τ_1) are 1.2 and 1.0 ns for CN and CdSe/CN, at 86.3% and 76.9%, respectively. Their radiative lifetimes are similar, but the percentage of photogenerated charge carriers on CdSe/CN is significantly reduced. This result implies that the recombination rate of photoinduced electron-hole pairs on the CdSe/CN composite are effectively suppressed after incorporating CdSe QDs with CN nanorods. Correspondingly, the nonradiative lifetimes (τ_2) of CN and CdSe/CN composites are 9.7 and 5.8 ns, at 13.7% and 23.1%, respectively (Table S1). The percentage of the long lifetimes (τ_2) for CdSe/CN composites is significantly higher than CN, showing a higher probability and priority of photogenerated charge carriers to participate in a series of photocatalytic reactions. The average lifetimes (τ_{av}) of CN and CdSe/CN composite samples are further calculated to be 2.3 and 2.1 ns, respectively. These PL results indicate the formation of hybrid structures lowers charge-carrier recombination and induces efficient photoinduced charge separation for improving photocatalytic efficiency.

2.5. Photocatalytic Water-Splitting Activities

The photocatalytic hydrogen evolution rates (HERs) of the prepared samples loaded with 3 wt% Pt using ascorbic acid (H_2A) as the sacrificial reagent at pH 4.0 are shown in Figure 6a. The CN nanorods exhibit a low hydrogen production rate (1.2 mmol g^{-1} h^{-1}). When CN nanorods are integrated with CdSe QDs, the hydrogen production rate of CdSe/CN is greatly improved. Specifically, when the weight percentage of CdSe QDs reached 10 wt%, the peak photocatalytic activity for CdSe/CN was achieved at 20.1 mmol g^{-1} h^{-1}. This value is 19-fold of CN nanorods and 4-fold of bare CdSe QDs. Nonetheless, when the amount of CdSe/CN increased to 20%, the photocatalytic activity of CdSe/CN was significantly reduced. This is because the light scattering effect and shadow effect of CdSe QDs can greatly block the absorption of incident light by CN materials, and the aggregation of excessive CdSe QDs could generate the recombination center of electron-hole pairs.

As can be seen in Figure 6b, four different electron sacrificial agents including ascorbic acid (H_2A), triethanolamine (TEOA), methanol, and lactic acid were chosen to investigate the photocatalytic hydrogen production of CdSe/CN. It is interesting to find that the rate of hydrogen production for CdSe/CN in H_2A is obviously advantageous over the other three systems. It is worth noting that the photocatalytic H_2 evolution activity in acidic condition (~pH 4) by ascorbic acid as the sacrificial reagent is much superior compared to the basic condition (~pH 11) by TEOA. This can be associated with the strong impact of pH value on the photocatalytic H_2 production activity of CdSe/CN composite.

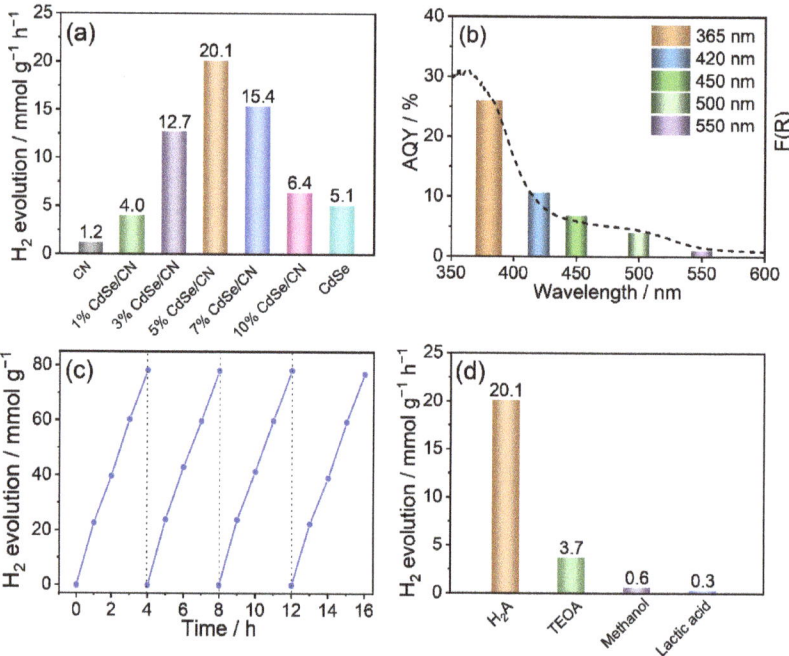

Figure 6. (**a**) Photocatalytic hydrogen evolution rates over CdSe/CN photocatalysts with different CdSe loading amount using ascorbic acid (H_2A) as sacrificial reagent at pH 4.0. (**b**) Hydrogen evolution rates evolved from 5% CdSe/CN by changing sacrificial reagent. (**c**) Wavelength-dependent hydrogen evolution rates of 5% CdSe/CN. (**d**) Time-dependent photocatalytic hydrogen evolution rates over 5% CdSe/ CN.

Furthermore, the effect of pH on the photocatalytic efficiency of CdSe/CN was studied at pH 2.0, 3.0, 4.0, 7.0, and 9.0 (Figure S7a). The photocatalytic H_2 evolution rate reached its highest value, 20.1 mmol g^{-1} h^{-1}, with pH 4.0. This is due to more efficient dissociation of H_2A toward HA$^-$ considering the pKa1 of H_2A as 4.0, which provides more HA$^-$ species acting as the sacrificial reductant to capture holes so that more photogenerated electrons can participate in proton reduction of hydrogen production. Moreover, the acidic reaction medium (~pH 4) can also help reduce the reduction potential of water, resulting in enhanced photocatalytic H_2 activity.

The apparent quantum yield (AQY) of H_2 production for 5% CdSe/CN hybrid loaded with 3 wt% Pt using ascorbic acid (H_2A) as the sacrificial reagent at 420 nm is 10.2%, surpassing the AQYs for most of the previously reported CN-based photocatalysts (Table S2). The AQY of 5% CdSe/CN under different wavelength range coincide well with its optical absorption feature (Figure 6c), suggesting that the photocatalytic reaction is initiated by the captured photons. Next, the relationship between the H_2 production rate and the amount of catalyst was studied (Figure S7b). With the increasing weight of catalyst, the AQY of CdSe/CN for photocatalytic hydrogen production increased first, and then reached the maximum value of 10.2% at 420 nm with the weight of 50 mg. When further increasing the weight of CdSe/CN catalyst above 50 mg, the AQY value slightly decreased and then remained stable.

The optimal 5% CdSe/CN photocatalyst was recycled for 16 h in four cycles in water-splitting arrays to explore the stability of the photocatalyst. Under light conditions, the hydrogen production rate on 5% CdSe/CN did not change significantly after four cycles of tests (Figure 6d). No noticeable changes were found in the XRD patterns, FTIR spectra, or Raman spectra of 5% CdSe/CN composite before and after photocatalytic hydrogen production, demonstrating the good stability of CdSe/CN composites (Figures S8 and S9).

2.6. Photocatalytic CO_2 Reduction Activities

Photocatalytic CO_2 reduction arrays of the samples were conducted and some reference experiments were carried out. No detectable amount of H_2 and CO was determined without catalyst or light (Table S3). No detectable amount of CO was noticed when replacing CO_2 with Ar gas, meaning that the decomposition of catalysts or organic additives (e.g., triethanolamine and 2,2′-bipyridyl) does not generate CO. The addition of cobalt ions (with organic ligands) cannot induce CO_2 conversion alone. These reference experiments proved that photoreduction reactions cannot occur without any component in the photosystem (e.g., photocatalyst, $Co(bpy)_3^{2+}$, triethanolamine, CO_2). Other products such as methane and methanol could be hardly generated in this photocatalytic CO_2 reduction system, in good accordance the results of previous work [48].

All CdSe/CN showed higher CO and H_2 yield than that of CN nanorods and CdSe. The highest CO yield of 5% CdSe/CN is 0.77 mmol g^{-1} h^{-1} with a turnover number of 23.7% and selectivity of 97.9% (Figure 7a and Table S4). The yields of CO and H_2 decrease with increasing illumination wavelength range, suggesting the photocatalytic CO_2 reduction is driven by the harvested photons (Figure 7b). The production amounts of CO and H_2 tend to increase gradually in a nonlinear model with the increasing reaction time for the photocatalytic CO_2 reduction system (Figure 7c). To test the photostability of the CdSe/CN mixture, the CO_2 reduction reaction was performed four times. No significant loss in CO_2 reduction activity was noticed (Figure 7d). XRD patterns, FTIR spectra, and Raman spectra of the CdSe/CN samples after photocatalytic reaction were monitored. The major chemical structure and morphology of CdSe/CN remained almost unchanged, which confirmed the stability of CdSe/CN during photocatalytic reactions (Figure S10).

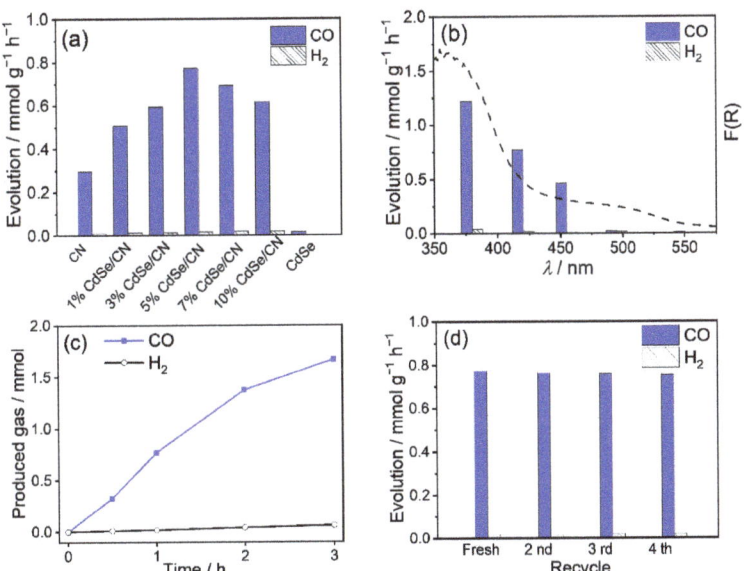

Figure 7. (a) Photocatalytic activity of CdSe/CN photocatalyst with different weight ratio in the conversion of CO_2 to CO, and (b) wavelength-dependent CO and H_2 production amount of 5% CdSe/CN composite. (c) Photocatalytic performance of CdSe/CN in CO_2-to-CO conversion. (d) Time-dependent CO production for 5% CdSe/CN, and (d) stability test for 5% CdSe/CN photocatalyst in CO_2 reduction.

2.7. Charge Transfer Process

The photoelectrochemical capability of pristine CN nanorods and CdSe/CN composites was evaluated. Transient photocurrent responses of CdSe/CN and CN nanorods for several on–off cycles were recorded (Figure 8a). At the end of irradiation, the photocurrent value quickly decreased to zero, revealing the photoexcitation properties of the process. Five percent CdSe/CN showed nearly 5-fold enhanced photocurrent higher than pristine CN nanorods, suggesting the enhanced mobility of photoexcited charge carriers. Moreover, electrochemical impedance spectroscopy (EIS) showed a significant decrease in the diameter of 5% CdSe/CN compared to CN nanorods, suggesting that CdSe/CN composites possess boosted charge-separation efficiency (Figure 8b). The obtained semicircle can be simulated by the electrical equivalent circuit model, as shown in the inset of Figure 8b. The diameter of EIS means a charge transfer resistance at the electrode/electrolyte interface (R_2) (Table S5). In comparison with CN, the smaller R_2 of 5% CdSe/CN indicates decreased charge-transfer resistance, higher electrical conductivity, and accelerated migration of photogenerated charges.

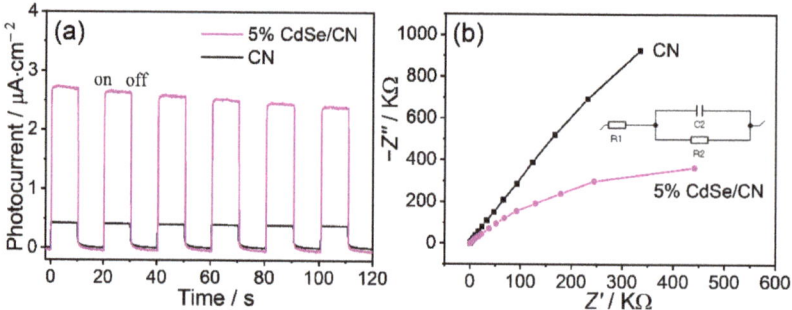

Figure 8. (a) Periodic on–off photocurrent response and (b) EIS Nyquist plots for 5% CdSe/CN composites and CN nanorods provided by drawing Z' versus $-Z''$. Z' and Z'' are defined by the real and imaginary part of impedance, respectively.

Mott–Schottky experiments were also performed to explore the relative position of the conduction band (CB) edges of CN nanorods and CdSe QDs (Figure S11). Because of the positive slope, CN nanorods, 5% CdSe/CN composites, and CdSe QDs possess the feature of n-type semiconductors. The flat band potentials of CN nanorods, 5% CdSe/CN, and CdSe QDs tests resulted in −0.95, −1.07, and −1.20 V vs. RHE at pH 7, respectively. The flat band potentials are in good accordance with the results of calculated E_{CB} for CdSe/CN composites.

From the slopes of the Mott–Schottky plots (Figure S12), carrier densities of CN and 5% CdSe/CN samples were calculated to be ~10^{20} and ~10^{21} cm^{-3}, respectively, using the Mott–Schottky relation [49]. CdSe/CN composite showed one order of magnitude increased carrier concentration compared with pristine CN, which is beneficial for boosting the photocatalytic activity.

Linear sweep voltamentary (LSV) curves for CdSe/CN and CN are shown in Figure S13. Under current density of −10 mA cm^{-2}, the overpotentials of CN and 5% CdSe/CN were found to be −210 and −150 mV, respectively. CdSe/CN presents a lower overpotential than CN, which demonstrates the construction of CdSe/CN hybrid is favorable for H_2 production in photocatalytic H_2 evolution.

The electronic band structure information of the CdSe/CN sample was further tested by electron paramagnetic resonance (EPR) at room temperature (Figure S14). Both CN and CdSe/CN presented one single Lorentzian line at 3515 G with a g value of 2.0034, which is ascribed to an unpaired electron on the carbon atoms of the aromatic rings within π-bonded nanosized clusters. In comparison with CN, the stronger spin intensity of CdSe/CN confirmed the promoted formation of unpaired electrons. A slightly enhanced

EPR intensity under visible light illumination of CdSe/CN suggested that photochemical formation of radical pairs was promoted in the CdSe/CN semiconductor.

2.8. Photocatalytic Mechanism

A further theoretical study by DFT calculation was conducted on the CdSe/CN composite heterojunction to understand the electron transfer process and the intrinsic photocatalytic mechanism.

The work functions (Φ) assigned to CN and CdSe were calculated according to Equation (3):

$$\Phi = E_{vac} - E_F \qquad (3)$$

where E_F and E_{vac} represent the Fermi level and the energy of stationary electrons in a vacuum, respectively.

Based on DFT calculation, the work functions of CN and CdSe are 4.32 and 2.67 eV, respectively (Figure 9a–d). Since CN possessed a higher work function than that of CdSe, the electrons would transfer from CdSe to CN until the E_F level reach the same levels, and the formed IEF at the heterointerface greatly facilitates the separation of photoinduced charge carriers.

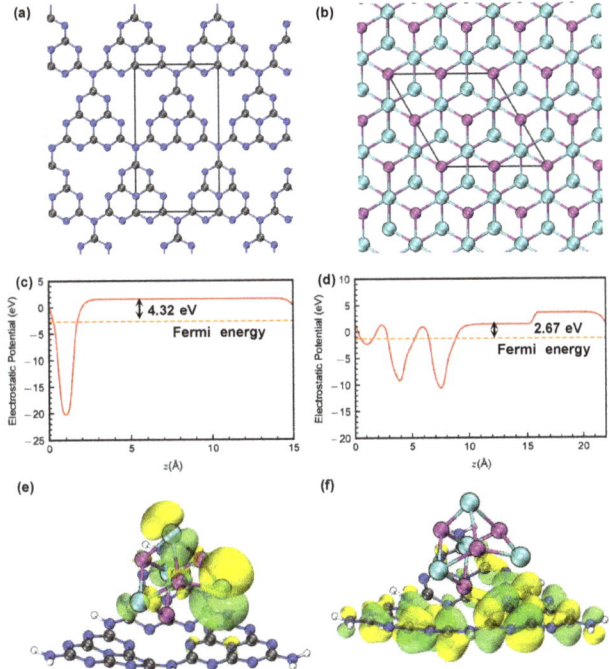

Figure 9. Structure of (**a**) CN and (**b**) CdSe. Electrostatic potential of (**c**) CN and (**d**) CdSe. (**e**) HOMO and (**f**) LUMO of CdSe/CN composites.

In addition, the calculation models based on CdSe/CN composites are presented in Figure 9e,f. For CdSe/CN composites, the lowest unoccupied molecular orbital (LUMO) and highest occupied molecular orbital (HOMO) are separately located at CN and CdSe, respectively. This suggests that CdSe and CN act as electron donor and acceptor, respectively; thus, the electrons could transfer from CdSe to CN.

A direct S-scheme photocatalytic reaction pathway based on calculation and experimental results is illustrated in Figure 10. Since CN has a higher work function than CdSe, the photogenerated electrons transfer from CdSe to CN until their E_F levels reach the same level, and IEF is produced at the contact interface due to the different electron densities [50–52]. Motivated by the IEF, the photogenerated electrons in the CB of CN combine with the holes in the VB of CdSe, similar to an S-path of charge transfer [53]. The electron transfer from the CB of CdSe to the CB of CN is then prohibited. In addition, the electrons in the CB of CdSe are transferred to its surface, thus increasing the electron density of CdSe. For water splitting, the electrons on the CB of CdSe migrate to the Pt nanoparticles and then take part in the water-splitting reactions. For CO_2 reduction, the electrons on the CB of CdSe initiate the redox reaction of the electron mediator $Co(bpy)_3^{2+}$, and then drive the reduction of CO_2 to CO. The photogenerated holes are consumed by sacrificial agents such as TEOA or H_2A. Thus, the accelerated charge separation and transfer rate is realized by constructing a 0D/1D S-scheme heterojunction of CdSe QDs and CN nanorods, thus significantly raising the photocatalytic efficiency for H_2 evolution and CO_2 reduction.

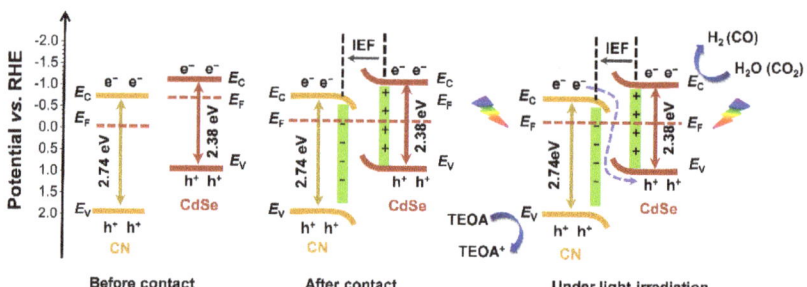

Figure 10. Schematic diagram of CdSe/CN composite for photocatalytic water splitting and CO_2 reduction using triethanolamine (TEOA) as a sacrificial agent and the formation of internal electric field (IEF). The potentials are relative to reversible hydrogen electrode (RHE).

3. Materials and Methods

3.1. Materials

Sodium hydroxide (NaOH, ≥96.0%), ammonium bifluoride (NH_4HF_2, ≥98.0%), cadmium chloride hemi(pentahydrate) ($CdCl_2 \cdot 2.5\ H_2O$, ≥99.0%), sodium borohydride ($NaBH_4$, 96%), chloroplatinic acid hexahydrate ($H_2PtCl_6 \cdot 6H_2O$, AR), cobalt(II) chloride hexahydrate ($CoCl_2 \cdot 6H_2O$, ≥99.0%), 2,2′-bipyridyl (bpy, ≥99.5%), sodium sulfide nonahydrate ($Na_2S \cdot 9H_2O$, 98.0%), mercaptoacetic acid (MPA, $C_2H_4O_2S$, ≥90.0%), sodium sulfate (Na_2SO_4, ≥99.0%), selenium (Se, ≥99.0%), hydrochloric acid (HCl, 36.0~38.0%), acetone (C_3H_6O, ≥99.5%), ethanol (C_2H_6O, ≥99.5%), petroleum ether (AR), triethanolamine (denoted as TEOA, $C_6H_{15}NO_3$, AR), acetonitrile (C_2H_3N, ≥99.8%), ascorbic acid (denoted as H_2A, $C_6H_8O_6$, ≥99.7%), N,N-dimethylformamide (C_3H_7NO, ≥99.5%) and Na_2SO_3 (sodium sulfite, ≥97.0%) were obtained from the China Sinopharm Chemical Reagent Co. Ltd. Tetradecanoyl chloride ($CH_3(CH_2)_8COCl$, 97%), tetraethoxysilane ($Si(OC_2H_5)_4$, 98%), 3-aminopropyl triethoxysilane ($H_2N(CH_2)_3Si(OC_2H_5)_3$, ≥98%), and cyanamide ($NCNH_2$, 98%), D-alanine ($C_3H_7NO_2$, ≥98%) were purchased from Sigma-Aldrich (Merck KGaA, Darmstadt, Germany). Carbon dioxide (super-grade purity, 99.999%), argon (super-grade purity, 99.999%), and nitrogen gas (99.99%) were obtained from Fujian Nanan Chenggong Gas Co. Ltd. (Fujian, China). Ultrapure water (18 mW cm^{-1}) was produced by a Millipore Milli-Q water purification system (Darmstadt, Germany). All reagents were utilized without purification.

3.2. Synthesis of N-Myristoylalanine (C_{14}-D-Ala)

C_{14}-D-Ala was synthesized according to [54]. D-alanine (0.24 mol, 21.4 g) was mixed with deionized water (140 mL), NaOH (19.2 g), and acetone (120 mL). Under vigorous stirring at 0 °C, the mixture was dropwise added to tetradecanoyl chloride (0.2 mol, 49.3 g). Additionally, 20 mL 0.2 mol L^{-1} NaOH solution was injected to maintain the pH at ~12. After reaction for 1 h, a certain amount of HCl solution was added to adjust the pH at 1.

The solids were washed with deionized water until neutral, cleaned with petroleum ether for several times, and vacuum dried at 50 °C. The yield of C_{14}-D-Ala was 30~35 g.

3.3. Synthesis of Chiral Mesoporous Silica Hard-Template

Chiral mesoporous silica was synthesized according to [55,56]. C_{14}-D-Ala (0.321 g, 1.0 mmol) surfactant was dissolved in water (22.1 g) and 0.01 M hydrochloric acid (10 g, 0.1 mmol), and then dropwise added to a mixture of tetraethoxysilane (1.40 g, 6.7 mmol) and 3-aminopropyl triethoxysilane (0.23 g, 1.0 mmol) with stirring at 400 rpm for 10 min. The mixture remained under static conditions at room temperature for 22 h, and then collected by filtration. The precipitates were dried at 80 °C for 12 h, and heated at 550 °C for 6 h in air. The yield of chiral mesoporous silica was 0.1~0.2 g.

3.4. Preparation of CN Nanorods

CN nanorods were prepared by a hard-templating method using chiral mesoporous silica nanorods as hard-templates based on [17]. Chiral mesoporous silica powder was dispersed in hydrochloric acid solution (1 mol L^{-1}) at 80 °C for 20 h, centrifuged, and dried at 80 °C for 10 h. The acidified chiral mesoporous silica powder (1.0 g) was mixed with cyanamide (6.0 g) in a flask, vacuum degassed for 5 h, and sonicated at 60 °C water bath for 5 h. The mixture was washed with water, stirred for 15 min, and centrifuged. The white solids were dried at 80 °C overnight, and heated at 550 °C for 240 min at a rate of 2.2 °C·min^{-1} with the flow of nitrogen. The yellow solids were mixed with ammonium bifluoride solution (4 mol L^{-1}) for 10 h, cleaned with water and ethanol 4 times, and finally vacuum dried at 80 °C for 10 h.

3.5. Preparation of Water-Soluble CdSe QDs

CdSe QDs were obtained based on [35,36]. In a three-necked flask, $CdCl_2·2.5 H_2O$ (5 mmol, 1.142 g) was dissolved in water (60 mL) and degassed with N_2 bubbles for 60 min. This solution was added to mercaptoacetic acid (0.85 mL), and dropwise added to sodium hydroxide (1 mol L^{-1}, 24 mL) solution to tune pH value to 7. NaHSe solution was prepared by mixing Se powder (0.19 g) with sodium borohydride (0.19 g) in water (12 mL), and then injected into the above solution under high-speed stirring. The mixture was refluxed at 80 °C for 240 min, and then added to ethanol (125 mL) and centrifuged. The precipitates were totally cleaned with water and methanol, and vacuum dried at 60 °C for 10 h to obtain the CdSe QDs powder.

3.6. Synthesis of CdSe/CN Composites

CdSe/CN photocatalysts were prepared by a linker-assisted hybridization approach. The binding of CdSe QDs to CN nanorods can be achieved via the assistance of mercaptoacetic acid, which is a stabilizer and a bifunctional linker of CdSe QDs. Two-tenths gram of CN nanorods powder was added to 5 mL of water and a suitable amount of CdSe QDs solution (10 mg mL^{-1}) and stirred at 80 °C for 12 h to acquire mixed solid by removing water. The as-prepared CdSe QDs modified CN nanorods sample was named as x% CdSe/CN, where x represents the weight percentage of the CdSe QDs to the CN nanorods (x = 1, 10, or 20).

3.7. Characterizations

Scanning emission microscope (SEM) analysis was carried out via an S4800 Field Emission Scanning Electron Microscope (Hitachi, Chiyoda, Tokyo, Japan). Transmission electron microscopy (TEM) analysis was performed via a Talos F200X (Thermo, Waltham, MA, USA) and TECNAI G2F20 instrument (FEI, Hillsboro, OR, USA). X-ray diffraction (XRD) patterns were obtained from a D/MAXRB diffractometer (Rigaku, Akishima-shi, Tokyo, Japan) with Cu-Kα radiation (λ = 1.54184 Å). Fourier transform infrared (FTIR) spectra were gathered from a Nicolet iS10 FTIR spectrometer (Thermo, Waltham, MA, USA). UV–Raman scattering tests were performed with a multichannel modular triple

Raman system (Renishaw Co., Wotton-under-Edge, Gloucestershire, UK) with confocal microscope at room temperature using a 325 nm laser. Solid-state ^{13}C cross-polarization nuclear magnetic resonance (^{13}C NMR) spectra were obtained using an Advance III 500 Spectrometer (Bruker, Billerica, MA, USA). X-ray photoelectron spectroscopy (XPS) was performed on an ESCALAB250 instrument with a monochromatized Al Kα line source (200 W) (Thermo Scientific, Waltham, MA, USA). All binding energies were referenced to the C 1s peak at 284.8 eV of surface adventitious carbon. The UV–vis diffuse reflectance spectra (DRS) were tested on a Shimadzu UV-2550 UV–vis–NIR system (Kyoto, Japan). Photoluminescence (PL) spectra were measured on a FLS-920 spectrophotometer (Edinburgh, Livingston, West Lothian, UK). Ultraviolet photoelectron spectroscopy (UPS) was performed on a PHI 5000 Versaprobe III instrument (Chigasaki, Kanagawa, Japan). UPS measurements were conducted with an unfiltered He I (21.22 eV) gas discharge lamp and a gold calibration. The width of binding energy (ΔE) was determined from the two intersections with the UPS spectrum baseline. The width value of He I UPS spectra (21.22 eV) was used as the standard. Since the reference standard 0 V vs. RHE (reversible hydrogen electrode) equals −4.44 eV vs. vacuum level, the calculated value in eV was converted to potentials in volts. Electron paramagnetic resonance (EPR) measurements were carried out on a Bruker model A300 spectrometer (Billerica, MA, USA).

3.8. Photoelectrochemical Measurement

To prepare the working electrode, 5 mg of photocatalyst and 5 mL DMF were firstly mixed by sonication for 1 h. Then, 40 μL of the suspension was spin-coated on an F-doped SnO$_2$ transparent conductive glass (FTO) slide with a specific round area (0.2826 cm^2), and naturally dried at room temperature. Subsequently, the sample was added to 10 μL of Nafion solution (0.05%) and naturally dried at room temperature. Photoelectrochemical measurements were conducted in a three-electrode cell in an aqueous Na$_2$SO$_4$ electrolyte (0.2 M, pH 6.6) using a VSP-300 (Biologic, Seyssinet-Pariset, France) electrochemical analyzer. The catalyst electrode, saturated calomel electrode, and Pt plate were utilized as the working electrode, the reference electrode, and the counter electrode, respectively. The electrolyte solution was 0.1 M Na$_2$SO$_4$ solution. Photocurrent densities were measured under a 300 W Xenon lamp (Perfect Light PLSSXE 300, Beijing, China) with a 420 nm cutoff filter. Electrochemical impedance spectroscopy (EIS) was tested at a 5 mV sinusoidal AC perturbation over the frequency range 0.1~10^5 Hz at −0.2 V. Linear sweep voltammetry (LSV) measurements were performed at a scan rate of 20 mV s^{-1} in the range −1.0~0.7 V. Carrier density was estimated through the Mott–Schottky relation according to Equation (4):

$$\frac{1}{C_{sc}^2} = \frac{2}{\varepsilon\varepsilon_0 A^2 e N_d}\left(V - V_{fb} - \frac{k_b T}{e}\right) \qquad (4)$$

where C_{sc} is the capacitance of the space charge region, e is the elementary charge of an electron (1.602 × 10^{-19} C), ε_0 is the permittivity of vacuum (8.854 × 10^{-14} F cm^{-1}), ε is the dielectric constant of carbon nitride polymer ($\varepsilon \approx 8$), A is the electrochemically active surface area (0.28 cm^2), V is the applied voltage, V_{fb} is the flatband potential, and N_d is the donor density (carrier concentration). T is the absolute temperature, k_b is the Boltzmann constant, and $k_b T/e$ is about 0.026 V at room temperature.

3.9. Photocatalytic Hydrogen Evolution

Hydrogen evolution was measured in a Ceaulight CEL-SPH2N-D5 closed gas-circulation–evacuation system. Photocatalyst powder (50 mg) was dispersed in an aqueous solution (100 mL) containing triethanolamine (10 mL) or 0.1 mol L^{-1} ascorbic acid (H$_2$A) with an adjusted pH of 3.5, and added into a Pyrex top-irradiation reaction vessel connected to a glass closed-gas system. The catalyst was stirred, loaded with 3 wt% Pt using an in situ photodeposition approach with H$_2$PtCl$_6$·6H$_2$O, and subjected to vacuum degassing for several times to completely remove air. Photocatalytic H$_2$ evolution array was conducted in a vacuum at 6 °C to avoid the evaporation of water, which interferes with light irradiation. Gas products were determined by a gas chromatograph with a 5A sieve column. The system was then irradiated under a Perfect Light PLSSXE 300 W Xe lamp (Beijing, China) equipped with an appropriate long-pass cutoff filter and maintained at room temperature by a flow of cooling water. The generated gases were detected by a Shimadzu GC-2014C gas chromatograph (Kyoto, Japan) equipped with a thermal conductive detector (5A sieve) and argon as the carrier gas.

3.10. Apparent Quantum Efficiency for H_2 Evolution Measurement

The apparent quantum efficiency (AQY) was calculated according to Equation (5):

$$AQY = \frac{2 \times \text{number of evolved hydrogen molecules}}{\text{number of incident photons}} \times 100\%$$

$$AQY = \frac{N_e}{N_p} \times 100\% = \frac{2 \times M \times N_A}{\frac{E_{total}}{E_{photon}}} \times 100\% = \frac{2 \times M \times N_A}{\frac{S \times P \times t}{h \times \frac{c}{\lambda}}} \times 100\% = \frac{2 \times M \times N_A \times h \times c}{S \times P \times t \times \lambda} \times 100\% \quad (5)$$

where M is the number of H_2 molecules (mol), N_A is the Avogadro constant (6.022×10^{23} mol), h is the Planck constant (6.626×10^{-34} J·s), c is the speed of light (3×10^8 m/s), S is the irradiation area (cm^2), P is the intensity of irradiation light (W/cm^2), t is the photoreaction time (s), and λ is the wavelength of the monochromatic light (m).

The AQY of H_2 generation was tested by using different band-pass filters. A PLSSXE 300 W Xe lamp with a band-pass filter (420 ± 15 nm) was used as the light source. The intensity of irradiation light was 5.369 mW/cm^2. The H_2 evolution amount of 5% CdSe/CN was 456.2 µmol for the 5 h reaction.

$$AQY = \frac{2 \times 456.2 \times 10^{-6} \times 6.02 \times 10^{23} \times 6.626 \times 10^{-34} \times 3.0 \times 10^8}{26.42 \times 5.369 \times 10^{-3} \times 5 \times 3600 \times 420 \times 10^{-9}} \times 100\% = 10.2\% \quad (6)$$

3.11. Photocatalytic CO_2 Reduction

Catalyst powder (30 mg), 2,2′-bipyridyl (15 mg), H_2O (1 mL), acetonitrile (3 mL), triethanolamine (1 mL), and $CoCl_2·6H_2O$ (1 µmol) were mixed under stirring in a Schlenk flask (80 mL). The mixture was subjected to vacuum degassing and backfilling with pure CO_2 gas for 3 times. The system was filled with CO_2 (1 atm) after the last cycle. The photocatalytic CO_2 reduction array was performed at 30 °C in an atmospheric system, free from obvious evaporation disturbance under the experimental conditions. The system was then irradiated with a Perfect Light PLSSXE 300 W Xe lamp (Beijing, China) with a 420 nm cutoff filter. The gas products were tested using a 7890B gas chromatograph (Agilent, Santa Clara, CA, USA) equipped with a methanizer, flame ionization detector, thermal conductivity detector, and TDX-1 packed column with argon as the carrier gas.

3.12. Computational Details

The slab calculations were performed with the Vienna ab initio package (VASP) [57–59]. In the calculations, within the framework of density functional theory (DFT), the PBE exchange-correlation functional [60] and the dispersion interaction corrected by the D3 scheme were considered [61]. The cutoff energy for the plane waves was 400 eV, and the atomic core region was described by PAW pseudopotentials [62]. Dipole correction along the z-direction was taken into consideration [63]. $5 \times 3 \times 1$ and $4 \times 4 \times 1$ k-point meshes were used for CN and CdSe (111) in DFT calculation, respectively. The vacuum layer of slab model was set to 15 Å.

The molecule-level calculations were performed using the Gaussian 09 programs [64]. The structures were fully optimized with the B3LYP [65,66] method and Ahlrichs' split-valence def2-SVP basis set [67]. Grimmes's DFT-D3 dispersion correction was used to describe the van der Waals interaction.

4. Conclusions

In summary, a zero-dimensional/one-dimensional S-scheme heterojunction of CdSe quantum dots coupled with polymeric carbon nitride nanorods were prepared via a chemical impregnation method. Five percent CdSe/CN composite showed the best photocatalytic efficiency in water splitting, with a hydrogen evolution rate of 20.1 mmol g^{-1} h^{-1} and apparent quantum yield of 10.2% at 420 nm irradiation, and exhibited a CO production rate of 0.77 mmol g^{-1} h^{-1} in CO_2 reduction. The superior reactivity of CdSe/CN heterojunction displays toward water splitting and carbon dioxide photoreduction are mainly attributed to the more efficient charge-carrier separation rate, stronger light absorption ability, and more abundant active sites. The higher work function value of polymeric carbon nitride than CdSe leads to the transfer of electrons from CdSe quantum dots to polymeric carbon nitride nanorods upon hybridization, and interfacial electric field is created at heterointerfaces. The photoinduced electrons in the conduction band of the polymeric carbon nitride nanorods then immigrate to the valence band of CdSe, confirming an S-path of charge transfer. This work demonstrates the possibility of employing zero-dimensional/one-dimensional S-scheme heterojunction photocatalysts for solar energy conversion.

Supplementary Materials: The following supporting information can be downloaded at: https://www.mdpi.com/article/10.3390/molecules27196286/s1, Figures S1–S14 and Tables S1–S5, with references [39,68–76]. Figure S1: SEM images of CN nanorods; Figure S2: SEM images of 5% CdSe/CN; Figure S3: (a) TEM image, (b) HRTEM image, (c) SAED image, and (d) EDX image of CdSe QD solutions; Figure S4: TEM-EDX image of 5% CdSe/CN hybrid; Figure S5: XPS survey spectra of CN, CdSe and 5% CdSe/CN hybrid; Figure S6: Tauc plots of (a) CN nanorods, CdSe/CN hybrids, and (b) CdSe QDs; Figure S7: (a) Amount of hydrogen evolved from 5% CdSe/CN using ascorbic acid (H_2A) as sacrificial reagent by changing pH value. (b) AQY of 5% CdSe/CN for photocatalytic H_2 evolution with different photocatalyst weight.; Figure S8: (a) XRD pattern, (b) FTIR spectra, and (c) UV–Raman spectra of 5% CdSe/CN before and after the hydrogen evolution reaction; Figure S9: SEM images of 5% CdSe/CN after photocatalytic reactions; Figure S10: (a) XRD pattern, (b) FTIR spectra, and (c) UV–Raman spectra of 5% CdSe/CN before and after the CO_2 conversion reaction; Figure S11: Mott–Schottky plots of (a) CN nanorods, (b) 5% CdSe/CN, and (c) CdSe QDs; Figure S12. Mott–Schottky plots of CN nanorods and 5% CdSe/CN to calculate the charge-carrier density; Figure S13. Linear sweep voltamentary for CN and 5% CdSe/CN; Figure S14. EPR spectra of CN and 5% CdSe/CN; Table S1: Fitted fluorescence decay components of CN nanorods and CdSe/CN (λ = 375 nm); Table S2. Literature values of AQY for CN-based photocatalysts in hydrogen evolution; Table S3: Various controlled experiments for CO_2 reduction; Table S4: The reactivity of photocatalysts for CO_2 reduction. Table S5. Calculated values of equivalent circuit elements for CN and 5% CdSe/CN samples.

Author Contributions: Data curation, Y.W. and H.W.; Funding acquisition, Y.Z. and M.Z.; Investigation, Y.W. and H.W.; Project administration, Y.L. and Y.Z.; Supervision, Y.L. and Y.Z.; Writing—original draft, Y.Z. and M.Z. All authors have read and agreed to the published version of the manuscript.

Funding: This research was funded by the National Natural Science Foundation of China (21902051 and 21902026), the Natural Science Foundation of Fujian Province (2019J05090), and the Fundamental Research Funds for the Central Universities (ZQN-807).

Institutional Review Board Statement: Not applicable.

Informed Consent Statement: Not applicable.

Data Availability Statement: Not applicable.

Acknowledgments: The authors thank Xinchen Wang, Yuanxing Fang and Sibo Wang from Fuzhou University for supporting this work with additional facilities. The authors thank the Instrumental Analysis Center of Huaqiao University for TEM analysis.

Conflicts of Interest: The authors declare no conflict of interest.

Sample Availability: Samples of the compounds of CdSe/CN hybrids are available from the authors.

References

1. Ghosh, S.; Kouamé, N.A.; Ramos, L.; Remita, S.; Dazzi, A.; Deniset-Besseau, A.; Beaunier, P.; Goubard, F.; Aubert, P.-H.; Remita, H. Conducting polymer nanostructures for photocatalysis under visible light. *Nat. Mater.* **2015**, *14*, 505–511. [CrossRef] [PubMed]
2. Vesali-Kermani, E.; Habibi-Yangjeh, A.; Diarmand-Khalilabad, H.; Ghosh, S. Nitrogen photofixation ability of g-C_3N_4 nanosheets/Bi_2MoO_6 heterojunction photocatalyst under visible-light illumination. *J. Colloid Interf. Sci.* **2020**, *563*, 81–91. [CrossRef] [PubMed]
3. He, Z.; Zhang, J.; Li, X.; Guan, S.; Dai, M.; Wang, S. 1D/2D heterostructured photocatalysts: From design and unique properties to their environmental applications. *Small* **2020**, *16*, 2005051. [CrossRef] [PubMed]
4. Bera, S.; Ghosh, S.; Maiyalagan, T.; Basu, R.N. Band edge engineering of BiOX/$CuFe_2O_4$ heterostructures for efficient water splitting. *ACS Appl. Energy Mater.* **2022**, *5*, 3821–3833. [CrossRef]
5. Wang, Y.; Wang, Q.; Zhan, X.; Wang, F.; Safdar, M.; He, J. Visible light driven type II heterostructures and their enhanced photocatalysis properties: A review. *Nanoscale* **2013**, *5*, 8326–8339. [CrossRef]
6. Pan, B.; Feng, M.; McDonald, T.J.; Manoli, K.; Wang, C.; Huang, C.H.; Sharma, V.K. Enhanced ferrate(VI) oxidation of micropollutants in water by carbonaceous materials: Elucidating surface functionality. *Chem. Eng. J.* **2020**, *398*, 125607. [CrossRef]
7. Pan, B.; Wu, Y.; Qin, J.; Wang, C. Ultrathin $Co_{0.85}$Se nanosheet cocatalyst for visible-light CO_2 photoreduction. *Catal. Today* **2019**, *335*, 208–213. [CrossRef]
8. Yang, C.; Li, R.; Zhang, K.A.I.; Lin, W.; Landfester, K.; Wang, X. Heterogeneous photoredox flow chemistry for the scalable organosynthesis of fine chemicals. *Nat. Commun.* **2020**, *11*, 1239. [CrossRef]
9. Wang, X.; Maeda, K.; Thomas, A.; Takanabe, K.; Xin, G.; Carlsson, J.M.; Domen, K.; Antonietti, M. A metal-free polymeric photocatalyst for hydrogen production from water under visible light. *Nat. Mater.* **2009**, *8*, 76–80. [CrossRef]
10. Zhao, Y.; Zheng, L.; Shi, R.; Zhang, S.; Bian, X.; Wu, F.; Cao, X.; Waterhouse, G.I.N.; Zhang, T. Alkali etching of layered double hydroxide nanosheets for enhanced photocatalytic N_2 reduction to NH_3. *Adv. Energy Mater.* **2020**, *10*, 2002199. [CrossRef]
11. Lin, L.; Lin, Z.; Zhang, J.; Cai, X.; Lin, W.; Yu, Z.; Wang, X. Molecular-level insights on the reactive facet of carbon nitride single crystals photocatalysing overall water splitting. *Nat. Catal.* **2020**, *3*, 649–655. [CrossRef]
12. Fang, Y.; Hou, Y.; Fu, X.; Wang, X. Semiconducting polymers for oxygen evolution reaction under light illumination. *Chem. Rev.* **2022**, *122*, 4204–4256. [CrossRef] [PubMed]
13. Volokh, M.; Peng, G.; Barrio, J.; Shalom, M. Carbon nitride materials for water splitting photoelectrochemical cells. *Angew. Chem. Int. Ed.* **2019**, *58*, 6138–6151. [CrossRef] [PubMed]
14. Chen, Z.; Fang, Y.; Wang, L.; Chen, X.; Lin, W.; Wang, X. Remarkable oxygen evolution by Co-doped ZnO nanorods and visible light. *Appl. Catal. B Environ.* **2021**, *296*, 120369. [CrossRef]
15. Zhang, B.; Hu, X.; Liu, E.; Fan, J. Novel S-scheme 2D/2D BiOBr/g-C_3N_4 heterojunctions with enhanced photocatalytic activity. *Chin. J. Catal.* **2021**, *42*, 1519–1529. [CrossRef]
16. Wang, J.; Yu, Y.; Cui, J.; Li, X.; Zhang, Y.; Wang, C.; Yu, X.; Ye, J. Defective g-C_3N_4/covalent organic framework van der Waals heterojunction toward highly efficient S-scheme CO_2 photoreduction. *Appl. Catal. B Environ.* **2022**, *301*, 120814. [CrossRef]
17. Zheng, Y.; Lin, L.; Ye, X.; Guo, F.; Wang, X. Helical graphitic carbon nitrides with photocatalytic and optical activities. *Angew. Chem. Int. Ed.* **2014**, *53*, 11926–11930. [CrossRef]
18. Chen, Y.; Su, F.; Xie, H.; Wang, R.; Ding, C.; Huang, J.; Xu, Y.; Ye, L. One-step construction of S-scheme heterojunctions of N-doped MoS_2 and S-doped g-C_3N_4 for enhanced photocatalytic hydrogen evolution. *Chem. Eng. J.* **2021**, *404*, 126498. [CrossRef]
19. Wang, J.; Wang, G.; Cheng, B.; Yu, J.; Fan, J. Sulfur-doped g-C_3N_4/TiO_2 S-scheme heterojunction photocatalyst for congo red photodegradation. *Chin. J. Catal.* **2021**, *42*, 56–68. [CrossRef]
20. Xu, Q.; Ma, D.; Yang, S.; Tian, Z.; Cheng, B.; Fan, J. Novel g-C_3N_4/g-C_3N_4 S-scheme isotype heterojunction for improved photocatalytic hydrogen generation. *Appl. Surf. Sci.* **2019**, *495*, 143555. [CrossRef]
21. Zheng, Y.; Yu, Z.; Ou, H.; Asiri, A.M.; Chen, Y.; Wang, X. Black phosphorus and polymeric carbon nitride heterostructure for photoinduced molecular oxygen activation. *Adv. Funct. Mater.* **2018**, *28*, 1705407. [CrossRef]
22. Zheng, Y.; Lin, L.; Wang, B.; Wang, X. Graphitic carbon nitride polymers toward sustainable photoredox catalysis. *Angew. Chem. Int. Ed.* **2015**, *54*, 12868–12884. [CrossRef] [PubMed]
23. Li, X.; Zhang, J.; Huo, Y.; Dai, K.; Li, S.; Chen, S. Two-dimensional sulfur- and chlorine-codoped g-C_3N_4/CdSe-amine heterostructures nanocomposite with effective interfacial charge transfer and mechanism insight. *Appl. Catal. B Environ.* **2021**, *280*, 119452. [CrossRef]
24. Yao, G.; Liu, Y.; Liu, J.; Xu, Y. Facile synthesis of porous g-C_3N_4 with enhanced visible-light photoactivity. *Molecules* **2022**, *27*, 1754. [CrossRef] [PubMed]
25. Zheng, Y.; Chen, Y.; Gao, B.; Lin, B.; Wang, X. Black phosphorus and carbon nitride hybrid photocatalysts for photoredox reactions. *Adv. Funct. Mater.* **2020**, *30*, 2002021. [CrossRef]
26. Ong, W.-J.; Tan, L.-L.; Ng, Y.H.; Yong, S.-T.; Chai, S.-P. Graphitic carbon nitride (g-C_3N_4)-based photocatalysts for artificial photosynthesis and environmental remediation: Are we a step closer to achieving sustainability? *Chem. Rev.* **2016**, *116*, 7159–7329. [CrossRef]
27. Li, X.; Wang, J.; Xia, J.; Fang, Y.; Hou, Y.; Fu, X.; Shalom, M.; Wang, X. One-pot synthesis of CoS_2 merged polymeric carbon nitride films for photoelectrochemical water splitting. *ChemSusChem* **2022**, *15*, e202200330. [CrossRef]

28. Zhang, Z.; Kang, Y.; Yin, L.-C.; Niu, P.; Zhen, C.; Chen, R.; Kang, X.; Wu, F.; Liu, G. Constructing CdSe QDs modified porous g-C_3N_4 heterostructures for visible light photocatalytic hydrogen production. *J. Mater. Sci. Technol.* **2021**, *95*, 167–171. [CrossRef]
29. Pan, J.; Liang, J.; Xu, Z.; Yao, X.; Qiu, J.; Chen, H.; Qin, L.; Chen, D.; Huang, Y. Rationally designed ternary CdSe/WS_2/g-C_3N_4 hybrid photocatalysts with significantly enhanced hydrogen evolution activity and mechanism insight. *Int. J. Hydrogen Energy* **2021**, *46*, 30344–30354. [CrossRef]
30. Li, C.; Zou, X.; Lin, W.; Mourad, H.; Meng, J.; Liu, Y.; Abdellah, M.; Guo, M.; Zheng, K.; Nordlander, E. Graphitic carbon nitride/CdSe quantum dot/iron carbonyl cluster composite for enhanced photocatalytic hydrogen evolution. *ACS Appl. Nano Mater.* **2021**, *4*, 6280–6289. [CrossRef]
31. Huo, Y.; Zhang, J.; Dai, K.; Liang, C. Amine-modified S-scheme porous g-C_3N_4/CdSe-diethylenetriamine composite with enhanced photocatalytic CO_2 reduction activity. *ACS Appl. Energy Mater.* **2021**, *4*, 956–968. [CrossRef]
32. Yang, H.; Zhang, J.; Dai, K. Organic amine surface modified one-dimensional $CdSe_{0.8}S_{0.2}$-diethylenetriamine/two-dimensional $SnNb_2O_6$ S-scheme heterojunction with promoted visible-light-driven photocatalytic CO_2 reduction. *Chin. J. Catal.* **2022**, *43*, 255–264. [CrossRef]
33. Kadi, M.W.; Mohamed, R.M.; Ismail, A.A.; Bahnemann, D.W. Construction of visible light responsive CdSe/g-C_3N_4 nanocomposites for H_2 production. *Nanosci. Nanotechnol. Lett.* **2019**, *11*, 1281–1291. [CrossRef]
34. Hu, T.; Li, Z.; Lu, L.; Dai, K.; Zhang, J.; Li, R.; Liang, C. Inorganic-organic CdSe-diethylenetriamine nanobelts for enhanced visible photocatalytic hydrogen evolution. *J. Colloid Interf. Sci.* **2019**, *555*, 166–173. [CrossRef] [PubMed]
35. Rogach, A.L.; Kornowski, A.; Gao, M.; Eychmüller, A.; Weller, H. Synthesis and characterization of a size series of extremely small thiol-stabilized CdSe nanocrystals. *J. Phys. Chem. B* **1999**, *103*, 3065–3069. [CrossRef]
36. Li, J.J.; Wang, Y.A.; Guo, W.; Keay, J.C.; Mishima, T.D.; Johnson, M.B.; Peng, X. Large-scale synthesis of nearly monodisperse CdSe/CdS core/shell nanocrystals using air-stable reagents via successive ion layer adsorption and reaction. *J. Am. Chem. Soc.* **2003**, *125*, 12567–12575. [CrossRef]
37. Zhong, Y.; Chen, W.; Yu, S.; Xie, Z.; Wei, S.; Zhou, Y. CdSe quantum dots/g-C_3N_4 heterostructure for efficient H_2 production under visible light irradiation. *ACS Omega* **2018**, *3*, 17762–17769. [CrossRef]
38. Raziq, F.; Hayat, A.; Humayun, M.; Mane, S.K.B.; Faheem, M.B.; Ali, A.; Zhao, Y.; Han, S.; Cai, C.; Li, W.; et al. Photocatalytic solar fuel production and environmental remediation through experimental and DFT based research on CdSe-QDs-coupled P-doped-g-C_3N_4 composites. *Appl. Catal. B Environ.* **2020**, *270*, 118867. [CrossRef]
39. Raheman, S.A.R.; Wilson, H.M.; Momin, B.M.; Annapure, U.S.; Jha, N. CdSe quantum dots modified thiol functionalized g-C_3N_4: Intimate interfacial charge transfer between 0D/2D nanostructure for visible light H_2 evolution. *Renew. Energy* **2020**, *158*, 431–443. [CrossRef]
40. Putri, L.K.; Ng, B.-J.; Ong, W.-J.; Lee, H.W.; Chang, W.S.; Mohamed, A.R.; Chai, S.-P. Energy level tuning of CdSe colloidal quantum dots in ternary 0D-2D-2D CdSe QD/B-rGO/O-gC_3N_4 as photocatalysts for enhanced hydrogen generation. *Appl. Catal. B Environ.* **2020**, *265*, 118592. [CrossRef]
41. Bao, Y.; Song, S.; Yao, G.; Jiang, S. S-scheme photocatalytic systems. *Solar RRL* **2021**, *5*, 2100118. [CrossRef]
42. He, F.; Zhu, B.; Cheng, B.; Yu, J.; Ho, W.; Macyk, W. 2D/2D/0D TiO_2/C_3N_4/Ti_3C_2 MXene composite S-scheme photocatalyst with enhanced CO_2 reduction activity. *Appl. Catal. B Environ.* **2020**, *272*, 119006. [CrossRef]
43. Cheng, C.; He, B.; Fan, J.; Cheng, B.; Cao, S.; Yu, J. An inorganic/organic S-scheme heterojunction H_2-production photocatalyst and its charge transfer mechanism. *Adv. Mater.* **2021**, *33*, 2100317. [CrossRef]
44. Wang, L.; Cheng, B.; Zhang, L.; Yu, J. In situ irradiated xps investigation on S-scheme TiO_2@$ZnIn_2S_4$ photocatalyst for efficient photocatalytic CO_2 reduction. *Small* **2021**, *17*, 2103447. [CrossRef] [PubMed]
45. Xia, P.; Cao, S.; Zhu, B.; Liu, M.; Shi, M.; Yu, J.; Zhang, Y. Designing a 0D/2D S-scheme heterojunction over polymeric carbon nitride for visible-light photocatalytic inactivation of bacteria. *Angew. Chem. Int. Ed.* **2020**, *59*, 5218–5225. [CrossRef]
46. Xu, F.; Meng, K.; Cheng, B.; Wang, S.; Xu, J.; Yu, J. Unique S-scheme heterojunctions in self-assembled TiO_2/$CsPbBr_3$ hybrids for CO_2 photoreduction. *Nat. Commun.* **2020**, *11*, 4613. [CrossRef]
47. Song, T.; Long, B.; Yin, S.; Ali, A.; Deng, G.-J. Designed synthesis of a porous ultrathin 2D CN@graphene@CN sandwich structure for superior photocatalytic hydrogen evolution under visible light. *Chem. Eng. J.* **2021**, *404*, 126455. [CrossRef]
48. Qin, J.; Wang, S.; Ren, H.; Hou, Y.; Wang, X. Photocatalytic reduction of CO_2 by graphitic carbon nitride polymers derived from urea and barbituric acid. *Appl. Catal. B Environ.* **2015**, *179*, 1–8. [CrossRef]
49. Resasco, J.; Zhang, H.; Kornienko, N.; Becknell, N.; Lee, H.; Guo, J.; Briseno, A.L.; Yang, P. TiO_2/$BiVO_4$ nanowire heterostructure photoanodes based on type II band alignment. *ACS Cent. Sci.* **2016**, *2*, 80–88. [CrossRef]
50. Xu, Q.; Zhang, L.; Cheng, B.; Fan, J.; Yu, J. S-scheme heterojunction photocatalyst. *Chem* **2020**, *6*, 1543–1559. [CrossRef]
51. Zhang, L.; Zhang, J.; Yu, H.; Yu, J. Emerging S-scheme photocatalyst. *Adv. Mater.* **2022**, *34*, 2107668. [CrossRef] [PubMed]
52. Li, X.; Kang, B.; Dong, F.; Zhang, Z.; Luo, X.; Han, L.; Huang, J.; Feng, Z.; Chen, Z.; Xu, J.; et al. Enhanced photocatalytic degradation and H_2/H_2O_2 production performance of S-pCN/$WO_{2.72}$ S-scheme heterojunction with appropriate surface oxygen vacancies. *Nano Energy* **2021**, *81*, 105671. [CrossRef]
53. Li, H.; Gong, H.; Jin, Z. Phosphorus modified Ni-MOF-74/$BiVO_4$ S-scheme heterojunction for enhanced photocatalytic hydrogen evolution. *Appl. Catal. B Environ.* **2022**, *307*, 121166. [CrossRef]
54. Takehara, M.; Yoshimura, I.; Takizawa, K.; Yoshida, R. Surface active N-acylglutamate: I. Preparation of long chain N-acylglutamic acid. *J. Am. Oil Chem. Soc.* **1972**, *49*, 157. [CrossRef]

55. Jin, H.; Liu, Z.; Ohsuna, T.; Terasaki, O.; Inoue, Y.; Sakamoto, K.; Nakanishi, T.; Ariga, K.; Che, S. Control of morphology and helicity of chiral mesoporous silica. *Adv. Mater.* **2006**, *18*, 593–596. [CrossRef]
56. Jin, H.; Qiu, H.; Sakamoto, Y.; Shu, P.; Terasaki, O.; Che, S. Mesoporous silicas by self-assembly of lipid molecules: Ribbon, hollow sphere, and chiral materials. *Chem. Eur. J.* **2008**, *14*, 6413–6420. [CrossRef]
57. Kresse, G.; Hafner, J. Ab initio molecular dynamics for liquid metals. *Phys. Rev. B Condens. Matter* **1993**, *47*, 558–561. [CrossRef]
58. Kresse, G.; Furthmuller, J. Efficient iterative schemes for ab initio total-energy calculations using a plane-wave basis set. *Phys. Rev. B Condens. Matter* **1996**, *54*, 11169–11186. [CrossRef]
59. Kresse, G.; Furthmüller, J. Efficiency of ab-initio total energy calculations for metals and semiconductors using a plane-wave basis set. *Comput. Mater. Sci.* **1996**, *6*, 15–50. [CrossRef]
60. Perdew, J.P.; Burke, K.; Ernzerhof, M. Generalized Gradient Approximation Made Simple. *Phys. Rev. Lett.* **1996**, *77*, 3865–3868. [CrossRef]
61. Grimme, S.; Ehrlich, S.; Goerigk, L. Effect of the damping function in dispersion corrected density functional theory. *J. Comput. Chem.* **2011**, *32*, 1456–1465. [CrossRef] [PubMed]
62. Kresse, G.; Joubert, D. From ultrasoft pseudopotentials to the projector augmented-wave method. *Phys. Rev. B* **1999**, *59*, 1758–1775. [CrossRef]
63. Neugebauer, J.; Scheffler, M. Adsorbate-substrate and adsorbate-adsorbate interactions of Na and K adlayers on Al(111). *Phys. Rev. B Condens. Matter* **1992**, *46*, 16067–16080. [CrossRef]
64. Frisch, M.J.; Trucks, G.W.; Schlegel, H.B.; Scuseria, G.E.; Robb, M.A.; Cheeseman, J.R.; Scalmani, G.; Barone, V.; Mennucci, B.; Petersson, G.A.; et al. *Gaussian 09, Revision D.01*; Gaussian Inc.: Wallingford, CT, USA, 2013.
65. Lee, C.; Yang, W.; Parr, R.G. Development of the Colle-Salvetti correlation-energy formula into a functional of the electron density. *Phys. Rev. B* **1988**, *37*, 785–789. [CrossRef] [PubMed]
66. Becke, A.D. A new mixing of Hartree–Fock and local density-functional theories. *J. Chem. Phys.* **1993**, *98*, 1372–1377. [CrossRef]
67. Weigend, F.; Ahlrichs, R. Balanced basis sets of split valence, triple zeta valence and quadruple zeta valence quality for H to Rn: Design and assessment of accuracy. *Phys. Chem. Chem. Phys.* **2005**, *7*, 3297–3305. [CrossRef]
68. Zhang, D.; Guo, Y.; Zhao, Z. Porous defect-modified graphitic carbon nitride via a facile one-step approach with significantly enhanced photocatalytic hydrogen evolution under visible light irradiation. *Appl. Catal. B Environ.* **2018**, *226*, 1–9. [CrossRef]
69. Wang, Y.; Liu, X.; Liu, J.; Han, B.; Hu, X.; Yang, F.; Xu, Z.; Li, Y.; Jia, S.; Li, Z.; et al. Carbon quantum dot implanted graphite carbon nitride nanotubes: Excellent charge separation and enhanced photocatalytic hydrogen evolution. *Angew. Chem. Int. Ed.* **2018**, *57*, 5765–5771. [CrossRef]
70. Fang, H.B.; Zhang, X.H.; Wu, J.; Li, N.; Zheng, Y.Z.; Tao, X. Fragmented phosphorus-doped graphitic carbon nitride nanoflakes with broad sub-bandgap absorption for highly efficient visible-light photocatalytic hydrogen evolution. *Appl. Catal. B Environ.* **2018**, *225*, 397–405. [CrossRef]
71. Iqbal, W.; Qiu, B.; Zhu, Q.; Xing, M.; Zhang, J. Self-modified breaking hydrogen bonds to highly crystalline graphitic carbon nitrides nanosheets for drastically enhanced hydrogen production. *Appl. Catal. B Environ.* **2018**, *232*, 306–313. [CrossRef]
72. Zhang, Y.; Zong, S.; Cheng, C.; Shi, J.; Guo, P.; Guan, X.; Luo, B.; Shen, S.; Guo, L. Rapid high-temperature treatment on graphitic carbon nitride for excellent photocatalytic H_2-evolution performance. *Appl. Catal. B Environ.* **2018**, *233*, 80–87. [CrossRef]
73. Han, Q.; Cheng, Z.; Wang, B.; Zhang, H.; Qu, L. Significant enhancement of visible-light-driven hydrogen evolution by structure regulation of carbon nitrides. *ACS Nano* **2018**, *12*, 5221–5227. [CrossRef] [PubMed]
74. Zou, Y.; Shi, J.-W.; Ma, D.; Fan, Z.; Cheng, L.; Sun, D.; Wang, Z.; Niu, C. WS_2/graphitic carbon nitride heterojunction nanosheets decorated with CdS quantum dots for photocatalytic hydrogen production. *ChemSusChem* **2018**, *11*, 1187–1197. [CrossRef] [PubMed]
75. Yang, P.; Ou, H.; Fang, Y.; Wang, X. A facile steam reforming strategy to delaminate layered carbon nitride semiconductors for photoredox catalysis. *Angew. Chem. Int. Ed.* **2017**, *56*, 3992–3996. [CrossRef]
76. Liu, J.; Liu, N.Y.; Li, H.; Wang, L.P.; Wu, X.Q.; Huang, H.; Liu, Y.; Bao, F.; Lifshitz, Y.; Lee, S.T.; et al. A critical study of the generality of the two step two electron pathway for water splitting by application of a C_3N_4/MnO_2 photocatalyst. *Nanoscale* **2016**, *8*, 11956–11961. [CrossRef] [PubMed]

Article

CdS Nanocubes Adorned by Graphitic C₃N₄ Nanoparticles for Hydrogenating Nitroaromatics: A Route of Visible-Light-Induced Heterogeneous Hollow Structural Photocatalysis

Zhi-Yu Liang [1,2], Feng Chen [1,2], Ren-Kun Huang [1,2], Wang-Jun Huang [3], Ying Wang [1,2,*], Ruo-Wen Liang [1,2,*] and Gui-Yang Yan [1,2,*]

[1] Province University Key Laboratory of Green Energy and Environment Catalysis, Ningde Normal University, Ningde 352100, China
[2] Fujian Provincial Key Laboratory of Featured Materials in Biochemical Industry, Ningde Normal University, Ningde 352100, China
[3] College of Environmental Science and Engineering, Fujian Normal University, Fuzhou 350117, China
* Correspondence: wy891203@163.com (Y.W.); t1629@ndnu.edu.cn (R.-W.L.); ygyfjnu@163.com (G.-Y.Y.); Tel.: +86-593-2965018 (Y.W.); +86-593-2954127 (R.-W.L.); +86-593-0593-2565503 (G.-Y.Y.)

Abstract: Modulating the transport route of photogenerated carriers on hollow cadmium sulfide without changing its intrinsic structure remains fascinating and challenging. In this work, a series of well-defined heterogeneous hollow structural materials consisting of CdS hollow nanocubes (CdS NCs) and graphitic C_3N_4 nanoparticles (CN NPs) were strategically designed and fabricated according to an electrostatic interaction approach. It was found that such CN NPs/CdS NCs still retained the hollow structure after CN NP adorning and demonstrated versatile and remarkably boosted photoreduction performance. Specifically, under visible light irradiation ($\lambda \geq 420$ nm), the hydrogenation ratio over 2CN NPs/CdS NCs (the mass ratio of CN NPs to CdS NCs is controlled to be 2%) toward nitrobenzene, p-nitroaniline, p-nitrotoluene, p-nitrophenol, and p-nitrochlorobenzene can be increased to 100%, 99.9%, 83.2%, 93.6%, and 98.2%, respectively. In addition, based on the results of photoelectrochemical performances, the 2CN NPs/CdS NCs reach a 0.46% applied bias photo-to-current efficiency, indicating that the combination with CN NPs can indeed improve the migration and motion behavior of photogenerated carriers, besides ameliorating the photocorrosion and prolonging the lifetime of CdS NCs.

Keywords: heterogeneous hollow structural materials; CdS nanocubes; graphitic carbon nitride; photoreduction catalysis

Citation: Liang, Z.-Y.; Chen, F.; Huang, R.-K.; Huang, W.-J.; Wang, Y.; Liang, R.-W.; Yan, G.-Y. CdS Nanocubes Adorned by Graphitic C₃N₄ Nanoparticles for Hydrogenating Nitroaromatics: A Route of Visible-Light-Induced Heterogeneous Hollow Structural Photocatalysis. *Molecules* 2022, 27, 5438. https://doi.org/10.3390/molecules27175438

Academic Editor: Munkhbayar Batmunkh

Received: 7 August 2022
Accepted: 25 August 2022
Published: 25 August 2022

Publisher's Note: MDPI stays neutral with regard to jurisdictional claims in published maps and institutional affiliations.

Copyright: © 2022 by the authors. Licensee MDPI, Basel, Switzerland. This article is an open access article distributed under the terms and conditions of the Creative Commons Attribution (CC BY) license (https://creativecommons.org/licenses/by/4.0/).

1. Introduction

Recently, micro- and nanomaterials with dimensions in the nanometer and submicron scales have attracted significant interest [1]. For example, in terms of the microstructure of materials, researchers have developed a myriad of micro- and nanomaterials with different morphologies, including nanoparticles [2], nanorods [3], nanosheets [4], and hollow structures [5]. Among the many morphologies, hollow structural micro- and nanomaterials possess a larger specific surface area, lower density, shorter charge transport distance, and higher loading capacity and have been successfully used in many fields, such as photocatalysis [6–8]. In particular, hollow CdS-based photocatalysts, which possess outstanding photocatalytic activity toward environmental remediation and energy production under visible light irradiation, have surpassed their solid counterparts in many aspects, such as larger surface area and more exposed active sites to trigger many redox reactions [9,10]. All of these are coupled with the high capacity for light absorption through light reflection and scattering in the inner cavity, as the hollow structure endows the CdS with enhanced photocatalytic properties. Therefore, hollow CdS-based photocatalysts have broad application prospects in the field of photocatalysis and are worthy of our further study.

Regardless of the structures of pure hollow CdS-based photocatalysts, the practical application remains restricted by the severe photocorrosion and high recombination rate of photogenerated charges [11]; thus, it is necessary to find a suitable co-catalyst to improve the shortcoming of severe photocorrosion. In the in-depth study of the morphology, size, and composition of hollow structural materials, heterogeneous hollow structural materials, which are formed by the combination of different compositions or different crystalline phases as units, were extremely fascinating due to the abundance of design approaches and the superiority of performance control [12–15]. Compared with conventional single-component hollow structural materials, heterogeneous hollow structural materials not only have the traditional advantages of hollow structures, but also show rich functionalities in energy conversion and storage, catalysis, and drug delivery due to the synergistic effect between different components [12,16–18].

Theoretically, the hollow structure can rapidly transport the generated electrons and holes to two counterparts by enhancing the built-in electric field at the shell level, avoiding the recombination of electrons and holes, and improving the photocatalytic activity [19–22]. Hitherto, multiple efforts have been devoted to developing heterogeneous hollow structural material-based photocatalysts. In particular, the effective transfer of photogenerated holes from the valence band of CdS would be an ideal strategy to deal with the photocorrosion problem and improve the stability of CdS. Fortunately, graphitic carbon nitride (g-C_3N_4) has a suitable energy level for constructing a heterostructure with CdS to afford an effective separation and transfer of photogenerated charges [23–28]. Thanks to its rigid heptazine ring structure, π-conjugated electronic structure, and high condensation degree, g-C_3N_4 possesses many advantages in building the heterostructure, including excellent physicochemical stability and attractive electronic structure combined with an appropriate band gap (~2.7 eV). In addition, g-C_3N_4 is readily available and can be easily prepared from cheap crude materials such as urea [29–31]. However, due to the irregular size and low photocatalytic activity caused by the buckling nature of bulk g-C_3N_4, it is necessary to design a favorable and facile morphology of g-C_3N_4 for constructing a hollow CdS-based heterogeneous photocatalyst.

In this work, we synthesized a series of heterogeneous hollow structural materials constructed by hollow CdS nanocubes (CdS NCs) and g-C_3N_4 nanoparticles (CN NPs), in which CN NPs with abundant active sites were employed to reduce the photocorrosion and enhance the photocatalytic performance of CdS NCs. Consequently, the resultant CN NP/CdS NC heterogeneous hollow structural materials demonstrated superior photoactivities to the single counterparts toward hydrogenation of nitroaromatics to amino derivatives under visible light irradiation. In addition, according to the variety of photoelectrochemical performances, the CN NPs worked as a competent and suitable co-catalyst to separate the photogenerated charges and improve the stability of CdS NCs; furthermore, they prolong the lifetime of the photogenerated charges. The possible mechanism for the enhancement of photocatalytic efficiency was studied in particular.

2. Results

2.1. Characterizations of CN NPs/CdS NCs

The detailed synthesis approach for CN NPs/CdS NCs is schematically demonstrated in Figure 1. Based on the results of the zeta potential test, the CdS NC aqueous solution is positively charged within the range of pH from 3 to 12 (Figure 1a). On the other hand, it is noteworthy that the CN NP solution (Figure 1b) shows a negatively charged status within the pH range from 6 to 12. In this regard, when the solution has a pH of more than 6, the CN NPs with a negative charge would be spontaneously and uniformly self-assembled on the surface of CdS NCs with a positive charge by electrostatic interaction.

X-ray diffraction (XRD) was performed to verify the crystalline phases. CdS NCs, CN NPs, and 2CN NPs/CdS NCs (the mass ratio of CN NPs to CdS NCs is controlled to be 2%) were selected for a concise comparison, as shown in Figure 2a. Hexagonal wurtzite CdS (JCPDS 41-1049) could be found in the pristine CdS NCs and the 2CN NPs/CdS

NCs [9,10,32–34]. The peak at 2θ = 27.2° in the XRD of the CN NPs (JCPDS 87-1526) was ascribed to the crystallographic plane (002) of graphite [31,35–37]. Note that this characteristic peak of g-C$_3$N$_4$ disappeared when the CN NPs hybridized with the CdS NCs, which may be ascribed to the rather low loading amount of CN NPs compared with CdS NCs in the composite; the peak of g-C$_3$N$_4$ was probably covered by the CdS peaks.

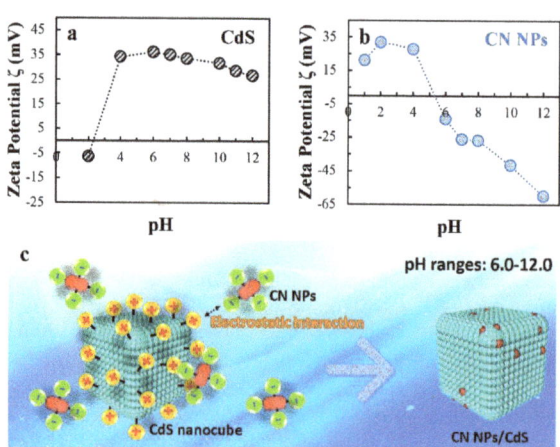

Figure 1. Zeta potential test of (a) CdS NCs and (b) CN NPs. (c) Schematic diagram of synthesis approach for CN NPs/CdS NCs by electrostatic interaction.

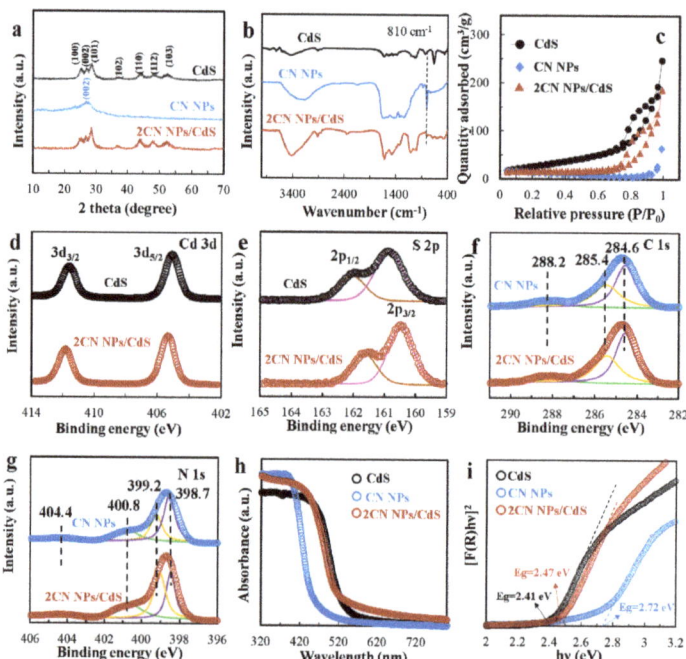

Figure 2. (a) XRD patterns, (b) FTIR spectra, and (c) BET measures of the CdS NCs, CN NPs, and 2CN/CdS NCs. Survey and high-resolution XPS spectra of (d) Cd 3d, (e) S 2p, (f) C 1s, and (g) N 1s for CdS NCs, CN NPs, and 2CN NPs/CdS NCs. (h) UV-Vis diffuse reflectance spectra of CdS NCs, CN NPs, and 2CN NPs/CdS NCs with (i) band gap determinations based on the Kubelka–Munk method.

The Fourier transform infrared spectroscopy (FTIR) spectra of different samples were inspected to verify the integration of CN NPs and CdS NCs in the composites. Figure 2b shows that the absorption bands at 625, 1078, and 1338 cm^{-1}, attributable to the vibration mode of Cd–S [38,39], could be observed both in the spectrum of CdS NCs and CN NPs/CdS NCs. Meanwhile, the absorption in the region of 1240–1680 cm^{-1}, corresponding to the stretching modes of the heterocycles in CN NPs, was also observed both in the CN NPs and the 2CN NP/CdS NC composite [40,41]. Therefore, in combination, the XRD and FTIR tests indicated that the integration of the CN NPs and CdS NCs was successful according to the electrostatic interaction.

Additionally, Brunauer–Emmett–Teller (BET) measurements were performed to evaluate the specific surface areas of CdS NCs, CN NPs, and 2CN NPs/CdS NCs. As illustrated in Figure 2c, specific surface areas of CdS NCs, CN NPs, and 2CN NPs/CdS NCs were 97.5, 6.3, and 45.3 m^2/g, respectively. It is noteworthy that after combining with CN NPs, the specific surface area of 2CN NPs/CdS NCs is dramatically decreased, which can be attributed to the integration of CN NPs on the CdS NC substrates shielding the pores of CdS NC frameworks.

Elemental chemical states of CN NPs, CdS NCs, and 2CN NPs/CdS NCs were determined by X-ray photoelectron spectroscopy (XPS). As shown in Figure 2d–g, the high-resolution spectrum of Cd 3d suggests the presence of Cd^{2+}. Additionally, the characterized peaks located at 161.3 eV for S 2p$_{3/2}$ and 162.7 eV for S (Figure 2e) can be ascribed to S^{2-} [42,43]. Peaks centered at 284.6, 285.4, and 288.2 eV in the C 1s spectrum corresponded to C-C/C-H, C-OH, and N-C=N, respectively. Finally, the peaks at 398.7, 399.2, and 400.8 eV of the N 1s spectrum are attributable to the nitrogen species of C-N=C, N-(C)$_3$, and C-N-H groups, respectively [44,45]. As such, XPS results signify again the successful combination of CN NPs and CdS NSs in the composites.

To determine the optical properties of as-prepared samples, diffuse reflectance spectroscopy (DRS) was performed. Figure 2h shows that the CdS NCs, CN NPs, and 2CN NPs/CdS NCs had substantial absorption in the visible region, which could be attributed to the intrinsic absorption properties of CdS and g-C$_3$N$_4$. Additionally, according to the Kubelka–Munk method [46,47], the band gap of CdS NCs, CN NPs, and 2CN NPs/CdS NCs could be roughly determined as 2.41, 2.72, and 2.47 eV by drawing a tangent line on the plots of $(\alpha h\nu)^2$ vs. $h\nu$, as shown in Figure 2i. The results of DRS further suggested that the energy band structures of as-prepared samples are suitable for absorption of visible light.

The morphologies of CdS NCs, CN NPs, and 2CN NPs/CdS NCs were investigated by FESEM. As demonstrated in Figure 3a,b, uniform and neat nanocubes were observed. Figure 3c,d show the SEM image of CN NPs, which ultrasmall nanoparticles with a mean diameter of ca. 5–20 nm. It is worth noting that the framework of CdS NCs had no change after combination with CN NPs, implying that electrostatic interaction can be considered as a mild and efficient assembly route, as shown in Figure 3e,f. Additionally, EDS mapping analysis was also performed (Figure 3g) to elucidate the elemental distribution in the nanocomposites, confirming the presence of the Cd, S, C, and N elements in the 2CN NPs/CdS NCs. Based on the SEM and EDS results, it is confirmed that the CN NPs were tightly integrated on the surface of CdS NCs after electrostatic self-assembly.

Transmission electron microscopy (TEM) measurements were performed to inspect the morphologies of the CdS NCs, CN NPs, and 2CN NPs/CdS NCs (Figure 4a). In the TEM images of CdS NCs, a polycrystalline structure of hollow nanocubes with a mean diameter of 100 nm could be observed. It was previously demonstrated that the complex hollow nanostructure of CdS possessed superior photocatalytic activity to the other forms of CdS. The TEM images of the CN NPs had many ultrasmall nanoparticles with a diameter of ca. 5–20 nm (Figure 4b). Therefore, given the surface charge properties of these assembly units, it could be reasonably speculated that the negatively charged CN NPs would spontaneously and uniformly self-assemble on the positively charged hollow CdS NCs to afford intimate interfacial contact under electrostatic interaction (Figure 4d). In 2CN NPs/CdS NCs, lattice spacing (0.359 nm) was accurately assigned to the (002) crystallographic plane of hexagonal

CdS [20,37,48]. The blue circles in Figure 4e mark some non-crystalline CN NPs that loaded on the CdS NCs surface uniformly and intimately, indicating the successful synthesis of heterogeneous hollow structural materials by combination of CdS NCs and CN NPs.

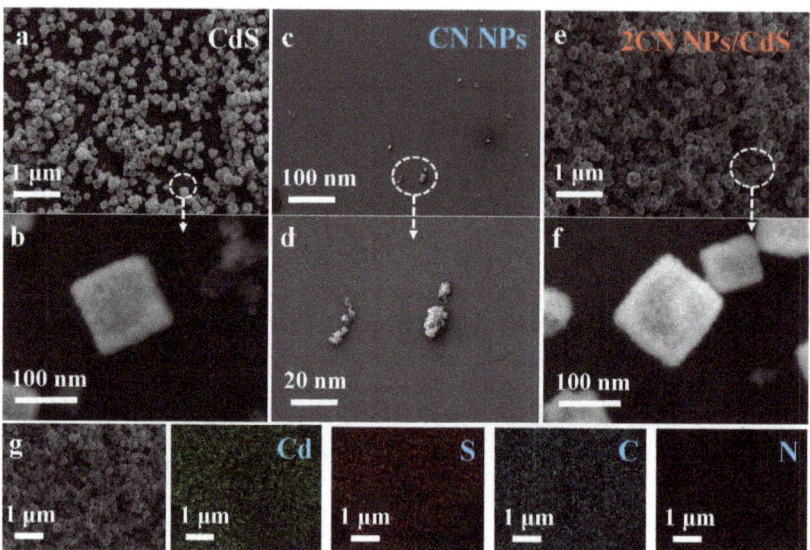

Figure 3. SEM images of the (a,b) CdS NCs, (c,d) CN NPs, and (e,f) 2CN NPs/CdS NCs. (g) EDS mapping analysis of Cd, S, C, and N elements over 2CN NPs/CdS NCs.

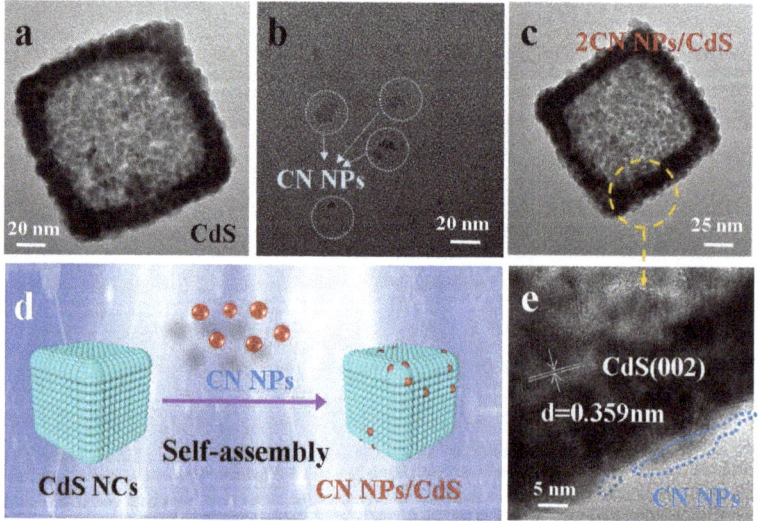

Figure 4. TEM images of (a) CdS NCs, (b) CN NPs, and (c) the 2CN NP/CdS NC heterogeneous hollow photocatalyst; (d) schematic illustration of the flowchart for fabricating the CN NPs/CdS NCs; (e) HRTEM image of the 2CN NP/CdS NC heterogeneous hollow photocatalyst.

2.2. Photocatalytic Performances of CN NPs/CdS NCs

The photocatalytic performance of the substrates was probed under visible light ($\lambda > 420$ nm) irradiation at ambient conditions by the selective anaerobic photocatalytic hydrogenation of aromatic nitro compounds to the corresponding amino compounds. An example is the photocatalytic hydrogenation of nitrobenzene to aniline. According to UV-Vis spectroscopy, the absorption peak at 273 nm was attributed to nitrobenzene, and the peaks at 238 nm were ascribed to aniline. Figure 5a shows that the CN NPs had a low photoactivity and reduced only ca. 9.4% of the nitrobenzene after irradiation for 15 min, whereas the CdS NCs reduced ca. 84.1% of the nitrobenzene in the same irradiation time. When the CN NPs and CdS NCs were combined, the CN NPs/CdS NC heterogeneous hollow photocatalysts exhibited strongly enhanced photocatalytic performances. Among the different CN NPs loading amounts, the photoactivity of 2CN NP/CdS heterogeneous hollow photocatalyst demonstrated the optimal photoactivity (~100%) in comparison with the other counterparts. Consequently, the results indicated that the integration of CdS NCs with CN NPs in a rational way contributes to enhancing performance. Furthermore, to demonstrate the critical role of photoelectrons in initiating the progress of reducing nitrobenzene, $K_2S_2O_8$ was used as an electron quench. Figure 5b shows that the photoactivity decreased remarkably for 2CN NPs/CdS under the same reaction conditions, therefore indicating the crucial role of in situ photoexcited electrons in initiating the reaction.

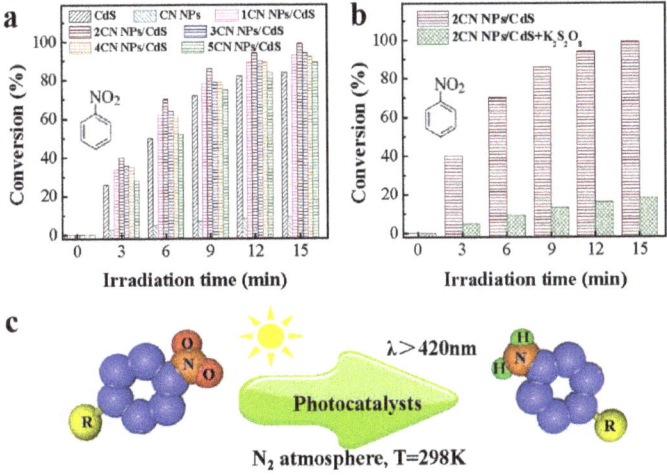

Figure 5. (a) Photocatalytic performance of CdS NCs, CN NPs, and xCN NP/CdS NC (x is the mass ratio of CN NPs to CdS NCs) heterogeneous hollow photocatalysts with different CN NP loading amounts under visible light irradiation with N_2 aeration. (b) Photoactivities of 2CN NPs/CdS NCs when adding $K_2S_2O_8$ as electron quench under visible light irradiation with N_2 aeration. (c) Typical reaction model of converting aromatic nitro compounds to aromatic amines under the experimental environment.

Apart from nitrobenzene, the selection of other aromatic nitro compounds such as *p*-nitroaniline, *p*-nitrotoluene, *p*-nitrophenol, and *p*-nitrochlorobenzene over CdS NCs, CN NPs, and 2CN NP/CdS NC heterogeneous hollow photocatalyst were also explored. As shown in Figure 6a–f, compared with CdS NCs and CN NPs, 2CN NPs/CdS still shows the optimal photocatalytic activity for hydrogenation of aromatic nitro compounds. The photocatalytic performance toward *p*-nitroaniline, *p*-nitrotoluene, *p*-nitrophenol, and *p*-nitrochlorobenzene over 2CN NPs/CdS was found to be 99.9%, 83.2%, 93.6%, and 98.2%, respectively. Hence, the integration of CdS NCs and CN NPs in an appropriate approach will be beneficial for enhancing the photocatalytic performances under visible light. Meanwhile, cycling experiments were carried out to demonstrate that CN NPs can improve the photostability of CdS. As shown in Figure 6b,c, the 2CN NPs/CdS NCs maintained excellent photostability (97.7%) for the photocatalytic hydrogenation of *p*-nitroaniline after four consecutive cycling experiments. However, the photocatalytic hydrogenation activity of CdS NCs was reduced to only 48.6%. It was clearly shown that CN NPs will ameliorate the photocorrosion of the CdS NCs after integration.

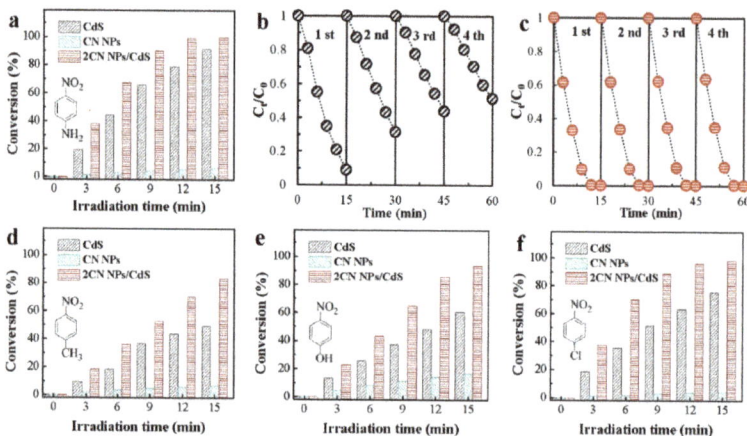

Figure 6. Photocatalytic performance toward reducing (**a**) *p*-nitroaniline, (**d**) *p*-nitrotoluene, (**e**) *p*-nitrophenol, and (**f**) *p*-nitrochlorobenzene over CdS NCs, CN NPs, and 2CN NPs/CdS under visible light irradiation with N_2 aeration. Cyclic experiments for photocatalytic reducing of *p*-nitroaniline over (**b**) CdS NCs and (**c**) 2CN NP/CdS NC heterogeneous hollow photocatalyst under the experimental environment.

Table 1 summarizes a comparative study of different photocatalysts reported in recent years toward the hydrogenation of nitrobenzene. It is clearly shown that the as-prepared samples showed excellent performance toward the hydrogenation of nitrobenzene.

Table 1. Comparison of the irradiation time and efficiency among various photocatalysts.

Materials	Irradiation Time (min)	Efficiency (%)	Ref.
2CN NPs/CdS NCs	15	100	This work
Ti_3C_2/Pd	150	99	[49]
1.2% PtO@Cr_2O_3	60	98	[50]
2% Bi_2O_3/Bi_2WO_6	40	93.1	[51]
20% rGO/$NiCo_2O_4$	60	100	[52]
1.2%PtO@$ZnCr_2O_4$	60	100	[53]

2.3. Photoelectrochemical (PEC) Measurements

Photoelectrochemical (PEC) measurements were performed to observe the PEC properties of CdS NCs, CN NPs, and CN NPs/CdS NC heterogeneous hollow photocatalysts. Figure 7a shows the transient photocurrent responses of CdS NCs, CN NPs, and the 2CN NP/CdS NC heterogeneous hollow photocatalyst under visible light, following the subsequence of 2CN NPs/CdS NCs > CdS NCs > CN NPs, which is in line with the photocatalytic performances. Additionally, electrochemical impedance spectroscopy (EIS) was performed to evaluate the separation efficiency of photogenerated charge carriers in the interfaces. Generally, a smaller Nyquist plot semicircle radius represents a faster interfacial charge transfer rate [54]. As shown in Figure 7b, the EIS Nyquist plot of the 2CN NP/CdS NC heterogeneous hollow photocatalysts exhibits a smaller semicircular arc radius under visible light than that of CdS NCs and CN NPs, implying that the interface of the 2CN NP/CdS NC heterogeneous hollow photocatalyst will facilitate the moving and transmission of photogenerated electrons. Furthermore, after 1 h of continuous irradiation, the 2CN CdS

NCs exhibit more photostability than CdS NCs (Figure 7c), indicating that the CN NPs can indeed ameliorate the photocorrosion of CdS NCs.

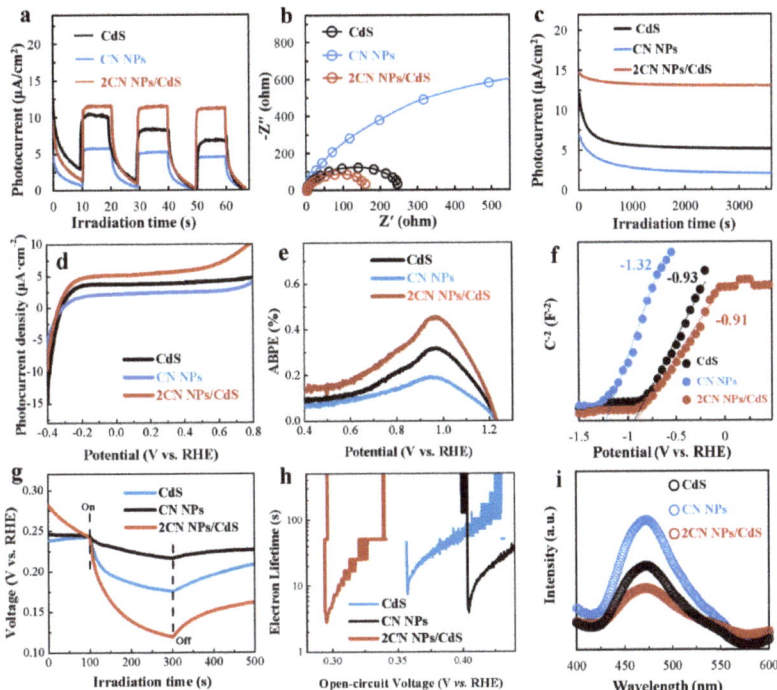

Figure 7. (a) Transient photocurrent responses with a 0.1 V vs. RHE bias voltage, (b) EIS results, (c) photostability, (d) LSV (scan rate: 5 mV/s), (e) ABPE curves based on the LSV results, (f) Mott–Schottky plots, and (g) OCVD curves of CdS NCs, CN NPs, and 2CN NP/CdS NC heterogeneous hollow photocatalyst under visible light irradiation. (h) Electron lifetime based on the results of OCVD curves and (i) PL spectra of the as-prepared materials with an excitation wavelength of 325 nm.

Additionally, linear sweep photovoltammetry (LSV) can be used to assess the kinetics of charge motion. As shown in Figure 7d, the photocurrent of the 2CN CdS NC heterogeneous hollow photocatalyst increases in the high bias voltage region, indicating that 2CN CdS NCs have a beneficial interface for charge movement [55]. For the results of LSV, the conversion efficiency can be estimated by calculating the applied bias photo-to-current efficiency (ABPE, η) with Equation (1).

$$\eta = \frac{I(1.23 - |V|)}{P} \tag{1}$$

where I is the photocurrent measured by LSV, V is the bias, and P is the intensity of irradiation light (about 20 mW·cm^{-2}). The results are reflected in Figure 7e, and the 2CN CdS NCs have the highest efficiency, about 0.46%. Notably, the results of LSV and ABPE are quite in line with the photoactivity performances, consistently confirming that the motion behavior of photogenerated charges will be significantly promoted after adorning the surface of CdS nanocubes with CN NPs.

Furthermore, the CB of semiconductors could be determined by Mott–Schottky (M-S) measurements, as shown in Figure 7f, flat-band potentials of CdS NCs and CN NPs are determined to be −0.93 and −1.22 eV (vs. RHE). Thus, combined with the results of

UV-Vis DRS, the VB location of CdS NSs and CN NPs can be roughly calculated as +1.48 V and +1.40 V, respectively. Such band structures are beneficial for the transmission of the photogenerated electrons from the CB of CN NPs to the CB of CdS NCs, as well as the photogenerated holes from the VB of CdS NCs to the VB of CN NPs, thus ameliorating the photocorrosion of CdS NCs.

Additionally, the recombination kinetics and lifetime of charge could be estimated by monitoring the open-circuit photovoltage decay (OCVD) tests [49]. The 2CN CdS NC heterogeneous hollow photocatalyst exhibits a larger photovoltage than CdS NCs and CN NPs, implying the lowest recombination of charge, as shown in Figure 7g. Based on the OCVD results, the logarithmic graph of electron lifetime vs. open-circuit voltage could be calculated by Equation (2),

$$\tau = \frac{-k_B T}{e}\left(\frac{dV_{oc}}{dt}\right)^{-1} \quad (2)$$

where electron lifetime, thermal energy, electron charge (~1.602 × 10^{-19} C), time-dependent photovoltage, and time are abbreviated as τ, $k_B T$, e, V_{oc}, and t. The calculated results are reflected in Figure 7h. The 2CN NP/CdS NC heterogeneous hollow photocatalyst shows a longer electron lifetime. Obviously, the result is in line with the other PEC tests. Such a conclusion can be further confirmed by the photoluminescence (PL) intensity test; as shown in Figure 7i, the lowest intensity of the 2CN NP/CdS NC heterogeneous hollow photocatalyst indicates an optimal charge separation efficiency.

Based on the above analysis, the photocatalytic mechanism of the CN/CdS NC heterogeneous hollow photocatalyst was proposed, as shown in Scheme 1. Under the irradiation of visible light, both CN NPs and CdS NCs were excited to produce the electron–hole pairs. The electrons in the conduction band of the CN NPs easily and immediately migrated into the conduction band of the CdS NCs because (1) CN NPs had a more negative electronic potential in their conduction band compared with CdS NCs and (2) the interfacial integration between CN NPs and CdS NCs was highly intimate. Meanwhile, the holes in the valence band of the CdS NCs spontaneously flew into the valence band of the CN NPs, which effectively prevented the photocorrosion of the CdS NCs and prolonged their lifetime. As for the photoreduction, the electrons in the conduction band of the CdS NCs efficiently reduced the aromatic nitro compounds to the corresponding amines under the N_2 atmosphere.

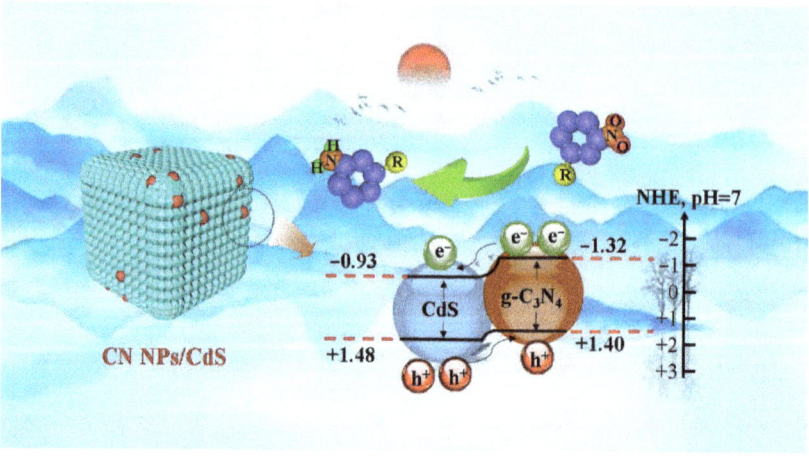

Scheme 1. Schematic illustration of the photocatalytic hydrogenation mechanism of the CN NP/CdS NC heterogeneous hollow photocatalysts.

3. Materials and Methods

3.1. Materials

Urea, sulfuric acid (H_2SO_4), nitric acid (HNO_3), sodium hydroxide (NaOH), hydrochloric acid (HCl), ammonium hydroxide ($NH_3 \cdot H_2O$), polyvinyl pyrrolidone K-30, silver nitrate ($AgNO_3$), sodium citrate, sodium borohydride ($NaBH_4$), ferric chloride ($FeCl_3$), sodium sulfide (Na_2S), cadmium nitrate ($Cd(NO_3)_2$), tributylphosphine, concentrated $NH_3\ H_2O$, potassium hexacyanoferrate (III) ($K_3[Fe(CN)_6]$), potassium hexacyanoferrate (II) ($K_4[Fe(CN)_6]$), sodium sulfate (Na_2SO_4), ethanol (C_2H_5OH), deionized water (DI H_2O, Millipore, 18.2 MΩ cm resistivity), ammonium formate (HCO_2NH_4), potassium persulfate (KPS, $K_2S_2O_8$), nitrobenzene, p-nitroaniline, p-nitrophenol, p-nitrotoluene, and p-nitrochlorobenzene were are analytical grade and used as received without further purification.

3.2. Synthesis

3.2.1. Synthesis of CdS Nanocubes (CdS NCs)

The CdS nanocubes were prepared as follows [10,32,33]: polyvinyl pyrrolidone K-30 (84 mg) and aqueous $AgNO_3$ (50 mmol/L, 1 mL) were added into a solution of aqueous sodium citrate (10 mmol/L, 40 mL) before the sequential addition of aqueous $NaBH_4$ (10 mmol/L, 1 mL), concentrated HCl (200 µL), and aqueous $FeCl_3$ (0.1 mol/L, 50 µL) under constant strong stirring. The mixed solution was maintained at 30 °C for 16 h to form a homogeneous suspension, to which aqueous Na_2S (0.1 mol/L, 1 mL) was added. The precipitates were collected, washed three times with methanol, and dispersed in methanol (4 mL). The solution of $Cd(NO_3)_2$ in methanol (50 mmol/L, 1 mL) and polyvinyl pyrrolidone K-30 (50 mg) were then added sequentially, and the mixture was stirred at 50 °C for 5 min. Finally, tributylphosphine (100 µL) was added and the mixture was kept at 50 °C for another 2 h. The precipitates were collected and washed three times with ethanol and deionized H_2O to give CdS nanocubes (CdS NCs).

3.2.2. Synthesis of g-C_3N_4 Nanoparticles (CN NPs)

Typically, urea (10 g) was heated from room temperature to 550 °C at 7 °C/min in a covered crucible for 2 h and then cooled to ambient temperature to obtain a yellow powder. The yellow powder (1 g) was added into a mixture solution including 20 mL H_2SO_4 and 20 mL HNO_3 and then stirred at room temperature for 2 h. The mixture was then diluted with 1 L of deionized water and washed several times to give a white residue. The white residue was finally dispersed in 30 mL concentrated $NH_3 \cdot H_2O$ and heated at 120 °C for 12 h in a 50 mL Teflon-lined autoclave, before being cooled to room temperature. The aqueous suspension was centrifuged at ~500 rpm to remove the precipitate and dialyzed to remove large-sized nanoparticles and NH_3 molecules to provide an aqueous solution of g-C_3N_4 nanoparticles (marked as CN NPs).

3.2.3. Preparation of g-C_3N_4 Nanoparticles/CdS Nanocubes (CN NPs/CdS NCs)

The g-C_3N_4 nanoparticles/CdS nanocubes (CN NPs/CdS NCs) were prepared by electrostatic self-assembly. CdS NCs (0.1 g) were firstly dispersed in ethanol (98 mL), and the aqueous g-CN NP solution was added dropwise; the pH of the mixture solution was adjusted to ~8.0 with aqueous NaOH (0.01 mol/L), and the mass ratio of CN NPs to CdS NCs was controlled to be 1%, 2%, 3%, 4%, and 5%. Finally, the precipitates were collected and washed three times with ethanol and deionized H_2O to give the g-C_3N_4 nanoparticles/CdS nanocubes (CN NPs/CdS NCs).

3.3. Characterizations

The zeta potential was monitored with a Zeta Potential Analyzer (Omin, BIC, America). An X-ray diffraction (XRD, X'Pert Pro MPD) instrument equipped with Cu Kα radiation under 36 kV and 30 mA was employed to analyze the crystal structures. A TJ270-30A spectrophotometer (Tianjin Tuopu Instrument, China) was used to monitor the Fourier transform infrared spectroscopy (FT-IR) in the range of 400–4000 cm^{-1}. The surface areas were determined from the nitrogen adsorption–desorption isotherm at 77.3 K by Brunauer–Emmett–Teller (BET, 3flex, Micromeritics, America). Field emission scanning electron microscopy (SEM, SUPRA 55, Carl Zeiss) was utilized to observe the morphologies. Transmission electron microscopy (TEM, FEI, America) was performed with an accelerating voltage of 200 kV. X-ray photoelectron spectroscopy (XPS, Escalab 250, Thermo Scientific, America) was analyzed with corrected by C 1s spectra of 284.6 eV. The UV-Vis diffuse reflectance spectroscopy (DRS) was performed on a Lambda 950 (PerkinElmer, America) by employing $BaSO_4$ as a reflectance substrate. The photofluorescence (PL) spectra were measured on a Varian Cary Eclipse spectrometer with an excitation wavelength of 325 nm. Finally, all the PEC measurements were monitored using a three-electrode configuration on an electrochemical workstation (CHI-660C, Chenhua, China). Working electrodes: added the as-prepared photocatalysts (5 mg) into 0.5 mL ethanol, dripped on a 1×1 cm^2 FTO glass, and finally dried the glass at 50 °C for 30 min to remove ethanol. Counter electrode: Pt foil. Reference electrode: Ag/AgCl. Electrolyte: Na_2SO_4 solution (1.0 mol·L^{-1}).

3.4. Photoactivity Evolution

The photoactivity of the as-prepared samples was evaluated by reducing a series of aromatic nitro compounds. Typically, the catalyst (5 mg) and ammonium formate (10 mg, for quenching photogenerated holes) were evenly dispersed into the solution including the target nitro compound (30 mg L^{-1}, 50 mL) with continuously bubbling N_2. After stirring for 0.5 h in dark to reach an adsorption–desorption equilibrium, the mixture system was irradiated by a 500 W Xe lamp equipped with an optical filter ($\lambda > 420$ nm). Based on the equation of Lambert–Beer (Equation (3)), the photoactivity of the photocatalysts could be evaluated as follows:

$$\text{Conversion} = \left(\frac{C_0 - C_t}{C_0}\right) \times 100\% \quad (3)$$

where C_0 and C_t represent the initial concentration and the time-dependent concentration, respectively.

4. Conclusions

In summary, we designed a series of well-fined heterogeneous hollow structural materials (CN NPs/CdS NCs) via a self-assembly approach by combining hollow CdS nanocubes and g-C_3N_4 nanoparticles. The intrinsically negatively charged CN NPs were uniformly and intimately interspersed on the positively charged hollow nanocubes of CdS NCs based on electrostatic interaction. The intimate integration of CN NPs with CdS NCs could effectively ameliorate the photocorrosion and enhance the performance of CdS NCs in photocatalytic hydrogenation. Under the irradiation of visible light, the CN NP/CdS NC heterogeneous hollow photocatalysts demonstrated markedly enhanced photocatalytic hydrogenation performance in the anaerobic photocatalytic hydrogenation of aromatic nitro compounds at ambient conditions; in particular, when the mass ratio of CN NPs to CdS NCs is 2%, the hydrogenation ratio over the heterogeneous hollow photocatalyst toward nitrobenzene, p-nitroaniline, p-nitrotoluene, p-nitrophenol, and p-nitrochlorobenzene can be increased to 100%, 99.9%, 83.2%, 93.6%, and 98.2%, respectively. The photogenerated electrons responsible for the photocatalytic hydrogenation process were unambiguously determined by systematic control experiments, based on which the photocatalytic mechanism was elucidated. Additionally, the 2CN NPs/CdS NCs reached the highest applied bias photo-to-current efficiency (0.46%) according to the PEC investigation. It is anticipated that our work may further inspire the rational design of heterogeneous hollow structural materials for highly efficient conversions making use of solar energy.

Author Contributions: Conceptualization, Y.W., R.-W.L. and G.-Y.Y.; investigation, Z.-Y.L., F.C., R.-K.H. and W.-J.H.; writing—review and editing, Z.-Y.L. All authors have read and agreed to the published version of the manuscript.

Funding: The support from the Natural Science Foundation of Fujian province (No. 2020J05224 and 2020H0050) is gratefully acknowledged. This work was financially supported by Education & Research Project for Young and Middle-aged Teachers of Fujian province (No. JAT200683), the Research Project of Ningde Normal University (No. 2019ZX410), and the Research Project of Fujian Provincial Key Laboratory of Featured Materials in Biochemical Industry (No. FJKL_FBCM201906). Moreover, we are also grateful to the Program of IRTSTFJ for the financial support.

Institutional Review Board Statement: Not applicable.

Informed Consent Statement: Not applicable.

Data Availability Statement: Not applicable.

Conflicts of Interest: The authors declare no conflict of interest.

Sample Availability: Not applicable.

References

1. Xu, C.; Anusuyadevi, P.R.; Aymonier, C.; Luque, R.; Marre, S. Nanostructured materials for photocatalysis. *Chem. Soc. Rev.* **2019**, *48*, 3868–3902. [CrossRef] [PubMed]
2. Chu, Y.; Guo, L.; Xi, B.; Feng, Z.; Wu, F.; Lin, Y.; Liu, J.; Sun, D.; Feng, J.; Qian, Y.; et al. Embedding MnO@Mn$_3$O$_4$ nanoparticles in an N-doped-carbon framework derived from Mn-organic clusters for efficient lithium storage. *Adv. Mater.* **2018**, *30*, 201704244. [CrossRef] [PubMed]
3. Zhang, T.; Wu, M.Y.; Yan, D.Y.; Mao, J.; Liu, H.; Hu, W.B.; Du, X.W.; Ling, T.; Qiao, S.Z. Engineering oxygen vacancy on NiO nanorod arrays for alkaline hydrogen evolution. *Nano Energy* **2018**, *43*, 103–109. [CrossRef]
4. Liu, Y.; Xiao, C.; Lyu, M.; Lin, Y.; Cai, W.; Huang, P.; Tong, W.; Zou, Y.; Xie, Y. Ultrathin Co$_3$S$_4$ Nanosheets that Synergistically Engineer Spin States and Exposed Polyhedra that Promote Water Oxidation under Neutral Conditions. *Angew. Chem. Int. Ed.* **2015**, *54*, 11231–11235. [CrossRef]
5. Lin, F.; Wang, H.; Wang, G. Facile synthesis of hollow polyhedral (cubic, octahedral and dodecahedral) NiO with enhanced lithium storage capabilities. *Electrochimi. Acta* **2016**, *211*, 207–216. [CrossRef]
6. Liang, M.; Borjigin, T.; Zhang, Y.; Liu, B.; Liu, H.; Guo, H. Controlled assemble of hollow heterostructured g-C$_3$N$_4$@CeO$_2$ with rich oxygen vacancies for enhanced photocatalytic CO$_2$ reduction. *Appl. Catal. B Environ.* **2019**, *243*, 566–575. [CrossRef]
7. Wang, Y.; Pan, A.; Zhang, Y.; Shi, J.; Lin, J.; Liang, S.; Cao, G. Heterogeneous NiS/NiO multi-shelled hollow microspheres with enhanced electrochemical performances for hybrid-type asymmetric supercapacitors. *J. Mater. Chem. A* **2018**, *6*, 9153–9160. [CrossRef]
8. Li, Q.; Xia, Y.; Yang, C.; Lv, K.; Lei, M.; Li, M. Building a direct Z-scheme heterojunction photocatalyst by ZnIn$_2$S$_4$ nanosheets and TiO$_2$ hollowspheres for highly-efficient artificial photosynthesis. *Chem. Eng. J.* **2018**, *349*, 287–296. [CrossRef]
9. Madhusudan, P.; Zhang, J.; Cheng, B.; Yu, J. Fabrication of CdMoO$_4$@CdS core-shell hollow superstructures as high performance visible-light driven photocatalysts. *Phys. Chem. Chem. Phys.* **2015**, *17*, 15339–15347. [CrossRef]
10. Kim, M.R.; Jang, D.J. One-step fabrication of well-defined hollow CdS nanoboxes. *Chem. Commun.* **2008**, *41*, 5218–5220. [CrossRef]
11. Liu, X.; Sayed, M.; Bie, C.; Cheng, B.; Hu, B.; Yu, J.; Zhang, L. Hollow CdS-based photocatalysts. *J. Mater.* **2021**, *7*, 419–439. [CrossRef]
12. Zou, Y.; Guo, C.; Cao, X.; Zhang, L.; Chen, T.; Guo, C.; Wang, J. Synthesis of CdS/CoP hollow nanocages with improved photocatalytic water splitting performance for hydrogen evolution. *J. Environ. Chem. Eng.* **2021**, *9*, 106270. [CrossRef]
13. Xia, C.; Xue, C.; Bian, W.; Liu, J.; Wang, J.; Wei, Y.; Zhang, J. Hollow Co$_9$S$_8$/CdS nanocages as efficient photocatalysts for hydrogen evolution. *ACS Appl. Nano Mater.* **2021**, *4*, 2743–2751. [CrossRef]
14. Wang, L.; Li, R.; Huang, M. Synthesis and adsorption properties of CdS-Au hybrid nanorings. *Mater. Lett.* **2021**, *304*, 130722. [CrossRef]
15. Liu, Y.; Huang, C.; Zhou, T.; Hu, J. Morphology-preserved transformation of CdS hollow structures toward photocatalytic H$_2$ evolution. *CrystEngComm* **2020**, *22*, 1057–1062. [CrossRef]
16. Wang, M.; Zhang, H.; Zu, H.; Zhang, Z.; Han, J. Construction of TiO$_2$/CdS heterojunction photocatslysts with enhanced visible light activity. *Appl. Surf. Sci.* **2018**, *455*, 729–735. [CrossRef]
17. Wang, Z.; Hou, J.; Yang, C.; Jiao, S.; Zhu, H. Three-dimensional MoS$_2$-CdS-gamma-TaON hollow composites for enhanced visible-light-driven hydrogen evolution. *Chem. Commun.* **2014**, *50*, 1731–1734. [CrossRef]
18. Hu, Z.S.; Song, C.X.; Wang, D.B.; Wang, H.T.; Fu, X. Preparation and characterization of hollow spheres consisting of CdS/TiO$_2$ composite. *Rare Metal Mat. Eng.* **2005**, *34*, 8–10.
19. Peng, J.; Zheng, Z.; Tan, H.; Yang, J.; Zheng, D.; Song, Y.; Lu, F.; Chen, Y.; Gao, W. Rational design of ZnIn$_2$S$_4$/CdIn$_2$S$_4$/CdS with hollow heterostructure for the sensitive determination of carbohydrate antigen 19-9. *Sensors Actuat. B Chem.* **2022**, *363*, 131863. [CrossRef]

20. Yuan, W.; Zhang, Z.; Cui, X.; Liu, H.; Tai, C.; Song, Y. Fabrication of hollow mesoporous CdS@TiO$_2$@Au microspheres with high photocatalytic activity for hydrogen evolution from water under visible light. *ACS Sustain. Chem. Eng.* **2018**, *6*, 13766–13777. [CrossRef]
21. Qiu, B.; Zhu, Q.; Du, M.; Fan, L.; Xing, M.; Zhang, J. Efficient solar light harvesting CdS/Co$_9$S$_8$ hollow cubes for Z-scheme photocatalytic water splitting. *Angew. Chem. Int. Ed.* **2017**, *56*, 2684–2688. [CrossRef] [PubMed]
22. Dai, Z.; Zhang, J.; Bao, J.; Huang, X.; Mo, X. Facile synthesis of high-quality nano-sized CdS hollow spheres and their application in electrogenerated chemiluminescence sensing. *J. Mater. Chem.* **2007**, *17*, 1087–1093. [CrossRef]
23. Liang, Z.-Y.; Wei, J.X.; Wang, X.; Yu, Y.; Xiao, F.X. Elegant Z-scheme-dictated g-C$_3$N$_4$ enwrapped WO$_3$ superstructures: A multifarious platform for versatile photoredox catalysis. *J. Mater. Chem. A* **2017**, *5*, 15601–15612. [CrossRef]
24. Liang, Z.; Wen, Q.; Wang, X.; Zhang, F.; Yu, Y. Chemically stable and reusable nano zero-valent iron/graphite-like carbon nitride nanohybrid for efficient photocatalytic treatment of Cr(VI) and rhodamine B under visible light. *Appl. Surf. Sci.* **2016**, *386*, 451–459. [CrossRef]
25. Shi, H.; Li, Y.; Wang, X.; Yu, H.; Yu, J. Selective modification of ultra-thin g-C$_3$N$_4$ nanosheets on the (110) facet of Au/BiVO$_4$ for boosting photocatalytic H$_2$O$_2$ production. *Appl. Catal. B Environ.* **2021**, *297*, 120414. [CrossRef]
26. Zhang, J.; Wu, M.; He, B.; Wang, R.; Wang, H.; Gong, Y. Facile synthesis of rod-like g-C$_3$N$_4$ by decorating Mo$_2$C co-catalyst for enhanced visible-light photocatalytic activity. *Appl. Surf. Sci.* **2019**, *470*, 565–572. [CrossRef]
27. Tzvetkov, G.; Tsvetkov, M.; Spassov, T. Ammonia-evaporation-induced construction of three-dimensional NiO/g-C$_3$N$_4$ composite with enhanced adsorption and visible light-driven photocatalytic performance. *Superlattices Microst.* **2018**, *119*, 122–133. [CrossRef]
28. Jiang, X.H.; Xing, Q.J.; Luo, X.B.; Li, F.; Zou, J.-P.; Liu, S.S.; Li, X.; Wang, X.-K. Simultaneous photoreduction of Uranium (VI) and photooxidation of Arsenic (III) in aqueous solution over g-C$_3$N$_4$/TiO$_2$ heterostructured catalysts under simulated sunlight irradiation. *Appl. Catal. B Environ.* **2018**, *228*, 29–38. [CrossRef]
29. Jian, J.; Sun, J. A review of recent progress on silicon carbide for photoelectrochemical water splitting. *Sol. RRL* **2020**, *4*, 2000111. [CrossRef]
30. Yang, Y.; Zhang, C.; Huang, D.; Zeng, G.; Huang, J.; Lai, C.; Zhou, C.; Wang, W.; Guo, H.; Xue, W.; et al. Boron nitride quantum dots decorated ultrathin porous g-C$_3$N$_4$: Intensified exciton dissociation and charge transfer for promoting visible-light-driven molecular oxygen activation. *Appl. Catal. B Environ.* **2019**, *245*, 87–99. [CrossRef]
31. Reddy, K.R.; Reddy, C.V.; Nadagouda, M.N.; Shetti, N.P.; Jaesool, S.; Aminabhavi, T.M. Polymeric graphitic carbon nitride (g-C$_3$N$_4$)-based semiconducting nanostructured materials: Synthesis methods, properties and photocatalytic applications. *J. Environ. Manag.* **2019**, *238*, 25–40. [CrossRef] [PubMed]
32. Liu, X.L.; Liang, S.; Li, M.; Yu, X.F.; Zhou, L.; Wang, Q.Q. Facile synthesis of Au nanocube-CdS core-shell nanocomposites with enhanced photocatalytic activity. *Chin. Phys. Lett.* **2014**, *31*, 064203. [CrossRef]
33. Han, L.L.; Kulinich, S.A.; Zhang, Y.Y.; Zou, J.; Liu, H.; Wang, W.H.; Liu, H.; Li, H.B.; Yang, J.; Xin, H.L.; et al. Synergistic synthesis of quasi-monocrystal CdS nanoboxes with high-energy facets. *J. Mater. Chem. A* **2015**, *3*, 23106–23112. [CrossRef]
34. Wu, C.; Jie, J.; Wang, L.; Yu, Y.; Peng, Q.; Zhang, X.; Cai, J.; Guo, H.; Wu, D.; Jiang, Y. Chlorine-doped n-type CdS nanowires with enhanced photoconductivity. *Nanotechnology* **2010**, *21*, 505203. [CrossRef] [PubMed]
35. Ran, J.; Guo, W.; Wang, H.; Zhu, B.; Yu, J.; Qiao, S.Z. Metal-Free 2D/2D Phosphorene/g-C$_3$N$_4$ Van der Waals Heterojunction for Highly Enhanced Visible-Light Photocatalytic H$_2$ Production. *Adv. Mater.* **2018**, *30*, 1800128. [CrossRef]
36. Fu, J.; Yu, J.; Jiang, C.; Cheng, B. g-C$_3$N$_4$-Based Heterostructured Photocatalysts. *Adv. Energy Mater.* **2018**, *8*, 1701503. [CrossRef]
37. Tonda, S.; Kumar, S.; Gawli, Y.; Bhardwaj, M.; Ogale, S. g-C$_3$N$_4$ (2D)/CdS (1D)/rGO (2D) dual-interface nano-composite for excellent and stable visible light photocatalytic hydrogen generation. *Int. J. Hydrog. Energy* **2017**, *42*, 5971–5984. [CrossRef]
38. Wei, R.B.; Huang, Z.L.; Gu, G.H.; Wang, Z.; Zeng, L.; Chen, Y.; Liu, Z.Q. Dual-cocatalysts decorated rimous CdS spheres advancing highly-efficient visible-light photocatalytic hydrogen production. *Appl. Catal. B Environ.* **2018**, *231*, 101–107. [CrossRef]
39. Ma, S.; Deng, Y.; Xie, J.; He, K.; Liu, W.; Chen, X.; Li, X. Noble-metal-free Ni$_3$C cocatalysts decorated CdS nanosheets for high efficiency visible-light-driven photocatalytic H$_2$ evolution. *Appl. Catal. B Environ.* **2018**, *227*, 218–228. [CrossRef]
40. Selvarajan, S.; Suganthi, A.; Rajarajan, M. Fabrication of g-C$_3$N$_4$/NiO heterostructured nanocomposite modified glassy carbon electrode for quercetin biosensor. *Ultrason. Sonochem.* **2018**, *41*, 651–660. [CrossRef]
41. Gong, Y.; Zhao, X.; Zhang, H.; Yang, B.; Xiao, K.; Guo, T.; Zhang, J.; Shao, H.; Wang, Y.; Yu, G. MOF-derived nitrogen doped carbon modified g-C$_3$N$_4$ heterostructure composite with enhanced photocatalytic activity for bisphenol A degradation with peroxymonosulfate under visible light irradiation. *Appl. Catal. B Environ.* **2018**, *233*, 35–45. [CrossRef]
42. Periasamy, P.; Krishnakumar, T.; Devarajan, V.P.; Sandhiya, M.; Sathish, M.; Chavali, M. Investigation of electrochemical supercapacitor performance of WO$_3$-CdS nanocomposites in 1 M H$_2$SO$_4$ electrolyte prepared by microwave-assisted method. *Mater. Lett.* **2020**, *274*, 127998. [CrossRef]
43. Zhang, X.; Meng, D.; Tang, Z.; Hu, D.; Geng, H.; Zheng, H.; Zang, S.; Yu, Z.; Peng, P. Preparation of radial ZnSe-CdS nano-heterojunctions through atomic layer deposition method and their optoelectronic applications. *J. Alloys Compd.* **2019**, *777*, 102–108. [CrossRef]
44. Liu, D.; Chen, D.; Li, N.; Xu, Q.; Li, H.; He, J.; Lu, J. Integration of 3D macroscopic graphene aerogel with 0D-2D AgVO$_3$-g-C$_3$N$_4$ heterojunction for highly efficient photocatalytic oxidation of nitric oxide. *Appl. Catal. B Environ.* **2019**, *243*, 576–584. [CrossRef]
45. Tan, Y.; Shu, Z.; Zhou, J.; Li, T.; Wang, W.; Zhao, Z. One-step synthesis of nanostructured g-C$_3$N$_4$/TiO$_2$ composite for highly enhanced visible-light photocatalytic H$_2$ evolution. *Appl. Catal. B Environ.* **2018**, *230*, 260–268. [CrossRef]

46. Jiang, K.Y.; Weng, Y.L.; Guo, S.Y.; Yu, Y.; Xiao, F.X. Self-assembly of metal/semiconductor heterostructures via ligand engineering: Unravelling the synergistic dual roles of metal nanocrystals toward plasmonic photoredox catalysis. *Nanoscale* **2017**, *9*, 16922–16936. [CrossRef]
47. Xiao, F.X.; Miao, J.; Liu, B. Layer-by-Layer Self-assembly of CdS quantum dots/graphene nanosheets hybrid films for photoelectrochemical and photocatalytic applications. *J. Am. Chem. Soc.* **2014**, *136*, 1559–1569. [CrossRef]
48. Garg, P.; Bhauriyal, P.; Mahata, A.; Rawat, K.S.; Pathak, B. Role of dimensionality for photocatalytic water splitting: CdS nanotube versus bulk structure. *Chemphyschem* **2019**, *20*, 383–391. [CrossRef]
49. Zhang, Y.; Chen, L.; Gui, Y.; Liu, L. Catalytic transfer hydrogenation of nitrobenzene over Ti_3C_2/Pd nanohybrids boosted by electronic modification and hydrogen evolution inhibition. *Appl. Surf. Sci.* **2022**, *592*, 153334. [CrossRef]
50. Ismail, A.A.; Albukhari, S.M.; Mahmoud, M.H.H. Highly efficient and accelerated photoreduction of nitrobenzene over visible-light-driven PtO@Cr_2O_3 nanocomposites. *Surf. Interfaces* **2021**, *27*, 101527. [CrossRef]
51. Hu, W.; Wu, F.; Liu, W. Facile synthesis of Z-scheme Bi_2O_3/Bi_2WO_6 composite for highly effective visible-light-driven photocatalytic degradation of nitrobenzene. *Chem. Phys.* **2022**, *552*, 111377. [CrossRef]
52. Mkhalid, I.A. Simple synthesis of $NiCo_2O_4$@rGO nanocomposites for conversion of nitrobenzene via its photoreduction to aniline using visible light. *J. Mater. Res. Technol.* **2021**, *12*, 1988–1998. [CrossRef]
53. Ismail, A.A.; Albukhari, S.M.; Mahmoud, M.H.H. Mesoporous $ZnCr_2O_4$ photocatalyst with highly distributed PtO nanoparticles for visible-light-induced photoreduction of nitrobenzene. *Opt. Mater.* **2021**, *122*, 111676. [CrossRef]
54. Lin, B.; Li, H.; An, H.; Hao, W.; Wei, J.; Dai, Y.; Ma, C.; Yang, G. Preparation of 2D/2D $g-C_3N_4$ nanosheet@$ZnIn_2S_4$ nanoleaf heterojunctions with well-designed high-speed charge transfer nanochannels towards high efficiency photocatalytic hydrogen evolution. *Appl. Catal. B Environ.* **2018**, *220*, 542–552. [CrossRef]
55. Fang, X.X.; Wang, P.F.; Yi, W.; Chen, W.; Lou, S.C.; Liu, G.Q. Visible-light-mediated oxidative coupling of vinylarenes with bromocarboxylates leading to gamma-ketoesters. *J. Org. Chem.* **2019**, *84*, 15677–15684. [CrossRef]

MDPI
St. Alban-Anlage 66
4052 Basel
Switzerland
Tel. +41 61 683 77 34
Fax +41 61 302 89 18
www.mdpi.com

Molecules Editorial Office
E-mail: molecules@mdpi.com
www.mdpi.com/journal/molecules